AF389435

Designer Biopolymers

Designer Biopolymers

Self-Assembling Proteins and Nucleic Acids

Special Issue Editors

Ayae Sugawara-Narutaki
Yukiko Kamiya

MDPI • Basel • Beijing • Wuhan • Barcelona • Belgrade • Manchester • Tokyo • Cluj • Tianjin

Special Issue Editors

Ayae Sugawara-Narutaki
Nagoya University
Japan

Yukiko Kamiya
Nagoya University
Japan

Editorial Office
MDPI
St. Alban-Anlage 66
4052 Basel, Switzerland

This is a reprint of articles from the Special Issue published online in the open access journal *International Journal of Molecular Sciences* (ISSN 1422-0067) (available at: https://www.mdpi.com/journal/ijms/special_issues/biopolymers_assembling).

For citation purposes, cite each article independently as indicated on the article page online and as indicated below:

LastName, A.A.; LastName, B.B.; LastName, C.C. Article Title. *Journal Name* **Year**, *Article Number*, Page Range.

ISBN 978-3-03936-370-4 (Hbk)
ISBN 978-3-03936-371-1 (PDF)

Cover image courtesy of Ayae Sugawara-Narutaki.

Contents

About the Special Issue Editors

Ayae Sugarawa-Narutaki completed her PhD at The University of Tokyo in 2004. After her postdoctoral studies at the Tokyo Medical and Dental University and the California Institute of Technology, she was appointed as an assistant professor at The University of Tokyo in 2008. She moved to Nagoya University in 2014 as an associate professor. She has been working as a professor at Nagoya University since 2020.

Yukiko Kamiya completed her PhD at Nagoya City University in 2008. She also completed a postdoctoral fellowship at the Institute for Molecular Science. Then, she was appointed as an assistant professor at Nagoya University in 2012. She is currently an associate professor at Nagoya University.

International Journal of
Molecular Sciences

Editorial

Designer Biopolymers: Self-Assembling Proteins and Nucleic Acids

Ayae Sugawara-Narutaki [1],* and Yukiko Kamiya [2]

1   Department of Energy Engineering, Graduate School of Engineering, Nagoya University, Furo-cho, Chikusa-ku, Nagoya 464-8603, Japan
2   Department of Biomolecular Engineering, Graduate School of Engineering, Nagoya University, Furo-cho, Chikusa-ku, Nagoya 464-8603, Japan; yukikok@chembio.nagoya-u.ac.jp
*   Correspondence: ayae@energy.nagoya-u.ac.jp; Tel.: +81-52-789-3602

Received: 1 May 2020; Accepted: 3 May 2020; Published: 6 May 2020

Nature has evolved sequence-controlled polymers such as DNA and proteins over its long history. The recent rapid progress of synthetic chemistry, DNA recombinant technology, and computational science, as well as the elucidation of molecular mechanisms in biological processes, drive us to design ingenious polymers that are inspired by naturally occurring polymers but surpass them in specialized functions. The term "designer biopolymers" refers to polymers consisting of biological building units such as nucleotides, amino acids, and monosaccharides in a sequence-controlled manner. They may contain non-canonical nucleotides/amino acids/monosaccharides, or they may be conjugated to synthetic polymers to acquire specific functions in vitro and in vivo.

This special issue, entitled "Designer Biopolymers: Self-Assembling Proteins and Nucleic Acids" particularly focuses on the self-assembling aspect of designer biopolymers. Self-assembly is one common feature in biopolymers used to realize their dynamic biological activities and is strictly controlled by the sequence of biopolymers. In a broad sense, the self-assembly of biopolymers includes a double-helix formation of DNA, protein folding, and higher-order protein assembly (e.g., viral capsids). Designer biopolymers are now going beyond what nature evolved: researchers have generated DNA origami, protein cages, peptide nanofibers, and gels. This special issue assembles three review papers and seven research articles on the latest interdisciplinary work on self-assembling designer biopolymers.

The review paper by Lee et al. covers design, self-assembly, and application of various designer peptides including dipeptides, amphiphilic peptides, and cyclic peptides [1]. These peptides are especially useful in drug delivery systems and tissue engineering. The in-cell self-assembly of peptides, termed "reverse engineering of peptide self-assembly," is highlighted as a new approach to deliver peptide-based nanostructures to cells. The protein-based self-assembly system is reviewed by Nesterenko et al. [2]. The building block, ZT, is a complex from two titin Z1Z2 domains and telethonin. The Z1Z2 double tandem proteins (Z1Z2–Z1Z2) and telethonins co-assemble into polymeric nanostructures. They are robust scaffolds that can be genetically functionalized with full-length proteins and bioactive peptides prior to self-assembly. Functionalized ZT polymers successfully sustain the long-term culturing of stem cells. The review paper by Pereira et al. focuses on designer polymers based on cyanobacterial extracellular polymeric substances (EPS) [3]. The cyanobacterial EPS, mainly composed of heteropolysaccharides, emerges as a valid alternative to address several biotechnological and biomedical challenges. The review covers the characteristics and biological properties of cyanobacterial EPS, approaches to improving the production of the polymers by metabolic engineering, strategies for their extraction, purification, and genetic/chemical functionalization, and their use in scaffolds and coatings.

Two research articles address the important self-assembly phenomena of natural peptides. Antimicrobial peptides (AMPs) are a diverse group of membrane-active peptides that can interact with target membranes and can cause cell death by disturbing the membrane structure. Petkov et al. report molecular dynamics simulations studies on the solution behaviour of an AMP, bombinin H2 [4]. The simulation results show that bombinin H2 rapidly self-associate when multiple peptide chains are present in the solution, and the aggregation promotes further folding of bombinin H2 towards the biologically active shape. This study suggests that AMPs reach the target membrane in a functional folded state and are able to effectively exert their antimicrobial action. Amyloidogenic peptides including $A\beta_{1-40}$, α-synuclein, and β2 microglobulin are regarded as hallmark peptides associated with key onset mechanisms of neurodegenerative diseases. Yokoyama et al. report pH-dependent adsorption of these peptides onto gold nanoparticles [5]. Nano-scale geometrical simulation with a simplified protein structure (i.e., prolate) represents peptide adsorption orientation on a gold colloid, indicating the presence of electrostatic intermolecular and gold-peptide interactions.

Two other articles use engineered peptides to control inorganic mineralization or peptide-cell interactions. Kojima et al. describe the effects of peptide secondary structures on hydroxyapatite (HAp) biomineralization [6]. HAp-peptide composites containing a β-sheet forming peptide show a higher adsorption ability for basic proteins than those containing an α-helix forming peptide, most likely due to higher carboxy group density at the surfaces of former composites. Nanofibers formed from antigenic peptides conjugating to β-sheet-forming peptides have been recognized as promising candidates for next-generation nanoparticle-based vaccines. Waku et al. demonstrate that the hydrophilic-hydrophobic balance of peptide nanofibers affects their cellular uptake, cytotoxicity, and dendritic cell activation ability, which will provide useful design guidelines for the development of effective nanofiber-based vaccines [7].

In nature, proteins are often designed to form filamentous and circular oligomers to play their function. The articles from Sekiguchi et al. and Satoh et al. provide mechanistic insights into an assembly system of 20S proteasome, which is a huge protein complex consisting of homologous subunits $\alpha1-\alpha7$ and $\beta1-\beta7$ [8,9]. The correct assembly of proteasome subunits is essential for the function. Sekiguchi et al. comprehensively characterize the oligomeric states of the $\alpha1-\alpha7$ [8]. The results provide potential mechanisms on how the assembly and disassembly of proteasomal α subunits are controlled. Assembly of some subunits are assisted by chaperones. Satoh et al. have created a model of PAC3-PAC4 associated with $\alpha4-\alpha5-\alpha6$ subcomplex based on their biophysical and biochemical analyses, providing functional mechanisms of the PAC3-PAC4 heterodimer as a molecular matchmaker underpinning the $\alpha4-\alpha5-\alpha6$ subcomplex during α-ring formation [9]. Their findings open up new opportunities for the creation of artificial protein-assembling machine and also design of inhibitors of proteasome biogenesis.

Creation of artificial nucleic acids and applications are key trends. Mercurio et al. use a peptide nucleic acid (PNA), which is the neutral pseudo-peptide backbone, based on N-(2-aminoethyl) glycine units for the downregulation of miRNA function in the ascidian *Ciona intestinalis*. They have evaluated the expression level of miR-7 in a developing stage dependent manner and inhibitory effect of anti-miR-7, which will provide potential usage of PNA for basic research and therapeutics [10].

As shown by this special issue, self-assembly of biopolymers has a great impact on a variety of research fields including molecular biology, neurodegenerative diseases, drug delivery, gene therapy, regenerative medicine, and biomineralization. Designer biopolymers will help researchers to better understand biological processes as well as to create innovative molecular systems. We believe that this issue will provide readers with new ideas in their molecular design strategies for frontier research.

Funding: This research received no external funding.

Conflicts of Interest: The authors declare no conflict of interest.

References

1. Lee, S.; Trinh, T.H.; Yoo, M.; Shin, J.; Lee, H.; Kim, J.; Hwang, E.; Lim, Y.-B.; Ryou, C. Self-Assembling Peptides and Their Application in the Treatment of Diseases. *Int. J. Mol. Sci.* **2019**, *20*, 5850. [CrossRef] [PubMed]
2. Nesterenko, Y.; Hill, C.J.; Fleming, J.R.; Murray, P.; Mayans, O. The ZT Biopolymer: A Self-Assembling Protein Scaffold for Stem Cell Applications. *Int. J. Mol. Sci.* **2019**, *20*, 4299. [CrossRef] [PubMed]
3. Pereira, S.B.; Sousa, A.; Santos, M.; Araújo, M.; Serôdio, F.; Granja, P.; Tamagnini, P. Strategies to Obtain Designer Polymers Based on Cyanobacterial Extracellular Polymeric Substances (EPS). *Int. J. Mol. Sci.* **2019**, *20*, 5693. [CrossRef] [PubMed]
4. Petkov, P.; Lilkova, E.; Ilieva, N.; Litov, L. Self-Association of Antimicrobial Peptides: A Molecular Dynamics Simulation Study on Bombinin. *Int. J. Mol. Sci.* **2019**, *20*, 5450. [CrossRef] [PubMed]
5. Yokoyama, K.; Brown, K.; Shevlin, P.; Jenkins, J.; D'Ambrosio, E.; Ralbovsky, N.; Battaglia, J.; Deshmukh, I.; Ichiki, A. Examination of Adsorption Orientation of Amyloidogenic Peptides Over Nano-Gold Colloidal Particle Surfaces. *Int. J. Mol. Sci.* **2019**, *20*, 5354. [CrossRef] [PubMed]
6. Kojima, S.; Nakamura, H.; Lee, S.; Nagata, F.; Kato, K. Hydroxyapatite Formation on Self-Assembling Peptides with Differing Secondary Structures and Their Selective Adsorption for Proteins. *Int. J. Mol. Sci.* **2019**, *20*, 4650. [CrossRef] [PubMed]
7. Waku, T.; Nishigaki, S.; Kitagawa, Y.; Koeda, S.; Kawabata, K.; Kunugi, S.; Kobori, A.; Tanaka, N. Effect of the Hydrophilic-Hydrophobic Balance of Antigen-Loaded Peptide Nanofibers on Their Cellular Uptake, Cellular Toxicity, and Immune Stimulatory Properties. *Int. J. Mol. Sci.* **2019**, *20*, 3781. [CrossRef] [PubMed]
8. Sekiguchi, T.; Satoh, T.; Kurimoto, E.; Song, C.; Kozai, T.; Watanabe, H.; Ishii, K.; Yagi, H.; Yanaka, S.; Uchiyama, S.; et al. Mutational and Combinatorial Control of Self-Assembling and Disassembling of Human Proteasome α Subunits. *Int. J. Mol. Sci.* **2019**, *20*, 2308. [CrossRef] [PubMed]
9. Satoh, T.; Yagi-Utsumi, M.; Okamoto, K.; Kurimoto, E.; Tanaka, K.; Kato, K. Molecular and Structural Basis of the Proteasome α Subunit Assembly Mechanism Mediated by the Proteasome-Assembling Chaperone PAC3-PAC4 Heterodimer. *Int. J. Mol. Sci.* **2019**, *20*, 2231. [CrossRef] [PubMed]
10. Mercurio, S.; Cauteruccio, S.; Manenti, R.; Candiani, S.; Scarì, G.; Licandro, E.; Pennati, R. miR-7 Knockdown by Peptide Nucleic Acids in the Ascidian Ciona intestinalis. *Int. J. Mol. Sci.* **2019**, *20*, 5127. [CrossRef] [PubMed]

Review

Self-Assembling Peptides and Their Application in the Treatment of Diseases

Sungeun Lee [1], Trang H.T. Trinh [1], Miryeong Yoo [1], Junwu Shin [1], Hakmin Lee [1], Jaehyeon Kim [1], Euimin Hwang [2], Yong-beom Lim [2] and Chongsuk Ryou [1,*]

[1]   Department of Pharmacy and Institute of Pharmaceutical Science and Technology, Hanyang University, Gyeonggi-do 15588, Korea; guranye@hanyang.ac.kr (S.L.); ds.trinhthihuyentrang@gmail.com (T.H.T.T.); sho_ymr0623@naver.com (M.Y.); dugalle1@naver.com (J.S.); gkrals92@naver.com (H.L.); rlawoguses@naver.com (J.K.)

[2]   Department of Materials Science and Engineering, Yonsei University, Seoul 03722, Korea; euimin92@naver.com (E.H.); yblim@yonsei.ac.kr (Y.-b.L.)

*   Correspondence: cryou2@hanyang.ac.kr; Tel.: +82-31-400-5811; Fax: +82-31-400-5958

Received: 30 September 2019; Accepted: 20 November 2019; Published: 21 November 2019

Abstract: Self-assembling peptides are biomedical materials with unique structures that are formed in response to various environmental conditions. Governed by their physicochemical characteristics, the peptides can form a variety of structures with greater reactivity than conventional non-biological materials. The structural divergence of self-assembling peptides allows for various functional possibilities; when assembled, they can be used as scaffolds for cell and tissue regeneration, and vehicles for drug delivery, conferring controlled release, stability, and targeting, and avoiding side effects of drugs. These peptides can also be used as drugs themselves. In this review, we describe the basic structure and characteristics of self-assembling peptides and the various factors that affect the formation of peptide-based structures. We also summarize the applications of self-assembling peptides in the treatment of various diseases, including cancer. Furthermore, the in-cell self-assembly of peptides, termed reverse self-assembly, is discussed as a novel paradigm for self-assembling peptide-based nanovehicles and nanomedicines.

Keywords: peptide; self-assembly; nanostructure; drug delivery; disease

---

1. Introduction

The development of effective drug delivery systems and patient-customized therapies has recently emerged as a popular research topic. The ability to control the production of functional materials at the nanometer level is currently being explored for various medical applications. Nanomedicines, in the forms of nanospheres, nanoparticles, and other nanostructures modified with antibodies, peptides, glycans, and carbon, offer an alternative approach to classical drugs through their potential selectivity for diseased cells. The Food and Drug Administration (FDA) has approved abraxane, a nanomedicine for metastatic breast cancer, which encapsulates the anticancer drug paclitaxel within protein (albumin) nanoparticles [1,2]. Other anticancer drugs such as doxorubicin [3], 5-fluorouracil [4], 10-hydroxycamptothecin [5], and methotrexate [6] were also used to fabricate nanomedicines with albumin. Alternatively, gelatin was used to fabricate protein-based nanomedicines to increase drug loading efficiency and extend the duration of drug release [7]; unlike albumin, which can only encapsulate hydrophilic compounds, gelatin can be used to encapsulate both hydrophobic and hydrophilic drugs. Despite the success in the controlled release of drugs from protein-conjugated nanostructures, cellular targeting and cellular delivery of drugs have remained challenging.

Based on the success of conjugation of drugs with biocompatible proteins, researchers developed peptide-based nanomedicines using small peptides designed for the control of drug release and targeting. The small peptides are biocompatible and biodegradable. Furthermore, these peptides can be easily modified, thus inducing various self-assembled structures with different shapes, depending on the biochemical environment. The lower occurrence of side effects and stable drug release are also advantages of peptide-based self-assembled structures [2,8]. Peptide self-assembly is a process in which peptides spontaneously form ordered aggregates [9]. Hydrogen bonding, hydrophobic interactions, electrostatic interactions, and van der Waals forces combine to maintain the peptide-based self-assembled structures in a stable low-energy state [8]. In addition to the building blocks of self-assembling peptides, the research has also focused on self-assembled nanostructures with different shapes [10], including micelles, vesicles, and fibrillar structures such as nanotubes and fibers [11–14]. Based on the characteristics of the self-assembling peptides, the self-assembled structures can be used for intracellular or targeted tissue delivery of various nucleotides and antibodies for therapy, and for the delivery of drugs that cannot be easily mobilized owing to their physicochemical characteristics or those that exhibit a rapid clearance rate. In addition, nanostructures composed of self-assembling peptides can be applied to the treatment of various diseases as peptide drugs.

In this review, we have described the types of self-assembling peptides and their associated characteristics, and have discussed the principles of peptide self-assembly. Furthermore, we have examined the applications of self-assembling peptides in disease treatment. Finally, the in-cell self-assembly of peptides, termed "reverse engineering of peptide self-assembly," is explored as a new approach to deliver peptide-based nanostructure to cells.

2. Self-Assembling Peptides: Structure and Characteristics

Self-assembling peptides comprise monomers of short amino acid sequences or repeated amino acid sequences that assemble to form nanostructures. Peptide assemblies show distinctive physicochemical and biochemical activities, depending on their morphology, size, and accessibility of the reactive surface area. In most cases, morphological control is the initial step in the design of functional peptide assemblies. For amphiphile molecules, the concept of the molecular packing parameter offers a simple and intuitive insight into morphological control. The molecular packing parameter, P, is calculated as $P = V_0/al_0$, where V_0 is the volume, l_0 is the length of the hydrophobic tail, and a is the surface area per molecule [15]. The relationship between P and the shape of molecular assemblies is as follows: $P < \frac{1}{3}$ for spherical micelles, $\frac{1}{3} < P < \frac{1}{2}$ for cylindrical micelles, $\frac{1}{2} < P < 1$ for flexible bilayers or vesicles, $P \approx 1$ for planar bilayers, and $P > 1$ for inverted micelles. In short, the morphology transitions from more highly curved assemblies to less curved structures as the packing parameter increases. This feature can also be found in amyloid fibrils, which have been linked to various diseases in their natural state and produce more stable and functional structures through various amino acid combinations. The building blocks described below can be used in the design of nanostructures by considering the molecular and chemical properties of amino acids and peptides.

2.1. Building Blocks

The building blocks of self-assembled peptide structures can be categorized by their different constituent amino acids and the various bound chains or motifs. The characteristics of some peptide building blocks are summarized in Table 1.

Table 1. Peptide building blocks that self-assemble.

Peptide Building Blocks	Characteristics	References
Dipeptides	Simple phenylalanine dipeptides with or without N-terminal modifications, such as N-fluorenylmethoxycarbonyl (Fmoc) and naphthyl	[16–19]
Surfactant-like peptides	Amphiphilic structure with both hydrophilic and hydrophobic amino acids included in the peptide head and tail	[20–22]
	Repeated sequence of hydrophobic amino acids	
Peptide amphiphiles with an alkyl group	An alkyl tail linked to the N- or C-terminus	[23,24]
	A hydrophilic functional region	
	Form a stable β-sheet, providing hydrogen bonds for self-assembly	
	Glycine linker residues support flexibility	
Bolaamphiphilic peptides	Two hydrophilic heads connected by a hydrophobic region that is generally composed of alkyls	[25–30]
Ionic-complementary self-assembling peptides	A hydrophobic tail promotes self-assembly in water	[31–34]
	A hydrophilic tail with charged amino acids residues forms an ionic bond	
	Classified by the number of repeated ion charges: Type I has a charge pattern of "+-+-+-", Type II has "++-++-", Type III has "+++—+++", and Type IV has "++++—-".	
Cyclic peptides	Even number of alternating D and L amino acids stacked by hydrogen bonding	[35–39]
	Other types of cyclic peptides are characterized by amphiphilic characteristics, i.e., one side of the cycle is hydrophilic, whereas the other side contains hydrophobic and/or aggregation-prone amino acids	

2.1.1. Dipeptides

Dipeptides are the simplest building block in peptide nanotechnology. The diphenylalanine peptide (L-Phe-L-Phe; FF) is the core recognition motif of the Alzheimer's β-amyloid peptide [16]. Many studies have indicated that the peptide and its derivatives can self-assemble into highly ordered structures and other forms with nanoscale order [16,17,40]. These building blocks are used in the production of functional peptide nanotubes for casting molds of metal nanowires or electrochemical biosensing platforms [41–43]. Aromatic interactions are suggested to play a key role in the tubular structures. Other tubular structures are also produced by N-terminal modification of diphenylalanine to a non-charged FF analog, such as Boc–F–F–COOH, Z-F–F–COOH and Fmoc–F–F–COOH (Boc: tert-butoxycarbonyl; F: phenylalanine; Z: N-Carbobenzoxy; Fmoc: 9-fluorenylmethoxycarbonyl) [44]. β-Amino acids, which provide notable structural diversity through their extra C–C bond, are also used in dipeptide self-assembly. The derivatives of β-amino acids form hydrogels by self-assembly and exhibit prolonged bioavailability relative to α-amino acid derivatives [18,45].

2.1.2. Surfactant-Like Peptides

Surfactant-like peptides are characterized by their large reductive effect on the surface tension of water and their solubility in both organic solvents and water. Their solubility stems from the amphiphilic structure of the peptide, with several consecutive hydrophobic residues that constitute the hydrophobic tail, and one or two hydrophilic charged residues that serve as the head [20]. Often, surfactant-like peptides include a hydrophilic head group of negatively charged aspartic acid at the C-terminus, thus containing two negative charges, and a lipophilic tail made of hydrophobic amino acids such as alanine (A), valine (V), or leucine (L); the acetylated N-terminus has no charge. When dissolved in water, these surfactant-like peptides tend to self-assemble to shield the hydrophobic tail from contact with water. As with lipids and fatty acids, the supramolecular structure is characterized by the formation of a polar interface that sequesters the hydrophobic tail from water. Aspartic acid (D) and glutamic acid (E) have hydrophilic characteristics with a negative charge. Lysine (K), histidine (H), and arginine (R) also have hydrophilic characteristics, but they are positively charged. In contrast, glycine (G), alanine (A), valine (V), leucine (L), and isoleucine (I) are hydrophobic. By directional organization, Ac-AAAAAAD (A6D), Ac-VVVVVVD(V6D), and positively charged Ac-AAAAAAK (A6K), or any other design, can be used for surfactant-like peptide design. Zhao [20], Wang et al. [22], and Vauthey et al. [21] suggest that, through self-assembly, nanotubes or nanovesicles are the main structures formed by surfactant-like peptide assembly, and that they can function in a manner similar to lipid detergent micelles on the lipid bilayer of cells.

2.1.3. Peptide Amphiphiles with an Alkyl Group

Most self-assembling peptides have very simple structures: A hydrophobic tail with a hydrophilic head. In this group of peptides, the link with the hydrophobic alkyl chains is the most common modification in the peptide building blocks. When the alkyl chain combined with a peptide block is exposed to aquatic solutions, the hydrophobic tail of the peptide adopts a three-dimensional (3D) structure, similar to protein folding. Usually, the peptides form nanofibers, micelles, vesicles, nanotapes, or nantotubes. Hargerink et al. [23] developed a mineralized self-assembling peptide, including an alkyl tail and phosphorylated serine residues, to interact with calcium. A C16 alkyl tail with a VVVAAAEEE (V3A3E3) peptide was reported to form a gel under pressure or through electrostatic interaction with divalent cations, and this gel functioned as a scaffold for mesenchymal stem cell or three-dimensional culture [23,46].

2.1.4. Bolaamphiphilic Peptides

The difference between surfactant-like peptides and bolaamphiphiles is the number of hydrophilic heads of the building block. The surfactant-like peptide building block has only one hydrophilic head, whereas the bolaamphiphile has two hydrophilic heads connected by a hydrophobic section [47]. The double-headed design results in special properties and a highly complex assembly phenomenon in bolaamphiphilic molecules. Notably, bolaamphiphilic molecules can possess different head groups at either end of the hydrophobic chain; these are called asymmetric bolas [48]. For example, one end of the bolaamphiphilic molecule can be functionalized with amine groups to bind negatively charged nucleotides, and it can assemble to form vesicles through amphiphilic properties [48]. Bolaamphiphilic peptides are related to amyloid-like aggregation. For example, K and R in KAAAAK (KA4K), KAAAAAAK (KA6K), and RAAAAAAR (RA6R) bolaamphiphilic peptides have hydrophilic character and are connected by the hydrophobic A residues. Their assembled product has a fibrous form [26]. Another bolaamphiphilic peptide, EFLLLLFE (EFL4FE), which contains an E residue, shows a flat membrane extension and forms peptide nanotubes by concentration differences [28]. As the charge of amino acids is altered in different pH conditions, based on their molecular properties, these bolaamphiphilic peptides may aggregate or disaggregate according to the environmental pH [26].

Bolaamphiphiles are a category of emerging nanomaterials with the ability to self-assemble into various valuable nanostructures [49–51].

2.1.5. Ionic-Complementary Self-Assembling Peptides

The study of ionic-complementary peptides began from research on the Z-DNA binding protein, which includes the unusual 16-amino acid sequence of AEAEAKAKAEAEAKAK (EAK16). This peptide shows a unique pattern of charge distribution and forms membrane-like structures [52]. The ionic-complementary peptides are characterized by an alternating arrangement of negatively and positively charged residues. According to their charge distribution, these peptides can be classified to three types: Type I, +- block; Type II, ++− block; Type III, +++—; and Type IV, ++++— block; in these, the charged amino acid repeats work like "molecular Lego" to assemble the structure [31]. To design other peptide blocks, additional ionic-complementary peptides can be combined and modified. The charge distribution is a major force determining the peptide structure; for example, −++−++ shows α-helical periodicity and -+-+ has β-strand periodicity. RADA16 is another ionic-complementary self-assembling peptide. RADA 16-I (RADARADARADARADA) has the charge distribution pattern of +-+-+-+-, whereas RADA16-II (RARADADARARADADA) has the charge distribution pattern of ++−++−, but both form β-sheets after assembly. Many alternative compositions of ionic complementary self-assembling peptides show α, β, or random coil structures after assembly, but transition between α and β has also been reported [53]

2.1.6. Cyclic Peptides

Cyclic peptides are easily explained by the stacking of amino acids to form a cylindrical structure. There is an intermolecular hydrogen bond between each amino acid, forming a β-sheet-like tubular structure. By stacking, the amino acid side chains are located outside the cylinder and the peptide backbone is located on the inner side of the cylinder [54]. The external surface properties and the internal diameter can be controlled by the appropriate choice of amino acid side chains and the number of amino acids employed in the cyclic peptide [35,55]. Cyclic peptides have advantages over linear peptides, owing to their stable conformations and the conformational stability of the exposed surface [37]. A recent study from Jeong et al. [38] suggested the use of a hybrid cyclic peptide for a more stable α-helical structure. From the results of α-helical stabilization with carbon nanotubes, the covalent linker peptide connected to the side chains decreased the conformational entropy of the unfolded state, resulting in α-helix stabilization between the target molecule and self-assembling peptide.

2.2. Formation of Nanostructures

Self-assembled peptide nanostructures are formed by the designed building blocks. The nanostructure of self-assembling peptides can be classified into several types based on their constructed results, such as fibers, cylinders, or flat forms. Micelles are also classified as self-assembled nanostructures. These differences arise from the hydrophobic interaction of peptides in aqueous solutions and are dependent on the building block designs. Here, we have summarized the classified nanostructures of self-assembling peptides.

2.2.1. Nanofibers

The self-assembling peptide EAK16 sequences with periodically repeating positive and negative charges form a stable structure by ionic-complementary forces in a checkerboard-like pattern and then assembles typical β-sheet structures, eventually forming a hydrogel network of nanofibers [52]. Generally, nanofibers have a diameter of less than 100 nm. Aqueous solutions, including ions with different pH, are generally used to produce nanofibers from self-assembling peptide building blocks. In recent research, a light-induced self-assembling peptide was developed by modification of the amino acid sequences, and nanofibers were produced [56]. Peptide amphiphiles with an alkyl group are the most renowned self-assembling peptides that form nanofibers. The designed peptides may include a specific sequence for RGD binding, fluorescence, or any other small molecule in their tails [57,58]

2.2.2. Nanotubes

The structure of nanotubes is similar to that of the nanofibers mentioned above. However, they are elongated nanostructures with a hole on the inner side of the capillary. Recent research has focused on the development of non-covalent nanotubes, owing to their advantages in self-organization and easy control of the nanotube diameters. The most commonly used materials for nanotubes are cyclic peptides. Cyclic peptide nanotubes are formed by the stacking of peptides with high stability compared with other peptide building blocks. Drugs can be loaded inside these tubes and can be conjugated or bound to the outside of tubes; therefore, nanotube-based peptide assemblies have a wide range of applications in drug delivery [59]. The cyclic peptide, cyclo[-(L-Gln-D-Ala- L -Glu- D -Ala)$_2$-] with an even number of alternating D- and L-amino acids, forms a distinct structure of nanotubes [35]. Amphiphilic and surfactant-like peptides also form nanotubes with lipidic or surfactant characteristics by self-assembly [21,60]. For example, the peptide diphenylalanine, which is the core recognition motif of Alzheimer's β-amyloid polypeptide, with an uncharged peptide can successfully produce nanotubes [44]. NH$_2$–F–F–COOH is efficiently self-assembled into a tubular structure that was most likely to have an antiparallel β-sheet conformation, but acetylation to form Ac–F–F–COOH resulted in a structure that did not dissolve in either water or fluoroalcohols. These results indicate that non-charged peptide blocks were better for nanotube synthesis.

2.2.3. Nanoparticles

Nanoparticles are diverse and are formed by different building blocks. The structures range from nanospheres with a hollow core to various solid structures [61]. The charged amphiphilic block co-polypeptides [poly(L-lysine)-b-poly(L-leucine)] self-assemble to form stable vesicles and micelles in aqueous solutions [62]. Their hydrophobicity contributes to their rigidity and stability. Moreover, the guanidine residue of arginine increases the cell-penetrating actions to facilitate the delivery of encapsulated materials such as drugs. The temperature-responsive self-assembling peptide, elastin-like polypeptide (ELP), is a linear di-block peptide, but in response to a temperature change, it forms spherical micelles upon drug loading. The sensitivity of ELP to temperature can be controlled by increasing the number of ELP units. Cyclic peptides also produce vesicle-forming nanostructures. Shirazi et al. reported that the [WR]4 peptide successfully functioned as a drug delivery vehicle with molecular cargo, with a circular vesicle-like structure ranging from 25 to 60 nm in size [63].

2.2.4. Nanotapes

β-Alanine-histidine dipeptide and lysine–threonine–threonine–lysine–serine pentapeptide each conjugated to C16 palmitoyl hydrophobic lipid chains (C16-βAH and C16-KTTKS) form the stacks of β-sheets structures, resulting in nanotapes [64]. C16-βAH self-assemble into fibrils due to the hydrophobicity of the lipid tail. Self-assembly of C16-KTTKS [65] is controlled by pH or temperature; if the pH decreases to 4, the morphological transition from tape to fibrils occurs, but if the pH decreases further to 3, the nanotaper structure reforms. Lipopeptides of bacterial origin also form nanotapes, as described by Hamley et al. [66]. *Bacillus subtilis* produces a lipopeptide comprising a cyclic peptide head with different alkyl chains. This bio-originated molecule forms either micelles or nanotapes. The nanotape structures of self-assembling peptides often interact with each other and form double-layers. If the concentration of these nanotapes exceeds a certain threshold, they tend to form hydrogels

2.2.5. Hydrogels

A hydrogel is a polymer network that is cross-linked or entangled. The properties of hydrogels formed from self-assembling peptides depend on pH, ionic strength, and temperature [67]. Some hydrogels can absorb large amounts of water and they can be designed to possess distinct structural elements with adjustable mechanical properties, similar to natural tissues. Peptide-based hydrogels are highly biocompatible, biodegradable, and simple [68–70]. The simplest dipeptide building block modified with Fmoc, diphenylalanine (Fmoc-FF), was found to form hydrogels comprised of nanofibril networks in aqueous solutions [71]. Modification of these peptides to Fmoc-FRGD and Fmoc-RGDF showed that the inclusion of the RGD motif also produced a hydrogel structure, but that it was not stable above pH 6.5. Hydrogel formation is not limited to a simple block structure. Alkyl chain peptide blocks form β-sheets by hydrophobic collapse and can also form aqueous gels [23]. Stable β-sheet structures produced by self-assembling peptides form hydrogels when the peptide block concentration is increased.

3. Factors for Peptide Self-Assembly

3.1. pH

pH is an important factor in determining the peptide structure. pH fluctuation results in changes in hydrogen bonds and salt bridges, which influence peptide structure [72]. A change in pH affects the charge of the side chains through protonation and deprotonation. These changes in amino acids result in disruption of the hydrogen bonds among amino acid residues and broken salt bridges, the ionic bonds formed between the positively and the negatively charged side chains of amino acids.

The peptide -ETATKAELLAKYEATHK- motif includes negatively charged amino acids towards the N-terminus and positively charged amino acids towards the C-terminus, conferring an α-helical structure. At pH 4, this peptide exhibits the clearest α-helical structure. However, the α-helical structure of the peptide turns to a structure similar to β-sheet at pH 8 [73]. The secondary structure of this peptide can be changed by pH fluctuation. Another self-assembling peptide, cyclic α, α-disubstituted α-amino acid (dAA), is a cyclic acetal that changes to acyclic dAA at low pH. The peptide, including the dAA side chain, is stabilized as an α-helix structure. However, structural changes in dAA induced by low pH affects the changes in the secondary structure of the peptide from an α-helix to a random coil [74]. The hydrophobic peptide -YVIFL- also demonstrates pH-dependent structural changes. The peptide forms an amorphous aggregate at pH 2. Protonation of -YVIFL- below pH 2 reduces the electrostatic and hydrogen bonds among peptides. This results in the formation of aggregates with an antiparallel stacking structure. At pH 9 and 11, electrostatic and hydrogen bonds are replaced, while the aggregates still maintain an antiparallel stacking structure [75]. Similarly, the glutamic acid of the -FKFEFKFEFKFE- peptide becomes hydrophobic by protonation at low pH. Thus, this peptide can aggregate through hydrophobic interaction and the aggregated structure can be maintained in the inner space of endosomes or lysosomes. When fused to the oligo-arginine R_{12} peptides, this hydrophobic peptide segment recruits itself and induces the formation of nanovesicle structures at low pH, which demonstrate stronger anti-prion activity and lower cytotoxicity than oligo-arginine R_{12} without the -FKFEFKFEFKFE- peptide [76]. These examples suggest the potential of using pH-dependently altered peptides to optimize peptide targeting inside the cells.

3.2. Temperature

Peptides that self-assemble by temperature variation initially exist as monomers, but when heated, they change forms to nanofibrils, micelles, and other peptide structures, such as non-specific aggregate-like networks, depending on the calcium concentration and pH of buffer [77]. Examples of such temperature-dependent self-assembling peptides include FF peptide [78], ELP (repeating VPGXG) [79], and the lauryl-VAGERGD peptide. FF forms a characteristic crystal nanowire structure during the process of heating to 90 °C and cooling to 25 °C. Heating FF to 90 °C reduces the ionization constant of $-NH_3^+$, making it highly soluble. However, if the temperature is lowered, the ionization constant increases and enhances the hydrogen bonds. As a result, self-assembly occurs, forming a nanowire structure [80]. ELP exists in a monomer form at temperatures below the transition temperature, but monomers are changed to micelles by heat energy. The increased heat energy to ELP micelles results in aggregates that form gels. By heating the ELP molecule, hydrophobic activity is increased, which results in the formation of a micellar structure in a polar solvent [78]. The backbone of the lauryl-VVVAGERGD peptide is stable at approximately 300 to 358 K and disintegrates at higher temperatures. However, rapid heating to high temperatures can also reshape the peptide backbone. This enables the system to bypass energy barriers and reach more thermodynamically stable configurations [81]. This process results in the peptide backbone, forming a thermodynamically stable nanofiber structure without disintegration [82].

Lim and coworkers reported that guest-associated ELP peptides with different additional nonpolar amino acid appendages showed disparate assembly behavior (Figure 1). They demonstrated the changes in thermo-responsive phase transition between miniaturized elastin-like peptides (MELPs) with different numbers of phenylalanine residues or different nonpolar amino acid composition. After establishing the tunable temperature-responsive system with short peptides, they investigated whether these peptides could be used as a platform for the development of thermo-responsive ELP amphiphiles with various functionalities. They coupled the MELP platform to a short linear RGD peptide and grafted an α-helical guest peptide to the platform in a lariat-type structure. The peptide platform devised by this group provides a new insight into the development of stimuli-responsive materials with wide ranging applications, including temperature-responsive drug delivery and controlled modulation of protein–protein interactions [83].

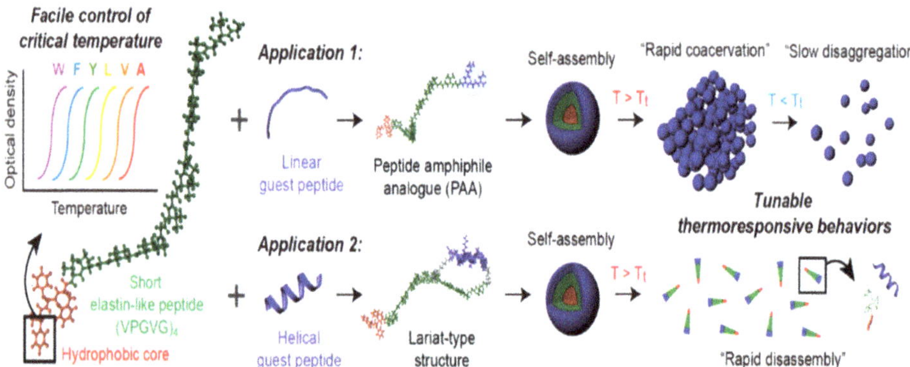

Figure 1. Peptide platform based on an miniaturized elastin-like peptide (MELP) for developing elastin-like peptide (ELP) amphiphiles with a thermo-responsive behavior that can be controlled by varying the types of conjugated guest–peptides, macromolecular topologies, and N-terminal amino acid residues. Reprinted with permission from [83].

3.3. Other Stimuli

Redox activity is another factor that mediates self-assembly of peptides. Phenylalanine derivatives conjugated to naphthalene diimide (NDI) form a nanofibril structure in high ionic strength aqueous solvents. This nanofibril changes to a non-fibril aggregate in a reductive environment; however, the nanofibrillar structure can reform in re-oxidation conditions [84]. Electrolytes are one of the factors that induce self-assembly in peptides. The peptide Ac–(AEAEAKAK)$_2$–CONH$_2$ is found in the yeast protein zuotin. This peptide has a β-sheet structure composed of hydrophilic and hydrophobic residues and exists in a monomer form. When an electrolyte is added, electrostatic repulsion between the peptides is reduced and hydrophobic interactions and hydrogen bonding increase, causing self-assembly and in-parallel alignment of the β-sheet structures [85].

Light can also be a factor regulating the behavior of peptide self-assembly. Lim and coworkers have developed an infrared (IR)-responsive self-assembling peptide–carbon nanotube hybrid material that enables spatiotemporal control of bioactive ligand multivalency and subsequent human neural stem cell (hNSC) differentiation (Figure 2). These hybrid materials exquisitely integrate the non-covalently assembled peptide ligands and the thermo-responsive dendrimers to remotely control multivalency. In this organic–inorganic hybrid, carbon nanotubes (CNTs) were used as a photothermal trigger. By IR-induced photothermal conversion of ligand density, these peptide–CNT hybrid materials organized the integrins into nanoscale clusters and subsequently induced functional neuronal differentiation of hNSCs [86].

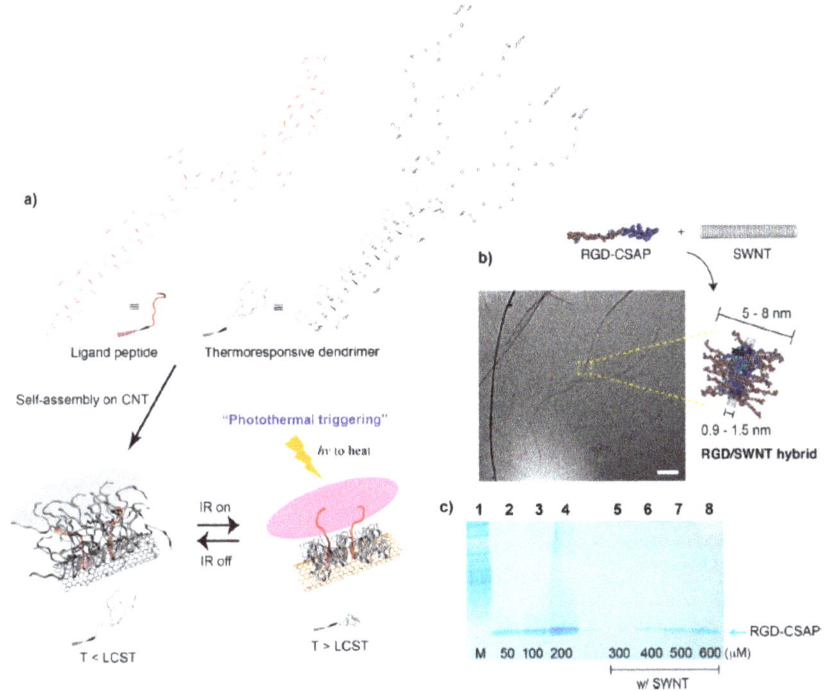

Figure 2. Photothermal control of multivalent ligand presentation. (**a**) Self-assembly of the ligand RGD peptide and thermoresponsive dendrimers on a carbon nanotube (CNT), followed by photothermal triggering of multivalent RGD ligands. (**b**) Transmission electron micrograph of an RGD/CNT hybrid. (**c**) SDS-PAGE analysis of the RGD complex that binds to CNT. Reprinted with permission from [86].

The enzymatic control of phosphorylation/dephosphorylation of the peptide can also regulate the formation of nanostructures. In the recent study by Shi and colleagues [87], enzymatic dephosphorylation of the amphiphilic peptide PP1 (VKVKVKVKV^{D}PPTK$_P$TEVKVKV), which contains a phosphorylated threonine, affected the folding of the peptide. As PP1 was dephosphorylated by phosphatase, the conformational equilibrium was shifted to the β-sheet conformation favoring the folded hairpin structure in the self-assembled fibril network. However, phosphorylated PP1 formed a gel with increased stiffness, resulting in broken fibril networks. This suggests that the use of enzymes can be a tool to control the self-assembly of peptides and affect the construction of nanostructure.

4. Application of Self-Assembling Peptide in Disease Treatment

The building blocks discussed above have been used to produce various types of structures based on material properties and environmental factors. Over the last few decades, the practical application of peptide-based self-assembled structures has been attempted widely. Table 2 summarizes nanostructures formed by particular self-assembling peptide blocks and their applications for specific biomedical uses.

Table 2. List of self-assembling peptide sequences and resultant nanostructures used for disease treatment.

Structure	Sequence	Applications	Reference
Nanofibers	VEVK9 (VEVKVEVKV) and VEVK12 (VEVKVEVKVEVK)/combined with RGD	Increase fibroblast migration	[88]
	V3A3E3 (VVVAAAEEE)	Stem cell culture and differentiation	[23,46]
Nanotubes	Heparin-binding peptide amphiphile (HBPA)	Hierarchical structure	[89,90]
	Q11 (QQKFQFQFEQQ)	Endothelial cell proliferation	[91]
Nano particle, vesicle, micelle, suspension	Lyp-1 (CGNKRTRGC)	Increase drug cellular uptake	[92]
	MAX8 (VKVKVKVKV^{D}PPTKVEVKVKV)	Drug delivery	[93]
	RADA16 with LRKKLGKA	Vascular endothelial growth factor (VEGF) delivery to the myocardium	[94]
	Tat/Tat combined with PEG/Cholesterol	Cross blood brain barrier (BBB)drug delivery	[95,96]
	cRGDfK	Drug targeting	[97]
	C16V2A2E2K(Hyd)	Drug stabilization	[98]
	V6K2(VVVVVVKK) combined with PLA	Drug delivery	[99]
	EAK16II (AEAEAKAKAEAEAKAK)	Drug stabilization	[100]
Hydrogel	RADA16I (RADARADARADARADA)	Controlled drug release	[101,102]
	RADA16I (RADARADARADARADA)	Hepatocyte regeneration	[103,104]
	RADA16 II (RARADADARARADADA)	Neuron regeneration	[105]
	RADA16-I combined with RGD motif	Neuron regeneration	[106,107]
	RADA16-I combined with RGD motif	Ligament regeneration	[108]
	KLD12 (KFDLKKDLKLDL)	Hepatocyte regeneration	[103]
	KLD12 (KFDLKKDLKLDL)	Chondrocyte regeneration	[69,109]
	KFE8 (FKFEFKFF)	Hepatocyte regeneration	[103]
	FEFEFKFK octarepeat	Extracellular matrix (ECM) accumulation	[110]

4.1. Application in Cancer Treatment

As a conventional cancer treatment, chemotherapy is fairly successful, but the off-target side effects causing damage to healthy cells and unavoidable development of multi-drug resistance are problematic [111,112]. The response of self-assembling peptides to environmental conditions may offer a means to prevent the aforementioned issues, because the tumor environment has a lower pH and higher temperatures than normal tissues. Thus, self-assembling peptides are well suited for controlled release or targeting of anticancer drugs to tumor sites [113].

4.1.1. Targeting

Chemotherapy in conventional cancer treatment is not only targeted to cancer cells, but also affects normal cells that are active in division and proliferation, such as bone marrow, hair, the mucous membrane of the gastrointestinal tract, and reproductive cells. To minimize the nonselective side effects, a specific peptide sequence or motif can be used in chemotherapy. Peptides designed as nanoparticles for targeting cancer cell surfaces or tumor vasculature in chemotherapy can be used to minimize systemic drug exposure and increase efficiency [114]. One of the cancer-targeting sequences, RGD, which binds to integrin, originates from a cell surface glycoprotein. RGD peptides can be linked to self-assembling peptides and increase the targeting effect of therapeutic drugs [115]. Furthermore, cyclic RGD increases binding affinity to integrins and is helpful in targeting drugs to cancer cells. Murphy et al. showed that cyclic RGDfK used to increase doxorubicin targeting suppressed growth of the primary tumor and prevented metastasis [97]. Another cancer-targeting peptide is the peptide Lyp-1, -CGNKRTRGC-, a nine-amino acid cyclic peptide that recognizes lymphatic metastatic tumors and exerts cytotoxic activity. LyP-1-conjugated PEG–PLGA nanoparticles (LyP-1-NPs) showed increased cellular uptake, by up to 4–8-fold in vitro and in vivo compared with PEG–PLGA nanoparticles without Lyp-1. LyP-1-NPs showed good targeting efficiency to cancer cells in vitro and to metastatic foci in vivo [92].

4.1.2. Drug Delivery

By modulating the properties of self-assembling peptides, drug release rates can be efficiently controlled. Ketone-containing drugs linked to peptide amphiphiles can be sustainably released at physiological pH. For example, doxorubicin or paclitaxel containing ketone functionality allows covalent tethering between the drug and peptide amphiphile via addition of a hydrazino acid to a lysine ε-amine [98]. In a study from the Stupp group, the peptide amphiphile C16V2A2E2K(Hyd) and its modifications were successfully bound to the ketone-containing fluorescent compound, Prodan. The release rate was dependent on the packing density, the order of the hydrophobic peptide amphiphilic core, and the β-sheet character of the peptide. It can also be controlled by chemical properties (Log P, pKa, pI, presence of aromatic rings, and steric hindrance) and by the solvent release [116]. RADA16 was also used in hydrophobic drug delivery and for self-assembling hydrogels. The diffusion properties of pindolol, quinine, and timolol maleate, through RADA16, showed sustained, controlled, reproducible, and efficient drug release [101,102]. RADA16-X controls the release of hydrophobic antitumor drugs and effectively inhibits the growth of a breast cancer cell line [7]. In this case, hydrogels with a higher peptide concentration have a longer release half-life and could block tumor cell proliferation more effectively.

As drug delivery carriers, self-assembling peptides offer many advantages, such as high efficiency of drug loading, a low ratio of drug loss, and high stability that avoids body clearance [117]. For example, EAK-16II can stabilize ellipticine, a hydrophobic anticancer drug, and form microcrystal suspensions in aqueous solutions. In particular, it is an ionic-complementary peptide that does not cause an immune response when applied to animals [100]. Curcumin, which has anti-inflammatory and antitumorigenic properties, has low water solubility and bioavailability, and is therefore difficult to use for therapeutic purposes. Self-assembling peptide hydrogels, such as the MAX8

(VKVKVKVKV^{D}PPTKVEVKVKV-NH2) peptide, can be an effective vehicle for the delivery of curcumin [93]. This peptide makes β-hairpin hydrogels for injectable agents to provide local curcumin delivery. The study with a medulloblastoma cell line confirmed that the encapsulation of curcumin with a hydrogel did not interfere with its drug activity. Peptide-based nanostructures made from polylactide (PLA) and VVVVVVKK (V6K2) are also used for the drug delivery of doxorubicin and paclitaxel [99]. The release of doxorubicin from PLA–V6K2 nanoparticles was slower than that from PLA–ethylene glycol nanoparticles. In addition, PLA–V6K2 nanoparticles showed a significantly increased cellular uptake rate with no induction of cytotoxicity in marrow stromal cells. However, it was more toxic to the 4T1 mouse breast carcinoma cell line than free doxorubicin. Moreover, PLA–V6K2 nanoparticles exhibited higher tumor toxicity and lower host toxicity in syngeneic breast cancer cells inoculated in mice, suggesting efficient drug delivery with selective toxicity.

4.2. Application in Regenerative Medicine

Regenerative medicine requires biocompatible scaffolds to increase cell engraftment and improve functionality, as well as to enhance cellular delivery processes. In recent years, engineering of nanostructures formed by self-assembling peptides can function as a scaffold in vivo.

4.2.1. Self-Assembling Peptides for Hepatocyte Regeneration

The liver is the only regenerative visceral organ in mammals. Self-assembled peptide scaffolds can assist cell growth with increased implantation rate, resulting in fully differentiated hepatocytes. RADA16-I peptide was used for the 3D culture of Lig-8 liver progenitor cells. The functional differentiation of hepatocytes by RADA16 scaffold was confirmed by the successful induction of CYP1A1 and CYP1A2 after hepatocyte cluster formation [103]. In addition, strongly hydrophobic peptide amphiphiles, such as KLD12 (AcN-KLDLKKDLKLDL-CNH2) or KFE8 (FKFEFKFE), resulted in greater stiffness of scaffold and increased cell growth, with a better differentiation rate compared with RADA16. From these results, the stiffness of self-assembled peptide scaffolds was shown to be one of the important factors for hepatocyte differentiation. An in vivo experiment with another hepatocellular carcinoma cell line, HepG2, also showed that RADA 16-I had a positive effect on cell growth and cluster formation [104]. RADA16 and other peptides combined with RADA16, such as RADA16-GRGDS, showed that myofibroblast replacement and hepatocyte cell proliferation were enhanced in vivo by RADA16 with an extended motif [118]. This result suggests that self-assembled peptide structures binding to other ECM motifs may be effective for liver tissue regeneration, although the enhancement of liver tissue regeneration requires various other forms of combined motifs.

4.2.2. Self-Assembling Peptides for Neuronal Regeneration

The regeneration or rehabilitation of neuronal tissue is difficult, owing to the characteristics of neuronal cells. Neuronal cells grow slowly, are difficult to differentiate, and have great networking complexity in vivo. Therefore, it is difficult to recover from ischemia or stroke damage in the brain, which manifest as neural function loss and cell death. Moreover, clinical conditions with chronic or idiopathic neuronal degeneration, such as Alzheimer's disease, Parkinson's disease, and prion disease, remain incurable [119].

Previous studies using RADA16-I and II as scaffolds in immature mouse cerebellum and rat hippocampus cells showed successful extensive neuron outgrowth. This growth was sustained for more than 4 weeks and resulted in the formation of an active synaptic connection [105]. The -IKVAV-penta-peptide combined with -Glu- and A4G3-alkyl residues is another example of using self-assembling peptides in neuronal cell regeneration, because the -IKVAV- peptide, a laminin epitope, was previously shown to promote neurite sprouting and growth. This peptide not only increased the fibril formation of self-assembled structures, but also improved cell growth, promoted neuronal cell differentiation, and inhibited astrocyte differentiation [120]. In addition, -SKPPGTSS, -PFSSTKT-, and RGD motifs combined with the RADA16-1 peptide increased the levels of nestin, β-tubulin, and other neuronal markers [106,107]. The self-assembling peptide nanofiber scaffold (SAPNS) composed of RADA16 resulted in reconnection of the injured spinal cord and facilitated axon regeneration and, eventually, locomotor functional recovery in animal models of spinal cord injury and acutely injured rat brain [121].

4.2.3. Self-Assembling Peptides for Cartilage Regeneration

Traumatic injury or degenerative articular cartilage defects require repair, with the deposition of ECM on the bone or previously existing cartilage. For successful cartilage regeneration, the newly assembled ECM must be combined with the remaining cartilage to achieve stable elastic restoration. The self-assembling peptide hydrogel scaffold is useful as a template for chondrocyte proliferation and ECM accumulation. In an experiment using the self-assembling KLD-12 peptides, chondrocytes showed increased proliferation and greater accumulation of cartilage-specific ECM molecules, such as proteoglycans, in the hydrogel scaffold [69,109], indicating that a highly porous hydrogel can be applied as a good scaffold for cartilage regeneration. In another report about alternating polar and non-polar amino acid residues, the -FEFEFKFK- octarepeat peptide formed β-sheet-rich nanofiber scaffolds and improved chondrocyte differentiation and ECM accumulation efficiency compared with previously studied RADA16 or KLD-12 [110].

4.2.4. Self-Assembling Peptides for Vascular Regeneration

Vascular regeneration aims to reshape blood vessels of various sizes and shapes, from microvascular to aortic [122]. Self-assembling peptides are also effective in vascular tissue regeneration and work as a scaffold to influence cell alignment, adhesion, and differentiation, and to promote better endothelization. In the study of Stupp et al. [89,90], peptide amphiphiles with a heparin-binding motif showed increased hierarchical structures and promoted rapid angiogenesis through heparin-binding growth factors involved in angiogenesis signaling. A Q11 (Ac-QQKFQFQFEQQ-Am) self-assembling peptide gel also promoted the proliferation of human umbilical vein endothelial cells and the expression of CD31 (PECAM), a vein endothelial cell marker, on the surface of these gels [91]. Both of these designed peptide structures helped the delivery and accumulation of angiogenic growth factors, such as VEGF and heparins, at the vessel regeneration site. RADA16 combined with the heparin binding domain motif sequence LRKKLGKA also increased VEGF delivery after myocardial infarction, and eventually improved cardiac function [94].

4.3. Other Applications

The BBB is the most challenging biological membrane encountered in drug delivery. Following the development of peptide science, researchers have tried to design self-assembling peptides to cross the BBB. The Tat (YGRKKRRQRRR) peptide found in a viral protein is known to penetrate the plasma membrane of cells, suggesting its function as a carrier to deliver drugs across the biological membrane. The Tat-polyethylene glycol-b-cholesterol (Tat-PEG-b-col) peptide, which forms micelles, allows antibiotics to migrate through the BBB [95]. Self-assembled polymersomes formed with lactoferrin-modified polyethylene glycol-poly ($_{D,L}$-lactic-co-glycolic acid)(PEG–PLGA) successfully deliver humanin, a neuroprotective peptide, across the BBB in a rat model of Alzheimer's disease [96].

Antimicrobial cationic peptides are used as peptide drugs to treat bacterial infection or prevent it. Antimicrobial peptides are amphipathic and destroy the cell membrane. For example, C16-KKK showed stronger antimicrobial activity than gentamicin in *E. coli* 25922 [123]. In another study, an antimicrobial cationic drug combined with RADA16 was shown to be released until 28 days after treatment [124]. Moreover, it also inhibited the growth of *Staphylococcus aureus*. Furthermore, cholesterol-conjugated Tat peptide (cholesterol–CG3R6TAT) formed nanoparticles that showed strong antimicrobial activity. It inhibited the growth of various types of Gram-positive bacteria, fungi, and yeast. These nanoparticles were able to cross the BBB and successfully treat *Staphylococcus aureus*-induced meningitis in rabbit [125].

The design of the self-assembling peptide scaffolds VEVK9 (Ac–VEVKVEVKV–CONH$_2$), VEVK12 (Ac–VEVKVEVKVEVK–CONH$_2$) and variants modified with RGD or a cell adhesion sequence resulted in a significant acceleration of fibroblast migration [88]. Periodontal ligament fibroblasts also significantly increased the production of Type I and Type II collagen upon culture with RADA16 combined with RGD and a laminin cell adhesion motif [108]. These results suggest that designed self-assembling peptides can regulate fibroblast cellular regeneration or reconstitution of the ECM. They can improve the skin or fibroblast tissue regeneration in scars or wound healing.

4.4. A New Paradigm of Nanostructure Formation with Reverse Self-Assembly

Based on the understanding of self-assembling peptides, a strong interest in optimizing the self-assembly of these peptides has emerged. In recent years, a new strategy by which self-assembling peptide monomers assemble inside a living cell has been proposed (Figure 3). The term "reverse self-assembly" has been coined to describe the new strategy, because this process is distinct from the conventional approach in which self-assembling peptides first form nanostructures in vitro, and then peptide nanostructure is provided to the cells. In reverse self-assembly, self-assembling peptides (or prodrugs) are first administered in a monomer form at the pre-structured states, specifically concentrated at target sites, and spontaneously convert to self-assembled nanodrugs. For instance, when the FF peptide conjugated to tri-phenyl-phosphonium, the mitochondria targeting motif, is provided to cells, self-assembled peptide structures are exclusively accumulated in mitochondria [126].

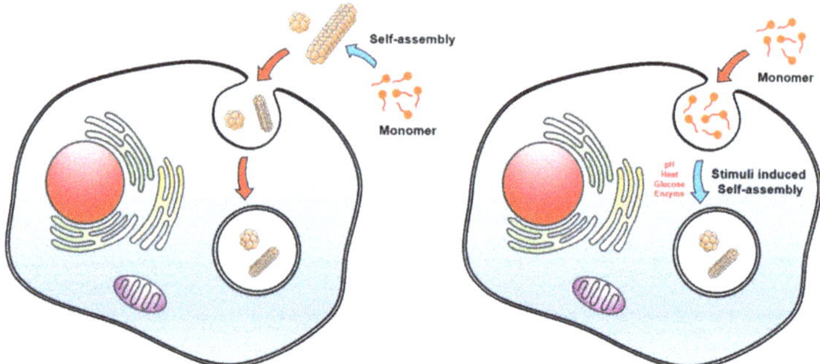

Figure 3. Concept of conventional forward (left) and novel reverse (right) self-assembly of peptides. Cellular environment and extrinsic stimuli such as endogenous changes of pH, glucose, enzyme activity, and heat drive in-cell self-assembly of peptides in the target sites.

Reverse self-assembly can be triggered by in situ stimuli, such as the induction of intracellular enzyme activity in a certain cell type or pH differences in a particular subcellular compartment. Enzyme-instructed self-assembly (EISA) is a useful strategy to accomplish intracellular reverse self-assembly. EISA is characterized by the self-assembly of monomers in cells or subcellular organelles, owing to endogenous enzymes targeted on-site. For instance, the hydrophilic precursor of hydrophobic FF peptide containing an esterase-cleavable bond provided to cancer cells is cleaved by intracellular esterase, resulting in formation of FF fibrils and gels within the cells, which induces cancer death [127]. Similarly, in-cell dephosphorylation of phosphorylated NDP1 (4-nitro-2,1,3-benzoxadiazole (NBD)-conjugated D-peptide) by intracellular alkaline phosphatase (AP) in some cancer cell lines, including HeLa cells, results in formation of non-diffusive nanofibrils [128]. Furthermore, self-assembled fibrils are formed in both pericellular space and endosomes, in which AP expresses abundantly. Compared with normal tissues, tumors exhibit a consequence of abnormal metabolism, such as increased temperature, decreased pH, induced expression of proteins and enzymes, and enhanced glycolysis. Because normal cells do not express AP as much as cancer cells do, reverse self-assembly of dephosphorylated NDP1 rarely occurs in normal cells, suggesting a feasibility of targeted drug delivery via in-cell reverse self-assembly. In addition, EISA is useful for controlled drug release in cells. Intracellular phosphatase activity facilitates in-cell formation of supramolecular hydrogels with Taxol-conjugated self-assembling peptide monomers, resulting in slow release of Taxol, a drug with low solubility, from the reverse assembled gels [129].

Another strategy of reverse self-assembly is the employment of pH responsiveness of self-assembling peptides in the acidic subcellular compartments. Waqas and colleagues demonstrated in-cell self-assembly of a -FKFEFKFEFKFE β-sheet-forming peptide motif combined with an oligo-arginine R_{12} peptide, which resulted in stronger anti-prion effects with improved cytotoxicity (Figure 4) [76]. In-cell self-assembled R_{12}-βUNS successfully decreased prions effectively to the level of anti-prion activity of the large oligo-arginine R_{60}, which was much improved from that of R_{12}-βSPN, the self-assembled nanostructure by "forward self-assembly". Similarly, reverse self-assembled R_{12}-βUNS improved cytotoxicity, exhibiting toxicity similar to that of the small oligo-arginine R_{12}, but lower than that of forward self-assembled R_{12}-βSPN. More interestingly, the R_{12}-FKFEFKFEFKFE peptide monomers (R_{12}-βUNS) provided to the cells formed nanostructures in the subcellular organelles (Figure 5). These result suggest that reverse-assembled peptide nanostructure can be used as an efficient therapeutic tool to circumvent the hurdles found in the application of forward-assembled peptide nanostructure in disease treatment.

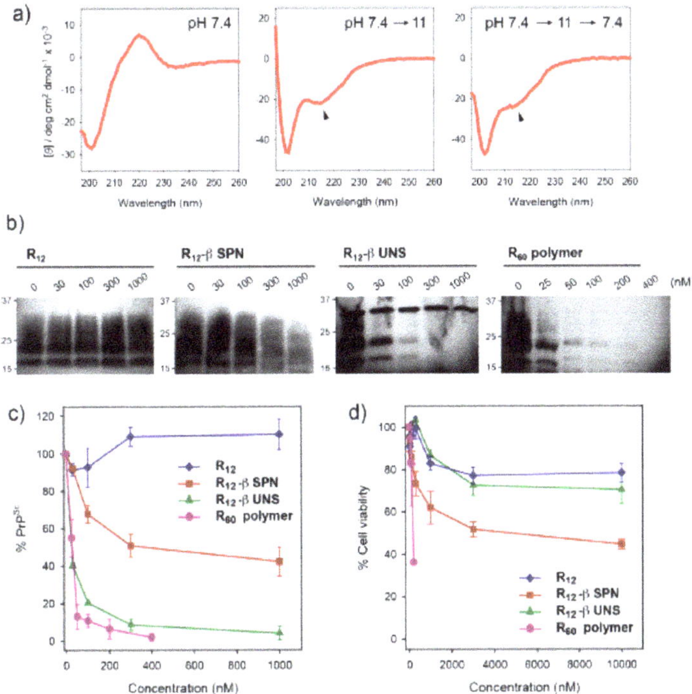

Figure 4. Improvement of anti-prion activity and cytotoxicity with reverse self-assembly of peptide blocks. (**a**) In vitro simulation of self-assembly of R_{12}-FKFEFKFEFKFE peptides as pH changes. (**b**) Western blots of prions of which level decreased by reverse self-assembly of R_{12}-FKFEFKFEFKFE peptides (R_{12}-βUNS). (**c**) Quantitative presentation of anti-prion activity of forward and reverse self-assembled R_{12}-FKFEFKFEFKFE peptides. (**d**) Cytotoxicity of forward and reverse self-assembled R_{12}-FKFEFKFEFKFE peptides. Reprinted with permission from [76].

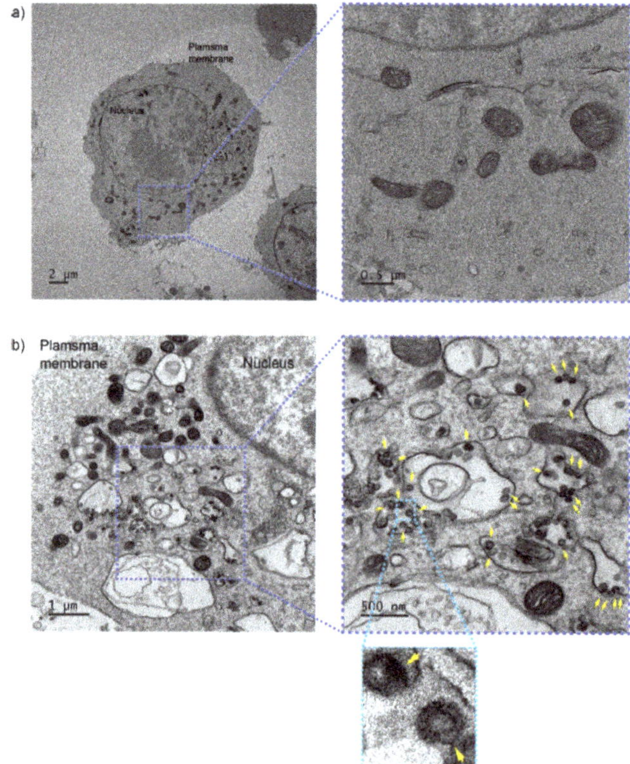

Figure 5. In-cell self-assembly of R_{12}–FKFEFKFEFKFE peptide (R_{12}-βUNS). Transmission electron micrographs of N2a neuroblastoma cells demonstrating reverse self-assembled nanostructure in the subcellular organelles. (**a**) Cells without incubation with the peptides. (**b**) Cells incubated with the peptides. Arrows indicate vesicle formed in the subcellular organelles. Reprinted with permission from [76].

5. Conclusions

In this review, we have summarized the characteristics, structures, and regulatory factors of self-assembled peptides and their self-assembled nanostructures. In addition, the application of self-assembling peptides to various diseases has been summarized in the context of cancer, regenerative medicine, and other diseases. Although self-assembling peptide drugs are rarely used clinically, drug delivery and motif design have been studied for several drugs. The biocompatibility and biodegradability of peptide materials are advantageous for targeted drug development and in vivo scaffold development. Thus, further developments in this field are expected in the future. Finally, the limitations of conventional forward self-assembled peptide structures associated with charge or size can be overcome through in-cell reverse self-assembly of peptides.

Author Contributions: S.L., T.H.T.T., M.Y., J.S., H.L., J.K., and E.H. wrote the first draft of the manuscript. S.L., Y.-b.L., and C.R. improved the final manuscript. C.R. organized the contents of manuscript.

Funding: This research was funded by the Korea Health Technology R&D Project grant number HI16C1085 through the Korea Health Industry Development Institute. And The APC was funded by Korea Health Technology R&D Project grant number HI16C1085.

Conflicts of Interest: Authors declare no conflicts of interest.

References

1. US Food and Drugs Administration. Drug Approval Package. Available online: https://www.accessdata.fda.gov/drugsatfda_docs/nda/2005/21660_AbraxaneTOC.cfm (accessed on 14 November 2019).
2. Cui, H.; Webber, M.J.; Stupp, S.I. Self-assembly of peptide amphiphiles: From molecules to nanostructures to biomaterials. *Biopolymers* **2010**, *94*, 1–18. [CrossRef] [PubMed]
3. Dreis, S.; Rothweiler, F.; Michaelis, M.; Cinatl, J., Jr.; Kreuter, J.; Langer, K. Preparation, characterisation and maintenance of drug efficacy of doxorubicin-loaded human serum albumin (HSA) nanoparticles. *Int. J. Pharm.* **2007**, *341*, 207–214. [CrossRef] [PubMed]
4. Maghsoudi, A.; Shojaosadati, S.A.; Farahani, E.V. 5-Fluorouracil-loaded BSA nanoparticles: Formulation optimization and in vitro release study. *AAPS PharmSciTech* **2008**, *9*, 1092–1096. [CrossRef] [PubMed]
5. Yang, L.; Cui, F.; Cun, D.; Tao, A.; Shi, K.; Lin, W. Preparation, characterization and biodistribution of the lactone form of 10-hydroxycamptothecin (HCPT)-loaded bovine serum albumin (BSA) nanoparticles. *Int. J. Pharm.* **2007**, *340*, 163–172. [CrossRef]
6. Burger, A.M.; Hartung, G.; Stehle, G.; Sinn, H.; Fiebig, H.H. Pre-clinical evaluation of a methotrexate-albumin conjugate (MTX-HSA) in human tumor xenografts in vivo. *Int. J. Cancer* **2001**, *92*, 718–724. [CrossRef]
7. Liu, J.; Zhang, L.; Yang, Z.; Zhao, X. Controlled release of paclitaxel from a self-assembling peptide hydrogel formed in situ and antitumor study in vitro. *Int. J. Nanomedicine* **2011**, *6*, 2143–2153. [CrossRef]
8. Edwards-Gayle, C.J.C.; Hamley, I.W. Self-assembly of bioactive peptides, peptide conjugates, and peptide mimetic materials. *Org. Biomol. Chem.* **2017**, *15*, 5867–5876. [CrossRef]
9. Whitesides, G.M.; Boncheva, M. Beyond molecules: Self-assembly of mesoscopic and macroscopic components. *Proc. Natl. Acad. Sci. USA* **2002**, *99*, 4769–4774. [CrossRef]
10. Ariga, K.; Hill, J.P.; Lee, M.V.; Vinu, A.; Charvet, R.; Acharya, S. Challenges and breakthroughs in recent research on self-assembly. *Sci. Tech. Adv. Mat.* **2008**, *9*, 014109. [CrossRef]
11. Stendahl, J.C.; Rao, M.S.; Guler, M.O.; Stupp, S.I. Intermolecular forces in the self-assembly of peptide amphiphile nanofibers. *Adv. Functi. Mat.* **2006**, *16*, 499–508. [CrossRef]
12. Dong, X.; Guo, X.; Liu, G.; Fan, A.; Wang, Z.; Zhao, Y. When self-assembly meets topology: An enhanced micelle stability. *Chem. Commun.* **2017**, *53*, 3822–3825. [CrossRef] [PubMed]
13. Sevink, G.J.A.; Zvelindovsky, A.V. Self-assembly of complex vesicles. *Macromolecules* **2005**, *38*, 7502–7513. [CrossRef]
14. Blau, W.J.; Fleming, A.J. Designer nanotubes by molecular self-assembly. *Science* **2004**, *304*, 1457. [CrossRef] [PubMed]
15. Israelachvili, J. Dynamic Biointeraction. In *Intermolecular and Surface Forces*, 3rd ed.; Israelachvili, J., Ed.; Elsevier Inc.: Waltham, MA, USA, 2011; pp. 617–633.
16. Reches, M.; Gazit, E. Casting metal nanowires within discrete self-assembled peptide nanotubes. *Science* **2003**, *300*, 625. [CrossRef] [PubMed]
17. Mahler, A.; Reches, M.; Rechter, M.; Cohen, S.; Gazit, E. Rigid, self-assembled hydrogel composed of a modified aromatic dipeptide. *Adv. Mat.* **2006**, *18*, 1365–1370. [CrossRef]
18. Yang, Z.; Liang, G.; Xu, B. Supramolecular hydrogels based on β-amino acid derivatives. *Chem. Commun.* **2006**, 738–740. [CrossRef]
19. Argudo, P.G.; Contreras-Montoya, R.; Álvarez de Cienfuegos, L.; Cuerva, J.M.; Cano, M.; Alba-Molina, D.; Martín-Romero, M.T.; Camacho, L.; Giner-Casares, J.J. Unravelling the 2D self-assembly of Fmoc-dipeptides at fluid interfaces. *Soft Matter* **2018**, *14*, 9343–9350. [CrossRef]
20. Zhao, X. Design of self-assembling surfactant-like peptides and their applications. *Curr. Opin. Colloid Interface Sci.* **2009**, *14*, 340–348. [CrossRef]
21. Vauthey, S.; Santoso, S.; Gong, H.; Watson, N.; Zhang, S. Molecular self-assembly of surfactant-like peptides to form nanotubes and nanovesicles. *Proc. Natl. Acad. Sci. USA* **2002**, *99*, 5355. [CrossRef]
22. Wang, J.; Han, S.; Meng, G.; Xu, H.; Xia, D.; Zhao, X.; Schweins, R.; Lu, J.R. Dynamic self-assembly of surfactant-like peptides A6K and A9K. *Soft Matter* **2009**, *5*, 3870–3878. [CrossRef]
23. Hartgerink, J.D.; Beniash, E.; Stupp, S.I. Self-assembly and mineralization of peptide-amphiphile nanofibers. *Science* **2001**, *294*, 1684. [CrossRef] [PubMed]
24. Hartgerink, J.D.; Beniash, E.; Stupp, S.I. Peptide-amphiphile nanofibers: A versatile scaffold for the preparation of self-assembling materials. *Proc. Natl. Acad. Sci. USA* **2002**, *99*, 5133. [CrossRef] [PubMed]

25. Claussen, R.C.; Rabatic, B.M.; Stupp, S.I. Aqueous self-assembly of unsymmetric peptide bolaamphiphiles into nanofibers with hydrophilic cores and surfaces. *J. Am. Chem. Soc.* **2003**, *125*, 12680–12681. [CrossRef] [PubMed]

26. Qiu, F.; Tang, C.; Chen, Y. Amyloid-like aggregation of designer bolaamphiphilic peptides: Effect of hydrophobic section and hydrophilic heads. *J. Peptide. Sci.* **2018**, *24*, e3062. [CrossRef] [PubMed]

27. Yao, L.; He, M.; Li, D.; Liu, H.; Wu, J.; Xiao, J. Self-assembling bolaamphiphile-like collagen mimetic peptides. *New J. Chem.* **2018**, *42*, 7439–7444. [CrossRef]

28. Da Silva, E.R.; Alves, W.A.; Castelletto, V.; Reza, M.; Ruokolainen, J.; Hussain, R.; Hamley, I.W. Self-assembly pathway of peptide nanotubes formed by a glutamatic acid-based bolaamphiphile. *Chem. Commun.* **2015**, *51*, 11634–11637.

29. Da Silva, E.R.; Walter, M.N.; Reza, M.; Castelletto, V.; Ruokolainen, J.; Connon, C.J.; Alves, W.A.; Hamley, I.W. Self-assembled arginine-capped peptide bolaamphiphile nanosheets for cell culture and controlled wettability surfaces. *Biomacromolecules* **2015**, *16*, 3180–3190. [CrossRef]

30. Dhasaiyan, P.; Prasad, B.L.V. Self-assembly of bolaamphiphilic molecules. *Chem. Rec.* **2017**, *17*, 597–610. [CrossRef]

31. Chen, P. Self-assembly of ionic-complementary peptides: A physicochemical viewpoint. *Colloids. Surf.* **2005**, *261*, 3–24. [CrossRef]

32. D'Auria, G.; Vacatello, M.; Falcigno, L.; Paduano, L.; Mangiapia, G.; Calvanese, L.; Gambaretto, R.; Dettin, M.; Paolillo, L. Self-assembling properties of ionic-complementary peptides. *J. Pept. Sci.* **2009**, *15*, 210–219. [CrossRef]

33. Zhang, H.; Park, J.; Jiang, Y.; Woodrow, K.A. Rational design of charged peptides that self-assemble into robust nanofibers as immune-functional scaffolds. *Acta. Biomater.* **2017**, *55*, 183–193. [CrossRef] [PubMed]

34. Kabiri, M.; Bushnak, I.; McDermot, M.T.; Unsworth, L.D. Toward a mechanistic understanding of ionic self-complementary peptide self-assembly: Role of water molecules and ions. *Biomacromolecules* **2013**, *14*, 3943–3950. [CrossRef] [PubMed]

35. Hartgerink, J.D.; Clark, T.D.; Ghadiri, M.R. Peptide nanotubes and beyond. *Chem. Eur. J.* **1998**, *4*, 1367–1372. [CrossRef]

36. Scott, C.P.; Abel-Santos, E.; Wall, M.; Wahnon, D.C.; Benkovic, S.J. Production of cyclic peptides and proteins in vivo. *Proc. Natl. Acad. Sci. USA* **1999**, *96*, 13638. [CrossRef] [PubMed]

37. Choi, S.J.; Jeong, W.J.; Kang, S.K.; Lee, M.; Kim, E.; Ryu du, Y.; Lim, Y.B. Differential self-assembly behaviors of cyclic and linear peptides. *Biomacromolecules* **2012**, *13*, 1991–1995. [CrossRef]

38. Jeong, W.J.; Choi, S.J.; Choi, J.S.; Lim, Y.B. Chameleon-like self-assembling peptides for adaptable biorecognition nanohybrids. *ACS Nano* **2013**, *7*, 6850–6857. [CrossRef] [PubMed]

39. Sun, L.; Fan, Z.; Wang, Y.; Huang, Y.; Schmidt, M.; Zhang, M. Tunable synthesis of self-assembled cyclic peptide nanotubes and nanoparticles. *Soft Matter* **2015**, *11*, 3822–3832. [CrossRef]

40. Reches, M.; Gazit, E. Formation of closed-cage nanostructures by self-assembly of aromatic dipeptides. *Nano Lett.* **2004**, *4*, 581–585. [CrossRef]

41. Song, Y.; Challa, S.R.; Medforth, C.J.; Qiu, Y.; Watt, R.K.; Pena, D.; Miller, J.E.; van Swol, F.; Shelnutt, J.A. Synthesis of peptide-nanotube platinum-nanoparticle composites. *Chem. Commun.* **2004**, *7*, 1044–1045. [CrossRef]

42. Yemini, M.; Reches, M.; Gazit, E.; Rishpon, J. Peptide nanotube-modified electrodes for enzyme-biosensor applications. *Anal. Chem.* **2005**, *77*, 5155–5159. [CrossRef]

43. Yemini, M.; Reches, M.; Rishpon, J.; Gazit, E. Novel electrochemical biosensing platform using self-assembled peptide nanotubes. *Nano Lett.* **2005**, *5*, 183–186. [CrossRef] [PubMed]

44. Reches, M.; Gazit, E. Self-assembly of peptide nanotubes and amyloid-like structures by charged-termini-capped diphenylalanine peptide analogues. *Isr. J. Chem.* **2005**, *45*, 363–371. [CrossRef]

45. Yang, Z.; Ho, P.-L.; Liang, G.; Chow, K.H.; Wang, Q.; Cao, Y.; Guo, Z.; Xu, B. Using β-lactamase to trigger supramolecular hydrogelation. *J. Am. Chem. Soc.* **2007**, *129*, 266–267. [CrossRef] [PubMed]

46. Zhang, S.; Greenfield, M.A.; Mata, A.; Palmer, L.C.; Bitton, R.; Mantei, J.R.; Aparicio, C.; de la Cruz, M.O.; Stupp, S.I. A self-assembly pathway to aligned monodomain gels. *Nat. Mater.* **2010**, *9*, 594–601. [CrossRef] [PubMed]

47. Fuhrhop, J.-H.; Wang, T. Bolaamphiphiles. *Chem. Rev.* **2004**, *104*, 2901–2938. [CrossRef]

48. Nuraje, N.; Bai, H.; Su, K. Bolaamphiphilic molecules: Assembly and applications. *Prog. Polym. Sci.* **2013**, *38*, 302–343. [CrossRef]
49. Sun, X.L.; Biswas, N.; Kai, T.; Dai, Z.; Dluhy, R.A.; Chaikof, E.L. Membrane-mimetic films of asymmetric phosphatidylcholine lipid bolaamphiphiles. *Langmuir* **2006**, *22*, 1201–1208. [CrossRef]
50. Maity, I.; Parmar, H.S.; Rasale, D.B.; Das, A.K. Self-programmed nanovesicle to nanofiber transformation of a dipeptide appended bolaamphiphile and its dose dependent cytotoxic behaviour. *J. Mat. Chem. B* **2014**, *2*, 5272–5279. [CrossRef]
51. Weissig, V.; Torchilin, V.P. Mitochondriotropic cationic vesicles: A strategy towards mitochondrial gene therapy. *Curr. Pharm. Biotechnol.* **2000**, *1*, 325–346. [CrossRef]
52. Zhang, S.; Holmes, T.; Lockshin, C.; Rich, A. Spontaneous assembly of a self-complementary oligopeptide to form a stable macroscopic membrane. *Proc. Natl. Acad. Sci. USA* **1993**, *90*, 3334–3338. [CrossRef]
53. Zhang, S.G.; Altman, M. Peptide self-assembly in functional polymer science and engineering. *React. Funct. Polym.* **1999**, *41*, 91–102.
54. Scanlon, S.; Aggeli, A. Self-assembling peptide nanotubes. *Nano Today* **2008**, *3*, 22–30. [CrossRef]
55. Chapman, R.; Danial, M.; Koh, M.L.; Jolliffe, K.A.; Perrier, S. Design and properties of functional nanotubes from the self-assembly of cyclic peptide templates. *Chem. Soc. Rev.* **2012**, *41*, 6023–6041. [CrossRef] [PubMed]
56. Muraoka, T.; Cui, H.; Stupp, S.I. Quadruple helix formation of a photoresponsive peptide amphiphile and its light-triggered dissociation into single fibers. *J. Am. Chem. Soc.* **2008**, *130*, 2946–2947. [CrossRef]
57. Bulut, S.; Erkal, T.S.; Toksoz, S.; Tekinay, A.B.; Tekinay, T.; Guler, M.O. Slow release and delivery of antisense oligonucleotide drug by self-assembled peptide amphiphile nanofibers. *Biomacromolecules* **2011**, *12*, 3007–3014. [CrossRef]
58. Ceylan, H.; Kocabey, S.; Tekinay, A.B.; Guler, M.O. Surface-adhesive and osteogenic self-assembled peptide nanofibers for bioinspired functionalization of titanium surfaces. *Soft Matter* **2012**, *8*, 3929–3937. [CrossRef]
59. Ghadiri, M.R.; Granja, J.R.; Milligan, R.A.; McRee, D.E.; Khazanovich, N. Self-assembling organic nanotubes based on a cyclic peptide architecture. *Nature* **1993**, *366*, 324–327. [CrossRef]
60. Zhao, X.; Pan, F.; Xu, H.; Yaseen, M.; Shan, H.; Hauser, C.A.; Zhang, S.; Lu, J.R. Molecular self-assembly and applications of designer peptide amphiphiles. *Chem. Soc. Rev.* **2010**, *39*, 3480–3498. [CrossRef]
61. Discher, D.E.; Eisenberg, A. Polymer vesicles. *Science* **2002**, *297*, 967. [CrossRef]
62. Bellomo, E.G.; Wyrsta, M.D.; Pakstis, L.; Pochan, D.J.; Deming, T.J. Stimuli-responsive polypeptide vesicles by conformation-specific assembly. *Nat. Mater.* **2004**, *3*, 244–248. [CrossRef]
63. Nasrolahi Shirazi, A.; Tiwari, R.K.; Oh, D.; Banerjee, A.; Yadav, A.; Parang, K. Efficient delivery of cell impermeable phosphopeptides by a cyclic peptide amphiphile containing tryptophan and arginine. *Mol. Pharm.* **2013**, *10*, 2008–2020. [CrossRef] [PubMed]
64. Aggeli, A.; Bell, M.; Boden, N.; Keen, J.N.; McLeish, T.C.B.; Nyrkova, I.; Radford, S.E.; Semenov, A. Engineering of peptide β-sheet nanotapes. *J. Mat. Chem.* **1997**, *7*, 1135–1145. [CrossRef]
65. Dehsorkhi, A.; Castelletto, V.; Hamley, I.W.; Adamcik, J.; Mezzenga, R. The effect of pH on the self-assembly of a collagen derived peptide amphiphile. *Soft Matter* **2013**, *9*, 6033–6036. [CrossRef]
66. Hamley, I.W.; Dehsorkhi, A.; Jauregi, P.; Seitsonen, J.; Ruokolainen, J.; Coutte, F.; Chataigne, G.; Jacques, P. Self-assembly of three bacterially-derived bioactive lipopeptides. *Soft Matter* **2013**, *9*, 9572–9578. [CrossRef]
67. Kopeček, J.; Yang, J. Peptide-directed self-assembly of hydrogels. *Acta Biomaterialia* **2009**, *5*, 805–816. [CrossRef]
68. Yu, Z.; Xu, Q.; Dong, C.; Lee, S.S.; Gao, L.; Li, Y.; D'Ortenzio, M.; Wu, J. Self-assembling peptide nanofibrous hydrogel as a versatile drug delivery platform. *Curr. Pharm. Des.* **2015**, *21*, 4342–4354. [CrossRef]
69. Kisiday, J.; Jin, M.; Kurz, B.; Hung, H.; Semino, C.; Zhang, S.; Grodzinsky, A.J. Self-assembling peptide hydrogel fosters chondrocyte extracellular matrix production and cell division: Implications for cartilage tissue repair. *Proc. Natl. Acad. Sci. USA* **2002**, *99*, 9996. [CrossRef]
70. Xing, R.; Li, S.; Zhang, N.; Shen, G.; Möhwald, H.; Yan, X. Self-assembled injectable peptide hydrogels capable of triggering antitumor immune response. *Biomacromolecules* **2017**, *18*, 3514–3523. [CrossRef]
71. Orbach, R.; Adler-Abramovich, L.; Zigerson, S.; Mironi-Harpaz, I.; Seliktar, D.; Gazit, E. Self-assembled fmoc-peptides as a platform for the formation of nanostructures and hydrogels. *Biomacromolecules* **2009**, *10*, 2646–2651. [CrossRef]
72. Khandogin, J.; Chen, J.; Brooks, C.L., 3rd. Exploring atomistic details of pH-dependent peptide folding. *Proc. Natl. Acad. Sci. USA* **2006**, *103*, 18546–18550. [CrossRef]

73. Cerpa, R.; Cohen, F.E.; Kuntz, I.D. Conformational switching in designed peptides: The helix/sheet transition. *Fold. Des.* **1996**, *1*, 91–101. [CrossRef]

74. Furukawa, K.; Oba, M.; Toyama, K.; Opiyo, G.O.; Demizu, Y.; Kurihara, M.; Doi, M.; Tanaka, M. Low pH-triggering changes in peptide secondary structures. *Org. Biomol. Chem.* **2017**, *15*, 6302–6305. [CrossRef] [PubMed]

75. Do, T.D.; LaPointe, N.E.; Economou, N.J.; Buratto, S.K.; Feinstein, S.C.; Shea, J.-E.; Bowers, M.T. Effects of pH and charge state on peptide assembly: The YVIFL model system. *J. Phys. Chem. B.* **2013**, *117*, 10759–10768. [CrossRef] [PubMed]

76. Waqas, M.; Jeong, W.-j.; Lee, Y.-J.; Kim, D.-H.; Ryou, C.; Lim, Y.-B. pH-dependent in-cell self-assembly of peptide inhibitors increases the anti-prion activity while decreasing the cytotoxicity. *Biomacromolecules* **2017**, *18*, 943–950. [CrossRef]

77. Ozkan, A.D.; Tekinay, A.B.; Guler, M.O.; Tekin, E.D. Effects of temperature, pH and counterions on the stability of peptide amphiphile nanofiber structures. *RSC Adv.* **2016**, *6*, 104201–104214. [CrossRef]

78. Huang, R.; Wang, Y.; Qi, W.; Su, R.; He, Z. Temperature-induced reversible self-assembly of diphenylalanine peptide and the structural transition from organogel to crystalline nanowires. *Nanoscale Res. Lett.* **2014**, *9*, 653. [CrossRef]

79. Weitzhandler, I.; Dzuricky, M.; Hoffmann, I.; Garcia Quiroz, F.; Gradzielski, M.; Chilkoti, A. Micellar self-assembly of recombinant resilin-/elastin-like block copolypeptides. *Biomacromolecules* **2017**, *18*, 2419–2426. [CrossRef]

80. Cao, M.; Shen, Y.; Wang, Y.; Wang, X.; Li, D. Self-assembly of short elastin-like amphiphilic peptides: Effects of temperature, molecular hydrophobicity and charge distribution. *Molecules* **2019**, *24*, 202. [CrossRef]

81. Hassouneh, W.; Fischer, K.; MacEwan, S.R.; Branscheid, R.; Fu, C.L.; Liu, R.; Schmidt, M.; Chilkoti, A. Unexpected multivalent display of proteins by temperature triggered self-assembly of elastin-like polypeptide block copolymers. *Biomacromolecules* **2012**, *13*, 1598–1605. [CrossRef]

82. Wang, J.; Liu, K.; Xing, R.; Yan, X. Peptide self-assembly: Thermodynamics and kinetics. *Chem. Soc. Rev.* **2016**, *45*, 5589–5604. [CrossRef]

83. Jeong, W.J.; Kwon, S.H.; Lim, Y.B. Modular self-assembling peptide platform with a tunable thermoresponsiveness via a single amino acid substitution. *Adv. Funct. Mat.* **2018**, *28*, 1803114. [CrossRef]

84. Liyanage, W.; Rubeo Paul, W.; Nilsson Bradley, L. Redox-sensitive reversible self-assembly of amino acid–naphthalene diimide conjugates. *Interface Focus* **2017**, *7*, 20160099. [CrossRef] [PubMed]

85. Piccoli, J.P.; Santos, A.; Santos-Filho, N.A.; Lorenzón, E.N.; Cilli, E.M.; Bueno, P.R. The self-assembly of redox active peptides: Synthesis and electrochemical capacitive behavior. *Pept. Sci.* **2016**, *106*, 357–367. [CrossRef] [PubMed]

86. Kim, H.W.; Yang, K.; Jeong, W.J.; Choi, S.J.; Lee, J.S.; Cho, A.N.; Chang, G.E.; Cheong, E.; Cho, S.W.; Lim, Y.B. Photoactivation of noncovalently assembled peptide ligands on carbon nanotubes enables the dynamic regulation of stem cell differentiation. *ACS Appl. Mater. Interfaces* **2016**, *8*, 26470–26481. [CrossRef] [PubMed]

87. Shi, J.; Fichman, G.; Schneider, J.P. Enzymatic control of the conformational landscape of self-assembling peptides. *Angew Chem. Int. Ed. Engl.* **2018**, *57*, 11188–11192. [CrossRef] [PubMed]

88. Kumada, Y.; Hammond, N.A.; Zhang, S.G. Functionalized scaffolds of shorter self-assembling peptides containing MMP-2 cleavable motif promote fibroblast proliferation and significantly accelerate 3-D cell migration independent of scaffold stiffness. *Soft Matter* **2010**, *6*, 5073–5079. [CrossRef]

89. Chow, L.W.; Bitton, R.; Webber, M.J.; Carvajal, D.; Shull, K.R.; Sharma, A.K.; Stupp, S.I. A bioactive self-assembled membrane to promote angiogenesis. *Biomaterials* **2011**, *32*, 1574–1582. [CrossRef]

90. Rajangam, K.; Behanna, H.A.; Hui, M.J.; Han, X.; Hulvat, J.F.; Lomasney, J.W.; Stupp, S.I. Heparin binding nanostructures to promote growth of blood vessels. *Nano Lett.* **2006**, *6*, 2086–2090. [CrossRef]

91. Jung, J.P.; Jones, J.L.; Cronier, S.A.; Collier, J.H. Modulating the mechanical properties of self-assembled peptide hydrogels via native chemical ligation. *Biomaterials* **2008**, *29*, 2143–2151. [CrossRef]

92. Luo, G.; Yu, X.; Jin, C.; Yang, F.; Fu, D.; Long, J.; Xu, J.; Zhan, C.; Lu, W. LyP-1-conjugated nanoparticles for targeting drug delivery to lymphatic metastatic tumors. *Int. J. Pharm.* **2010**, *385*, 150–156. [CrossRef]

93. Altunbas, A.; Lee, S.J.; Rajasekaran, S.A.; Schneider, J.P.; Pochan, D.J. Encapsulation of curcumin in self-assembling peptide hydrogels as injectable drug delivery vehicles. *Biomaterials* **2011**, *32*, 5906–5914. [CrossRef] [PubMed]

94. Guo, H.D.; Cui, G.H.; Yang, J.J.; Wang, C.; Zhu, J.; Zhang, L.S.; Jiang, J.; Shao, S.J. Sustained delivery of VEGF from designer self-assembling peptides improves cardiac function after myocardial infarction. *Biochem. Biophys. Res. Commun.* **2012**, *424*, 105–111. [CrossRef] [PubMed]

95. Liu, L.; Venkatraman, S.S.; Yang, Y.Y.; Guo, K.; Lu, J.; He, B.; Moochhala, S.; Kan, L. Polymeric micelles anchored with TAT for delivery of antibiotics across the blood-brain barrier. *Biopolymers* **2008**, *90*, 617–623. [CrossRef] [PubMed]

96. Yu, Y.; Jiang, X.; Gong, S.; Feng, L.; Zhong, Y.; Pang, Z. The proton permeability of self-assembled polymersomes and their neuroprotection by enhancing a neuroprotective peptide across the blood-brain barrier after modification with lactoferrin. *Nanoscale* **2014**, *6*, 3250–3258. [CrossRef]

97. Murphy, E.A.; Majeti, B.K.; Barnes, L.A.; Makale, M.; Weis, S.M.; Lutu-Fuga, K.; Wrasidlo, W.; Cheresh, D.A. Nanoparticle-mediated drug delivery to tumor vasculature suppresses metastasis. *Proc. Natl. Acad. Sci. USA* **2008**, *105*, 9343–9348. [CrossRef]

98. Matson, J.B.; Stupp, S.I. Drug release from hydrazone-containing peptide amphiphiles. *Chem. Commun.* **2011**, *47*, 7962–7964. [CrossRef]

99. Jabbari, E.; Yang, X.; Moeinzadeh, S.; He, X. Drug release kinetics, cell uptake, and tumor toxicity of hybrid VVVVVVKK peptide-assembled polylactide nanoparticles. *Eur. J. Pharm. Biopharm.* **2013**, *84*, 49–62. [CrossRef]

100. Fung, S.Y.; Yang, H.; Bhola, P.T.; Sadatmousavi, P.; Muzar, E.; Liu, M.; Chen, P. Self-assembling peptide as a potential carrier for dydrophobic anticancer drug ellipticine: Complexation, release and in vitro delivery. *Adv. Funct. Mater.* **2009**, *19*, 74–83. [CrossRef]

101. Briuglia, M.L.; Urquhart, A.J.; Lamprou, D.A. Sustained and controlled release of lipophilic drugs from a self-assembling amphiphilic peptide hydrogel. *Int. J. Pharm.* **2014**, *474*, 103–111. [CrossRef]

102. Nagai, Y.; Unsworth, L.D.; Koutsopoulos, S.; Zhang, S. Slow release of molecules in self-assembling peptide nanofiber scaffold. *J. Control. Release* **2006**, *115*, 18–25. [CrossRef]

103. Semino, C.E.; Merok, J.R.; Crane, G.G.; Panagiotakos, G.; Zhang, S. Functional differentiation of hepatocyte-like spheroid structures from putative liver progenitor cells in three-dimensional peptide scaffolds. *Differentiation* **2003**, *71*, 262–270. [CrossRef] [PubMed]

104. Wu, M.; Yang, Z.; Liu, Y.; Liu, B.; Zhao, X. The 3-D culture and in vivo growth of the human hepatocellular carcinoma cell line HepG2 in a self-assembling peptide nanofiber scaffold. *J. Nanomater.* **2010**, *2010*, 1–7. [CrossRef]

105. Holmes, T.C.; de Lacalle, S.; Su, X.; Liu, G.; Rich, A.; Zhang, S. Extensive neurite outgrowth and active synapse formation on self-assembling peptide scaffolds. *Proc. Natl. Acad. Sci. USA* **2000**, *97*, 6728–6733. [CrossRef] [PubMed]

106. Gelain, F.; Bottai, D.; Vescovi, A.; Zhang, S. Designer self-assembling peptide nanofiber scaffolds for adult mouse neural stem cell 3-dimensional cultures. *PLoS ONE* **2006**, *1*, e119. [CrossRef]

107. Cunha, C.; Panseri, S.; Villa, O.; Silva, D.; Gelain, F. 3D culture of adult mouse neural stem cells within functionalized self-assembling peptide scaffolds. *Int. J. Nanomed.* **2011**, *6*, 943–955. [CrossRef]

108. Kumada, Y.; Zhang, S. Significant type I and type III collagen production from human periodontal ligament fibroblasts in 3D peptide scaffolds without extra growth factors. *PLoS ONE* **2010**, *5*, e10305. [CrossRef]

109. Kisiday, J.D.; Jin, M.; DiMicco, M.A.; Kurz, B.; Grodzinsky, A.J. Effects of dynamic compressive loading on chondrocyte biosynthesis in self-assembling peptide scaffolds. *J. Biomech.* **2004**, *37*, 595–604. [CrossRef]

110. Mujeeb, A.; Miller, A.F.; Saiani, A.; Gough, J.E. Self-assembled octapeptide scaffolds for in vitro chondrocyte culture. *Acta Biomater.* **2013**, *9*, 4609–4617. [CrossRef]

111. Senapati, S.; Mahanta, A.K.; Kumar, S.; Maiti, P. Controlled drug delivery vehicles for cancer treatment and their performance. *Signal Transduct. Target Ther.* **2018**, *3*, 7. [CrossRef]

112. Zhang, X.X.; Eden, H.S.; Chen, X. Peptides in cancer nanomedicine: Drug carriers, targeting ligands and protease substrates. *J. Control. Release* **2012**, *159*, 2–13. [CrossRef]

113. Torchilin, V. Multifunctional and stimuli-sensitive pharmaceutical nanocarriers. *Eur. J. Pharm. Biopharm.* **2009**, *71*, 431–444. [CrossRef] [PubMed]

114. Zhao, F.; Ma, M.L.; Xu, B. Molecular hydrogels of therapeutic agents. *Chem. Soc. Rev.* **2009**, *38*, 883–891. [CrossRef] [PubMed]

115. Sadatmousavi, P.; Soltani, M.; Nazarian, R.; Jafari, M.; Chen, P. Self-assembling peptides: Potential role in tumor targeting. *Curr. Pharm. Biotechnol.* **2011**, *12*, 1089–1100. [CrossRef] [PubMed]

116. Matson, J.B.; Newcomb, C.J.; Bitton, R.; Stupp, S.I. Nanostructure-templated control of drug release from peptide amphiphile nanofiber gels. *Soft Matter* **2012**, *8*, 3586–3595. [CrossRef] [PubMed]

117. Fan, T.; Yu, X.; Shen, B.; Sun, L. Peptide self-assembled nanostructures for drug delivery applications. *J. Nanomater.* **2017**, *2017*, 1–16. [CrossRef]

118. Cheng, T.Y.; Wu, H.C.; Huang, M.Y.; Chang, W.H.; Lee, C.H.; Wang, T.W. Self-assembling functionalized nanopeptides for immediate hemostasis and accelerative liver tissue regeneration. *Nanoscale* **2013**, *5*, 2734–2744. [CrossRef] [PubMed]

119. Ryou, C. Prions and prion diseases: Fundamentals and mechanistic details. *J. Microbiol. Biotechnol.* **2007**, *17*, 1059–1070.

120. Silva, G.A.; Czeisler, C.; Niece, K.L.; Beniash, E.; Harrington, D.A.; Kessler, J.A.; Stupp, S.I. Selective differentiation of neural progenitor cells by high-epitope density nanofibers. *Science* **2004**, *303*, 1352–1355. [CrossRef]

121. Guo, J.; Leung, K.K.; Su, H.; Yuan, Q.; Wang, L.; Chu, T.H.; Zhang, W.; Pu, J.K.; Ng, G.K.; Wong, W.M.; et al. Self-assembling peptide nanofiber scaffold promotes the reconstruction of acutely injured brain. *Nanomedicine* **2009**, *5*, 345–351. [CrossRef]

122. Chung, E.; Ricles, L.M.; Stowers, R.S.; Nam, S.Y.; Emelianov, S.Y.; Suggs, L.J. Multifunctional nanoscale strategies for enhancing and monitoring blood vessel regeneration. *Nano. Today* **2012**, *7*, 514–531. [CrossRef]

123. Makovitzki, A.; Baram, J.; Shai, Y. Antimicrobial lipopolypeptides composed of palmitoyl Di- and tricationic peptides: In vitro and in vivo activities, self-assembly to nanostructures, and a plausible mode of action. *Biochemistry* **2008**, *47*, 10630–10636. [CrossRef] [PubMed]

124. Yang, G.; Huang, T.; Wang, Y.; Wang, H.; Li, Y.; Yu, K.; Dong, L. Sustained release of antimicrobial peptide from self-assembling hydrogel enhanced osteogenesis. *J. Biomater. Sci. Polym. Ed.* **2018**, *29*, 1812–1824. [CrossRef] [PubMed]

125. Liu, L.; Xu, K.; Wang, H.; Tan, P.K.; Fan, W.; Venkatraman, S.S.; Li, L.; Yang, Y.Y. Self-assembled cationic peptide nanoparticles as an efficient antimicrobial agent. *Nat. Nanotechnol.* **2009**, *4*, 457–463. [CrossRef] [PubMed]

126. Jeena, M.T.; Palanikumar, L.; Go, E.M.; Kim, I.; Kang, M.G.; Lee, S.; Park, S.; Choi, H.; Kim, C.; Jin, S.-M.; et al. Mitochondria localization induced self-assembly of peptide amphiphiles for cellular dysfunction. *Nat. Commun.* **2017**, *8*, 26. [CrossRef]

127. Yang, Z.M.; Xu, K.M.; Guo, Z.F.; Guo, Z.H.; Xu, B. Intracellular enzymatic formation of nanofibers results in hydrogelation and regulated cell death. *Adv. Mat.* **2007**, *19*, 3152–3156. [CrossRef]

128. Zhou, J.; Du, X.; Berciu, C.; He, H.; Shi, J.; Nicastro, D.; Xu, B. Enzyme-instructed self-assembly for spatiotemporal profiling of the activities of alkaline phosphatases on live cells. *Chem* **2016**, *1*, 246–263. [CrossRef] [PubMed]

129. Gao, Y.; Kuang, Y.; Guo, Z.-F.; Guo, Z.; Krauss, I.J.; Xu, B. Enzyme-instructed molecular self-assembly confers nanofibers and a supramolecular hydrogel of taxol derivative. *J. Am. Chem. Soc.* **2009**, *131*, 13576–13577. [CrossRef]

International Journal of
Molecular Sciences

Review

The ZT Biopolymer: A Self-Assembling Protein Scaffold for Stem Cell Applications

Yevheniia Nesterenko [1,†], Christopher J. Hill [2,†], Jennifer R. Fleming [1,†], Patricia Murray [2] and Olga Mayans [1,*]

1   Department of Biology, University of Konstanz, 78457 Konstanz, Germany
2   Department of Cellular and Molecular Physiology, Institute of Translational Medicine, University of Liverpool, Liverpool L69 3BX, UK
*   Correspondence: olga.mayans@uni-konstanz.de; Tel.: +49-7531-8822-12
†   These authors contributed equally to this work.

Received: 29 July 2019; Accepted: 30 August 2019; Published: 3 September 2019

Abstract: The development of cell culture systems for the naturalistic propagation, self-renewal and differentiation of cells ex vivo is a high goal of molecular engineering. Despite significant success in recent years, the high cost of up-scaling cultures, the need for xeno-free culture conditions, and the degree of mimicry of the natural extracellular matrix attainable in vitro using designer substrates continue to pose obstacles to the translation of cell-based technologies. In this regard, the ZT biopolymer is a protein-based, stable, scalable, and economical cell substrate of high promise. ZT is based on the naturally occurring assembly of two human proteins: titin-Z1Z2 and telethonin. These protein building blocks are robust scaffolds that can be conveniently functionalized with full-length proteins and bioactive peptidic motifs by genetic manipulation, prior to self-assembly. The polymer is, thereby, fully encodable. Functionalized versions of the ZT polymer have been shown to successfully sustain the long-term culturing of human embryonic stem cells (hESCs), human induced pluripotent stem cells (hiPSCs), and murine mesenchymal stromal cells (mMSCs). Pluripotency of hESCs and hiPSCs was retained for the longest period assayed (4 months). Results point to the large potential of the ZT system for the creation of a modular, pluri-functional biomaterial for cell-based applications.

Keywords: protein design; protein self-assembly; polymer functionalization; biomimetic material; cell culture substrate; stem cells

---

1. Introduction

The study of fundamental processes in cell biology, as well as biomedical and technological cell-based applications, require the culturing of cells ex vivo. For this purpose, culture systems that mimic the native extracellular matrix (ECM) microenvironment by providing biochemical (cell adhesion sites, growth factors) and biophysical (mechanical stiffness) cues are in high demand. Cell culture substrates exist that are derived from natural sources as well as produced synthetically. Prominent examples of naturally derived substrates are Matrigel™ and Geltrex™. The well-studied Matrigel™ is a heterogeneous glycoprotein mixture secreted by mouse sarcoma cells whose composition is comparable to that of an embryonic basement membrane, containing a variety of ECM proteins (particularly laminin-111, collagen IV, entactin, and heparan sulfate proteoglycan) and active growth factors (e.g., fibroblast growth factor, epidermal growth factor, transforming growth factor-β, insulin-like growth factor and platelet-derived growth factor) [1,2]. Although Matrigel™ is widely used for in vitro and in vivo applications [3], it is xenogenic and has a poorly defined, complex composition that suffers from batch-to-batch variability and offers limited experimental control. Simpler functional substrates of better defined composition are constituted by isolated ECM proteins. To this effect, full-length human fibronectin and vitronectin can either be extracted from plasma as native proteins or recombinantly produced (R&D Systems), while laminin isoforms can be expressed recombinantly in human cell lines (BioLamina). However, the production yield of these proteins is limited and their use can be troubled by the presence of impurities, uncontrollable degradation, and, in in vivo applications, possible immunogenicity [4]. Fragments from ECM proteins can also be bioactive, recapitulating the activity of the full-length protein with reasonable efficiency, to the point of supporting the long-term culturing of demanding cell types such as human pluripotent stem cells (hPSCs). Examples of bioactive ECM-protein fragments are: Laminin-511 (a combination of α5, β1 and γ1 laminin chains) produced recombinantly in human embryonic kidney cells [5]; the N-terminal somatomedin B domain of vitronectin expressed in *Escherichia coli* [6]; and the Fn7-Fn14 fragment from fibronectin expressed in mouse myeloma cells [7]. Fragments have the advantage of easing recombinant production and yielding substrates of improved purity. However, they are incomplete mimics of the ECM and proper fragment choice is critical to achieve suitable performance [7].

There is considerable interest in overcoming the limitations of cell substrates based on ECM-components through the engineering of materials that are xeno-free, feeder-free, of controlled composition, and with tailored biological functionalities. To this effect, synthetic substrates have been developed, largely in the form of peptide conjugates [8,9]. A leading example of a synthetic cell substrate is Synthemax™. In this, an acrylate base carries carboxylic acid groups to which short linear peptides (instead of folded proteins) are conjugated using chemical linkers [10]. Synthemax™ fulfills the requirements listed above and successfully supports cell proliferation. However, its usage for long-term cell culturing is not widely spread as some concerns exist related to a possible higher propensity for spontaneous cell differentiation [11] and the potential induction of karyotypic abnormalities [12].

Fulfilling the requirement of improved biodegradability and biocompatibility, materials based on self-assembling peptides have also been developed [13–17]. Peptides self-assemble to form hydrogels and can be designed to be responsive to different physical parameters, like pH and temperature. In addition, peptides produced by chemical synthesis or recombinant technologies offer optimized homogeneity as required to standardize cell-based applications. The molecular design of self-assembling peptides is usually inspired on naturally occurring, self-assembling fibrous proteins, such as collagens, elastin, silk, keratins, amyloids, and coiled-coils. Self-assembling peptides carrying bioactive sequence motifs have yielded excellent achievements in cell-based applications and are now well established [9]. Despite their demonstrated potential, the functionalization limits of peptidic systems in regards to bulky, three-dimensional bioactive components are still uncertain, as the peptidic building block is small and largely invested in mediating the polymeric assembly of the scaffold.

Compared to existing substrates, biomaterials based on self-assembling, full-length proteins can offer economy of production via recombinant methodologies, controlled composition, scalable yields, high purity with low batch variability, biodegradability, full encodability, and ease of "bottom-up" functionalization through genetic engineering—thereby bridging the advantages of synthetic designer peptides and those of natural systems based on ECM-components. Importantly, protein-based materials are ideally suited to exploit modularity by combining multiple protein domains with different functions into a single, multi-block polymer. The prospect of independent tunability of individual domain functions holds high promise for achieving fine control over multiple material properties, which is expected to lead to the development of complex multifunctional matrices of clinical significance [18–20]. Yet, designing controlled self-assembly in full-length globular proteins is highly challenging and, currently, designer protein polymers are mostly based on small and well-characterized protein subunits, where a rational molecular design is feasible. A notable example is the consensus tetratricopeptide repeat protein (CTPR), a de novo designed protein system composed of small, independently folded super-secondary helical motifs that can form a range of supramolecular assemblies, including nanofibers, nanotubes, films, and ordered monolayers [21]. Interestingly, strategies for the fusion of symmetric globular proteins and the re-design of proteins that naturally assemble into nano-objects (e.g., viral capsids) are also proving successful, with a variety of assemblies in the form of filaments, molecular layers and 3D-crystals being possible [22,23]. The resulting molecular designs are directed to applications in encapsulation and drug delivery, multivalent epitope display and in synthetic biology. However, such multi-domain protein systems are rarely conceived as microenvironments for cell culturing applications.

We have developed a functionalized, modular biomaterial formed by the controlled self-assembly of two proteins. The polymer, termed ZT, efficiently supports the long-term self-renewal of pluripotent stem cells. Here, we review the molecular design of this new cell substrate, the principles of its assembly and functionalization, and its current application to cell culturing. Its robustness, ease and economy of recombinant production and its high versatility make it a promising system to support complex functionalities in biological applications.

2. The ZT Building Block: A Unique Palindromic Assembly of Titin Z1Z2 with its Binding Partner Telethonin

The ZT polymer is composed of proteins of human origin produced recombinantly in bacteria. It is based on the naturally occurring complexation of the protein telethonin (Tel) with the two N-terminal immunoglobulin (Ig) domains, Z1Z2, from titin in the Z-disc of the muscle sarcomere. Tel is an insoluble, intrinsically disordered protein that becomes structured upon binding to Z1Z2. Specifically, its 90 N-terminal amino-acid residues become "sandwiched" between two antiparallel Z1Z2 doublets, forming a wing-like hairpin structure that builds an intermolecular β-sheet across the three components (Figure 1a) [24]. The resulting complex is strong and displays a unique palindromic structure. Interestingly, protein complexation through the β-sheet augmentation mechanism is not greatly dependent on sequence as it largely rests on main-chain hydrogen bonding [25,26], a fact that has also been demonstrated for the Z1Z2/Tel association [24]. In this fashion, Tel acts as a sturdy biological glue that joins the N-termini of two titin molecules in the sarcomere.

Z1Z2 and the truncated Tel (residues 1–90) can be expressed recombinantly in bacteria as stable protein products in high yield (>50 mg and >10 mg of pure protein per liter of bacterial culture, respectively). For improved recombinant production, Tel is produced as a Cys-null variant where the five native cysteine residues are exchanged for serines to avoid sample aggregation due to oxidation [27,28]. The Z1Z2/Tel complex displays a high tensile resistance to mechanical forces applied along its molecular axis [29] and its Ig constituent domains are highly thermostable, e.g., the melting temperature of Z1 was 72.6 ± 0.16 °C when monitored using differential scanning fluorimetry [30] and 69.4 ± 0.1 °C when monitored by circular dichroism [31]. Thereby, this two-component protein "sandwich" displayed excellent robustness for biomaterial development.

3. Design of the Self-Assembling ZT Polymer

The capability of self-assembling into a high-order polymer was engineered into the Z1Z2/Tel unit by genetically duplicating the coding sequence for Z1Z2 into a double tandem, Z1Z2-Z1Z2 (Z_{1212}; Figure 1b). The larger Z_{1212} tandem (41.6 kDa) did not compromise recombinant production in bacterial systems, with high yields being maintained (in excess of 65 mg of pure protein per gram of *E. coli* wet cell mass [30]). The Z_{1212} tandem underwent propagative assembly via sequential Tel-mediated cross-linking [30] and micrographs obtained by transmission electron microscopy (TEM) confirmed the formation of a polymer (Figure 1c). Computational assembly simulations suggested that the observed fibrous formations corresponded to two primary assembly modes: A longitudinal assembly resulting in curly fibers and a transversal assembly leading to a tape-like polymer. Both polymeric formations were thin (the diameter of curly fibers was 7 ± 1.6 nm, while tape-like formations were 13.4 ± 1.6 nm). In routine preparations, the thinner curly formations predominate, being observed with higher frequency and in higher yield.

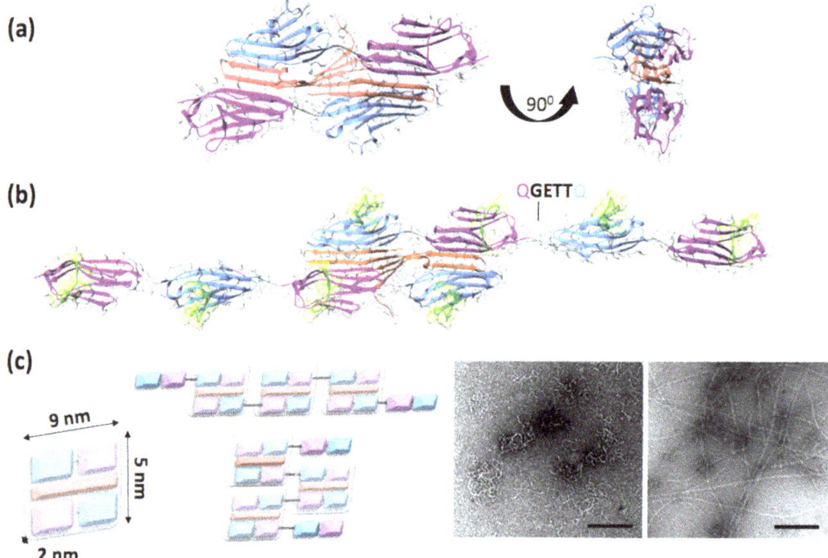

Figure 1. (**a**) Crystal structure of the naturally occurring Z1Z2/Tel complex (Protein Data Bank ID: 1YA5). Domain Z1 is shown in blue, Z2 in purple, and Tel is orange; (**b**) structural model of the Z1Z2-Z1Z2 tandem (Z_{1212}) generated by genetically fusing two Z1Z2 pairs using a QGETTQ engineered linker sequence. The model shows the proposed cross-linking of two Z_{1212} units by a Tel molecule at the initial polymerization event. The CD loop serving as internal functionalization site for the grafting of exogenous peptide sequences onto the Z1 and Z2 Ig domains is highlighted in green; (**c**) schematic representation of the Z1Z2/Tel complex with its molecular dimensions (left), its predicted assembly modes (centre), and electron micrographs of the assembled polymer (right). Thin, curly fibers allegedly resulting from sequential tandem assembly can be seen interwoven with thicker, straighter formations possibly corresponding to perpendicular assemblies. Scale bars correspond to 500 nm (left image reproduced from [30] with permission).

The high flexibility and overall morphology of the ZT polymer is determined by the linker engineered in the Z_{1212} fusion tandem: A linker sequence Gln-Gly-Glu-Thr-Thr-Gln (QGETTQ) introduced in the coding DNA between Z1Z2 pairs, which is the only tethering of Z1Z2/Tel blocks in the polymer (Figure 1b,c). In order to avoid the common problem of proteolytically unstable linkers in poly-proteins, this joining sequence was inspired on the Lys-Ala-Glu-Thr (KAET) natural linker between Z1 and Z2 in titin, which was known to be stable. In addition, as polymerization requires Tel to bind sequentially to domain pairs in Z_{1212}, the engineered linker was designed to promote extended domain conformations in the tandem Ig. To this effect, the linker included a glutamate residue, which is highly conserved in titin linkers and supports extended domain arrangements through hydrodynamic effects [32,33]. The conformational dynamics of the native KAET hinge in Z1Z2 has been characterized experimentally through the elucidation of the 3D-structure of Z1Z2 using X-ray crystallography and nuclear magnetic resonance (NMR) [34] as well as by analyzing the energetics of hinge closure using molecular dynamics simulations [35,36]. The data showed that the natural Z1Z2 linker was flexible and allowed for a wide range of interdomain conformations (from linearly extended to compact V-shape arrangements) permitting domain motions of large amplitude (Figure 2). The tight V-shape arrangements are short-lived in solution however, as the Z1 and Z2 domains do not form stable direct contacts [34]. The recently available crystal structure of the Z_{1212} tandem (PDB ID: 6FXW) allows comparing the properties of the native (KAET) and engineered (QGETTQ) linker sequences. The structure shows the QGETTQ linker free of interactions and in a bent conformation induced by the tight lattice packing of neighboring domains in the crystal. This indicates that this hydrophilic linker is highly flexible and that it permits a broad range of loose domain conformations, as expected. Further, the high conformational freedom of this linker explains the coiled and very collapsible structure of the polymer, as it constitutes the only junction between adjacent Z1Z2/Tel rigid blocks (Figure 1b,c). Engineering stiffer, extended linkers may allow for the formation of ZT polymers of diverse morphologies. The natural variability of domain junctions in titin could serve as a source of new linkers with defined control on interdomain conformations. Harvesting such natural linkers for polymer design would profit from the existing molecular characterization of titin [37]. Therefore, the linker engineered between Z1Z2/Tel units represents an additional functionalization point of the system, which could be altered to tune the polymer properties, namely its morphology and stiffness.

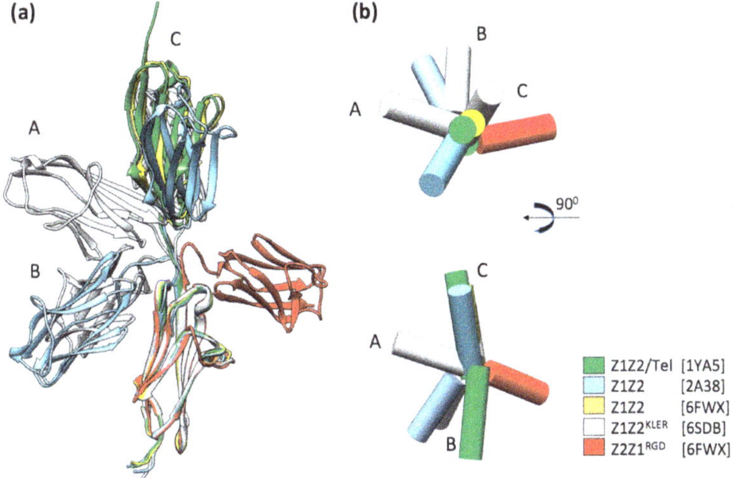

Figure 2. Conformational dynamics in the natural linker sequence of Z1Z2 from titin. (**a**) Conformations of the Z1Z2 Ig-doublets observed experimentally in crystal structures. All pairs are superimposed onto Z1 with the exception of the engineered Z2Z1 pair, where its Z2 domain is superimposed onto Z1 in the other pairs as to allow for comparison. The natural KAET linker sequence that connects Z1 with Z2 is flexible and permits inter-domain motions of large amplitude. In the Z1Z2/Tel pair, the Z1Z2 molecule displayed has been extracted from the structure of the complex illustrated in Figure 1a. The structure of Z1Z2 in isolation (blue) was observed in two conformations: Extended and a tight V-shape. The structure of Z1Z2 carrying the KLER motif in the CD-loop (white) was observed in three inter-domain conformations, here labeled A, B, and C to ease identification. The structure of the Z2–Z1 pair (red) shows that the flexibility of the KAET sequence is reproduced by the engineered linker sequence; (**b**) schematic representation of the domain arrangements shown in (a) where Ig domains are displayed as cylinders (lateral and top views are provided).

The loose structure of the ZT polymer appears to facilitate the incorporation and accessibility of bulky functionalization groups in the biomaterial, such as full-length proteins (described in Section 5). Additionally, the avoidance of a compact peptidic aggregate in this case opens the attractive possibility of using ZT as a possibly safe ECM substitute for cells that are destined for in vivo applications. Dense proteolysis-resistant peptidic aggregates (such as amyloidic fibers) cause human disease [38] and the introduction of biomaterials based on dense fibrous aggregates in living systems is a cause for concern. The loose, accessible fibrillar formations of the ZT polymer and the natural existence of the Z1Z2/Tel building block in humans leads to the expected good biodegradability of the material and clearance by the body. Future studies will be required to investigate this aspect.

4. Functionalization of the ZT Polymer Using Genetic Engineering: Display of Exogenous Peptide Sequences

The ZT polymer and its components are non-cytotoxic and biologically inert. However, biological functionality can be introduced in the polymer using genetic approaches, such as site-directed mutagenesis or gene fusion, which modify the protein building block pre-assembly. These genetic methods enable the grafting of exogenous peptide sequences onto the ZT scaffold as well as the incorporation of full-length proteins. Contrary to chemical approaches, "bottom-up" engineering provides full control on the amount and stoichiometry of functional groups displayed on the material.

The display of peptidic sequences on the ZT polymer is established, with sequences having been introduced both in Tel and Z_{1212} components without preventing polymerization [30,39,40]. Both components permit fusing peptidic sequences to their N- and C-termini and, in addition, Z_{1212} permits the introduction of motifs in four internal positions, namely the CD loop of each of its four Ig domains (Figure 1b). The Ig fold of Z1 and Z2 belongs to the I (intermediate)-type, which lacks developed hyper-variable loops as those of the V (variable)-type that is the archetypal binder fold in nature [41]. Z1 and Z2, as commonly observed in Ig domains from titin, have short loops with partly conserved sequences [32]. However, they possess a surface exposed loop, the CD loop, which is a region of low sequence conservation. We have demonstrated that the CD loop can tolerate drastic diversification in length and composition and have established its suitability as a locus for peptide grafting. Specifically, we replaced the native loop in Z1 for a highly charged FLAG affinity tag (DYKDDDDK) [39], a native PxxP SH3-interaction motif (EAMPPTLPHRDWKD) [39], a sequence carrying the integrin-attachment motif GRGDS from fibronectin (SSGRGDSS) [40], and a decorin-mimetic KLER motif (PDB ID: 6SDB; Figure 3a). All variants could be produced recombinantly in bacteria in yields equivalent to those of the wild-type. The crystal structures of RGD (Z_{1212}^{RGD}; PDB: 6FWX) and KLER (Z_{12}^{KLER}; PDB: 6SDB) modified variants as well as the NMR analysis of the FLAG single-domain variant ($Z1^{FLAG}$) [39] confirmed that Z1 retained its native fold and structural integrity after the modifications and that the modified loop did not affect the long-range conformation of the Z_{1212} tandem or its capability to polymerize through interaction with Tel. Furthermore, functional studies demonstrated that FLAG and SSGRGDSS sequences remained accessible and functional upon grafting, respectively supporting molecular and cell-based interactions with good efficiency [39,40]. These results established these Ig domains as successful scaffolds amenable to extensive protein engineering without detriment to their structural integrity, stability or assembly.

The grafting of peptides onto stable protein scaffolds is an established methodology. However, achieving the constrainment of the introduced peptide in its bioactive conformation remains a highly challenging task. The RGD tripeptide is a common cell binding motif present in various ECM proteins including fibronectin, vitronectin, and laminin [42]. Thus, RGD-containing sequences are often grafted on scaffolds designed to support cell attachment [43,44]. However, the conformation of the RGD motif and its direct flanking groups significantly affect integrin selectivity and cell response [44,45]. The crystal structure of Z_{1212}^{RGD} proved that the RGD motif incorporated in the CD loop of the third tandem Ig mimicked well the flexibility and the local conformation of the RGD motif within the 10th cell binding domain of native fibronectin, Fn10 (Figure 3b). Size-exclusion chromatography (SEC), SEC-multi angle light scattering, native-polyacrylamide gel electrophoresis (PAGE), and TEM micrographs showed that the introduction of the RGD carrying-sequence at the CD locus did not alter the interaction with Tel or the ensuing polymerization [40]. Subsequent cell assays (below) showed the Z_{1212}^{RGD} variant to be functional.

In the Z1Z2 variant carrying the decorin motif KLER in the CD loop of Z1 (Z_{12}^{KLER}), the motif replaced five native residues so as to become inserted between residues V45 and L51 (Figure 3b). This insertion point was chosen as the V and L residues in Z1 corresponded to the natural flanking residues of KLER in decorin (VKLERL). Decorin is a protein associated with cartilage homeostasis that binds to collagens and plays a role in their stability, organization, and fibrillogenesis [46]. In decorin, the KLER motif is located within a loop of the leucine-rich repeat 3, the region that binds collagen [47] and the incorporation of this sequence in a PEG-peptide co-polymer promoted chondrogenesis of human mesenchymal stromal cells [48]. It was envisioned that the CD loop would constitute a structurally suitable area to incorporate the KLER sequence. The crystal structure of Z_{12}^{KLER} (PDB: 6SDB) confirmed that the overall Z1 fold remained intact, but showed that the KLER motif in Z1 (RMSD = 0.34 Å for four non-crystallographic symmetry molecular copies) did not adopt a native conformation as in decorin (PDB: 1XCD). Interestingly, the overall pattern of exposed Lys, Glu, and Arg side chains with a buried Leu sidechain is preserved in Z_{12}^{KLER}, but the main chain adopts a slight helical turn that causes the side chains to be positioned closer together producing a compacter KLER motif than that of decorin (Figure 3b). Given that loop sequences can undergo changes induced by their binding targets, it remains to be established whether the observed difference affects the biological function of the engineered motif. Nonetheless, a Z_{1212} variant carrying the KLER motif in the CD loop of Z1 domain in third position within the tandem (Z_{1212}^{KLER}) was created. This was found to maintain the polymerization capability upon mixing with Tel, with its electrophoretic mobility profile being nearly identical to that of wild-type ZT in native-PAGE. In conclusion, data on the various variants establish the suitability of the CD loop as a locus for peptide grafting, capable of supporting the display of bioactive peptidic motifs on the ZT polymer.

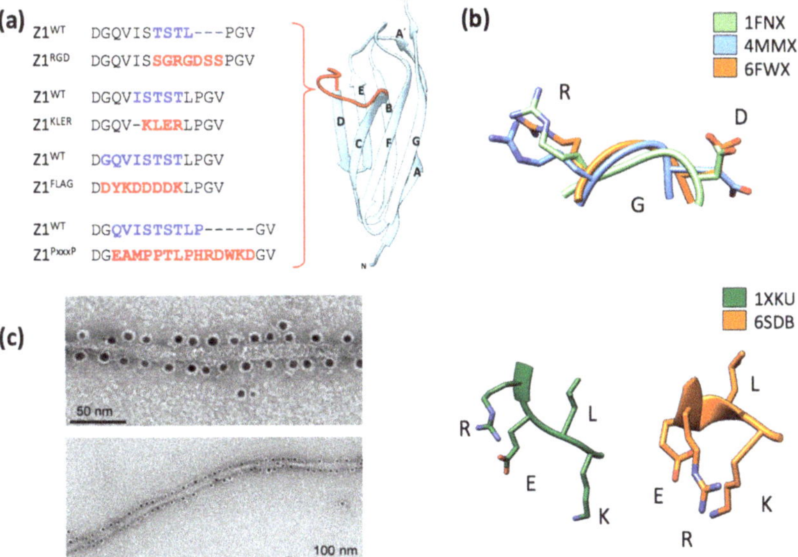

Figure 3. Exogenous peptide display on ZT components. (**a**) Overview of sequences grafted to date onto the CD loop of Z1. The crystal structure of Z1 (extracted from PDB 2A38) is shown and its CD loop is highlighted in red. For each grafted variant, native residues removed from Z1 are indicated in blue and the introduced exogenous residues in red. In the largest substitution, 14 new residues were introduced; (**b**) local structure of the RGD and KLER sequences grafted onto Z1 analyzed using X-ray crystallography and compared to the native sequences in their natural protein environments (only bioactive residues are shown). PDB accession codes for crystal structures are given; (**c**) TEM images of ZT fibers with repetitively attached gold nanoparticles functionalized with Ni^{2+}-NTA at different magnifications. The recruitment of the gold nanoparticles (AuNPs) to the ZT polymer was mediated by a His_6-tag fused N-terminally to Tel. (Images in (**c**) reproduced from [30] with permission).

Peptide display on the Tel component has also been demonstrated. In this case, polyethylene glycol (PEG)-passivated gold nanoparticles (AuNPs) functionalized with Ni^{2+}-NTA were specifically bound to a poly-histidine (His_6) tag fused to the N-terminus of Tel via an 18 residues linker sequence [30].

The recruitment of AuNPs to the ZT polymer via this affinity tag in Tel resulted in the repetitive decoration of the polymer at the nanoscale (~5 nm inter-particle distance), proving that Tel can display functional and accessible peptides without compromising the polymer structure [30] (Figure 3c).

5. Functionalization Through Domain Fusion: Protein Chimeras

A second approach to the functionalization of the ZT polymer is through the addition of full-length proteins or protein domains to the termini of either Z_{1212} or Tel components via gene fusion. Such incorporations can comprise natural integrin attachment domains, fluorescent proteins, or even processive enzymes. Examples of such functional chimeras are the fusion of the Fn10 domain from fibronectin (Fn10; 10.5 kDa) [40] and the fusion of yellow fluorescent protein (mCitrineFP; 26.9 kDa) to the C-terminus of Z_{1212} (Z_{1212}^{Fn} and Z_{1212}^{mYFP}, respectively; Figure 4a). SEC showed that the chimeric Z_{1212} tandems remained monomeric and highly monodisperse (Figure 4b) [40]. TEM and native-PAGE proved that the chimeric Z_{1212} remained able to polymerize upon complexation with Tel (Figure 4c) [40]. The Z_{1212}^{mYFP} chimera polymerized by Tel (ZT^{mYFP}) was subsequently used for the coating of polystyrene plates (commonly used in monolayer cell culture) by dispensing on the plastic surface. Fluorescence imaging confirmed that the polymer had adhered to the polystyrene by simple deposition with good homogeneous distribution, demonstrating effective surface coating (Figure 4d). The images revealed that the protein formed concentration-dependent clusters on the polystyrene surface, possibly reflecting the coiled morphology of the polymer. All data considered, we concluded that in the ZT system, the region involved in polymerization is well segregated from the loci of functionalization and that the incorporation of bulky moieties (such as full-length globular proteins) does not disrupt its self-assembly.

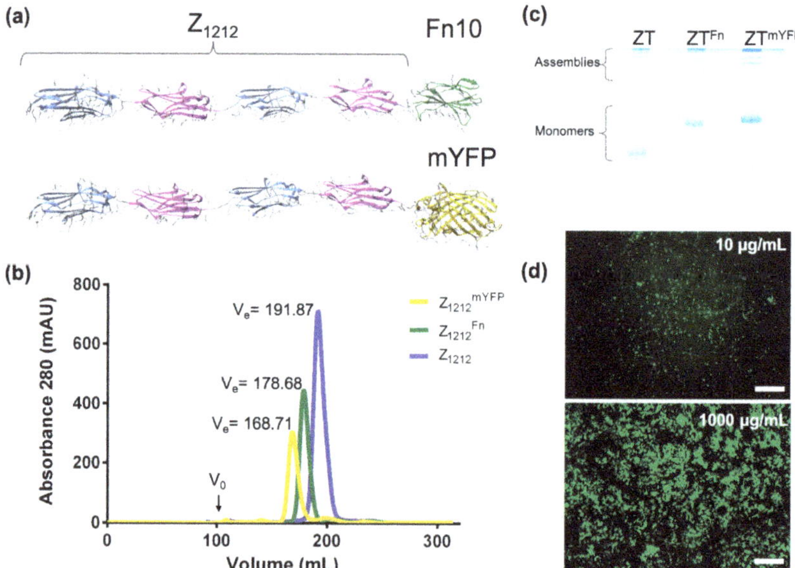

Figure 4. ZT functionalization through genetic fusion of proteins. (**a**) Molecular models of functional chimeric variants of Z_{1212} tandems; (**b**) size exclusion chromatograms of unmodified and chimeric tandems performed on a Superdex S200 26/60 column (GE Healthcare) that demonstrate the monodispersity of the samples (molecular mass: Z_{1212} = 41.76 kDa; Z_{1212}^{Fn} = 52.27 kDa; and Z_{1212}^{mYFP} = 68.78 kDa); (**c**) native-PAGE of Z-tandems and Tel mixtures 24 h post-assembly; (**d**) fluorescence images of polymeric ZT^{mYFP} adhered onto a polystyrene surface (96 well-plate, COSTAR, Corning) acquired using an Axi Zeiss Zoom V16 fluorescence microscope (Scale bar = 100 µm).

6. The Functionalized ZT Polymer is an Effective Substratum for Stem Cell Culturing

The ZT polymer is commonly produced as a light hydrogel that can be essentially dispensed as a liquid and efficiently adsorbed onto polystyrene surfaces by simple pipetting, acting as a biocoating (Figure 4d). Importantly, the polymer is highly stable in cell culture media and does not become eroded by cell populations. Interestingly, the Ig domains that compose the Z_{1212} tandem have similar molecular dimensions and a related topography to the Fn domains forming native fibronectin. This suggests that the ZT polymer has a coarse molecular granularity that resembles that of components naturally present in the ECM [49]. Upon functionalization by fusion of the Fn10 domain from fibronectin (ZT^{Fn}), the polymer has been shown to successfully sustain the long-term culturing of human embryonic stem cells (hESCs), murine mesenchymal stromal cells (mMSCs) [40], and human induced pluripotent stem cells (hiPSCs) [50] (Figures 5 and 6). The ZT^{RGD} polymer variant was shown to elicit mMSC adhesion and spreading, although with lower efficiency than ZT^{Fn} (Figure 5a) [40]. mMSCs are known to be responsive to RGD sequences presented in peptide format [51], but hESCs are more demanding and require more complex substrates with folded ECM components for optimal viability [7]. The hESC line (HUES7) assayed attached and spread on ZT^{Fn} comparably to reference substrates fibronectin, vitronectin, and Matrigel^{TM} (hESCs; Figure 5b) [40]. However, the cytoskeletal structures and cell morphology of cells grown on ZT^{Fn} more closely resembled those cultured on vitronectin. An analysis of integrin engagement upon hESC attachment to ZT^{Fn} showed that $\alpha V \beta 5$, an integrin that primarily acts as a vitronectin receptor [52], was engaged along with the canonical fibronectin receptor $\alpha 5 \beta 1$ [40] (Figure 5c). This finding adds to early observations, where cells cultured on large fibronectin fragments containing the RGD site were shown to engage preferentially the $\alpha 5 \beta 1$ integrin, whereas a small fragment consisting of the Fn10 domain, no longer withheld this preference and bound instead better to the $\alpha V \beta 3$ vitronectin receptor [53]. It is now known that $\alpha V \beta 3$ and $\alpha 5 \beta 1$ are closely related integrins. $\alpha V \beta 3$ is somewhat promiscuous and can bind to several ECM proteins including vitronectin, fibronectin, osteopontin, and bone sialoprotein, whilst the $\alpha 5 \beta 1$ integrin primarily recognizes fibronectin as a consequence of the presence of the synergistic amino acid sequence PHSRN in the cell attachment site of the protein [54,55]. It must also be considered that RGD motifs are present in many ECM proteins, including fibronectin, vitronectin, fibrinogen, von Willebrand factor, thrombospondin, laminin, entactin, and tenascin, amongst others [42] and that RGD-binding integrin receptors can recognize more than one specific RGD motif [42,52,56]. RGD motifs are individualized by flanking sequence groups as well as unique 3D-structural contexts in their containing proteins, and they often act in unison with other synergy motifs, thereby engaging integrins with differing efficacy. The engagement of prototypical $\alpha V \beta 5$ vitronectin integrin receptors by stem cells grown on the fibronectin-based ZT^{Fn} illustrates the plasticity of cell response, where not only the type of bioreactive moiety but also its mode of presentation to the cell and the specific proteomic constitution of a cell type jointly dictate the biological outcome.

In addition to evaluating viability, the long-term stability of stem cells cultured on ZT^{Fn} was assessed by monitoring cell morphology and the expression of transcription factors in hESCs. It was confirmed that the cells retained their embryonic stem cell phenotype, expressing the OCT4 and NANOG transcription factors. Long-term culturing, for up to 10 cell passages proved that hESCs were able to proliferate and maintain a pluripotent phenotype over time, as evidenced by the fact that they could generate embryoid bodies that comprise derivatives of the three embryonic germ layers [40]. These results have been recently extended by our laboratory using cells cultured for 18 cell passages (c.a. 4 months), demonstrating the long-term stability of cells grown on the ZT polymer.

To further validate ZT^{Fn} as a substrate for stem cells, we extended our studies to the culture of hiPSCs [50]. hiPSCs were cultured for up to 10 passages on ZT^{Fn} (under conditions equivalent to those used for the HUES7 hESC line; [40]). Cells retained a canonical PSC morphology typified by colony formation and a high nuclear-cytoplasmic ratio (Figure 6a). Cells also retained nuclear expression of pluripotency markers OCT4 and NANOG (Figure 6b). At the population level, expression of nuclear OCT4 and surface markers SSEA-4 and TRA-1-60 in hiPSCs cultured on ZT^{Fn} for 10 passages were found to be comparable to cells cultured on recombinant vitronectin (from ThermoFisher Scientific)

(Figure 6d). Additionally, *OCT4*, *NANOG*, and *SOX2* transcript levels in cells cultured on ZT^{Fn} were not significantly different to those cultured on vitronectin (Figure 6e). Taken together, these data show that ZT^{Fn} is comparable to vitronectin for hiPSC self-renewal. To ensure that culture on ZT^{Fn} did not induce genetic abnormalities, hiPSCs at passage 10 were karyotyped (Cell Guidance Systems). A normal human female karyotype (46, XX) was observed, confirming that the ZT^{Fn} polymer does not induce coarse genetic aberrations (Figure 6f). Finally, an in vitro differentiation assay was employed to confirm maintenance of pluripotency (as previously described for HUES7 cells; [40]). Derivatives of all three germ layers were observed (Figure 6c), demonstrating that hiPSCs remained pluripotent following 10 passages on ZT^{Fn}. These results further validate the claim that ZT^{Fn} can be employed for the propagation of hPSCs.

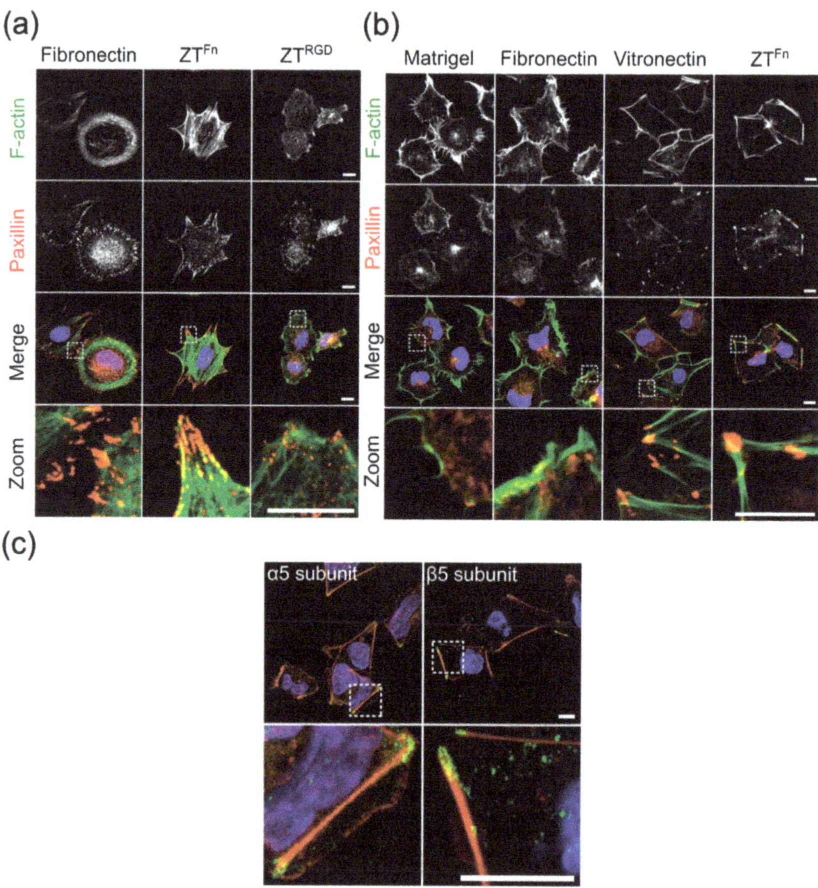

Figure 5. Culturing of murine mesenchymal stromal cells (mMSCs) and human embryonic stem cells (hESCs) on the ZT^{Fn} polymer. Focal adhesion formation and cytoskeletal organization in mMSCs and hESCs cultured on different substrates. Immunofluorescence micrographs show representative z-series projections of mMSCs (**a**) and hESCs (**b**) cultured on ZT variants and control substrates. Cells were stained for F-actin (green), paxillin (red), and DAPI (blue). Zoomed views of the boxed areas in the upper panels are shown to highlight focal adhesions and cytoskeletal structures. (**c**) hESC integrin engagement on ZT^{Fn}. Immunofluorescence micrographs show hESCs cultured on ZT^{Fn} and stained for F-actin (red), α5 or β5 integrin subunits (green), and DAPI (blue). Zoomed views of the boxed areas in the upper panels are shown to highlight integrin staining. Scale bars = 10 μm.

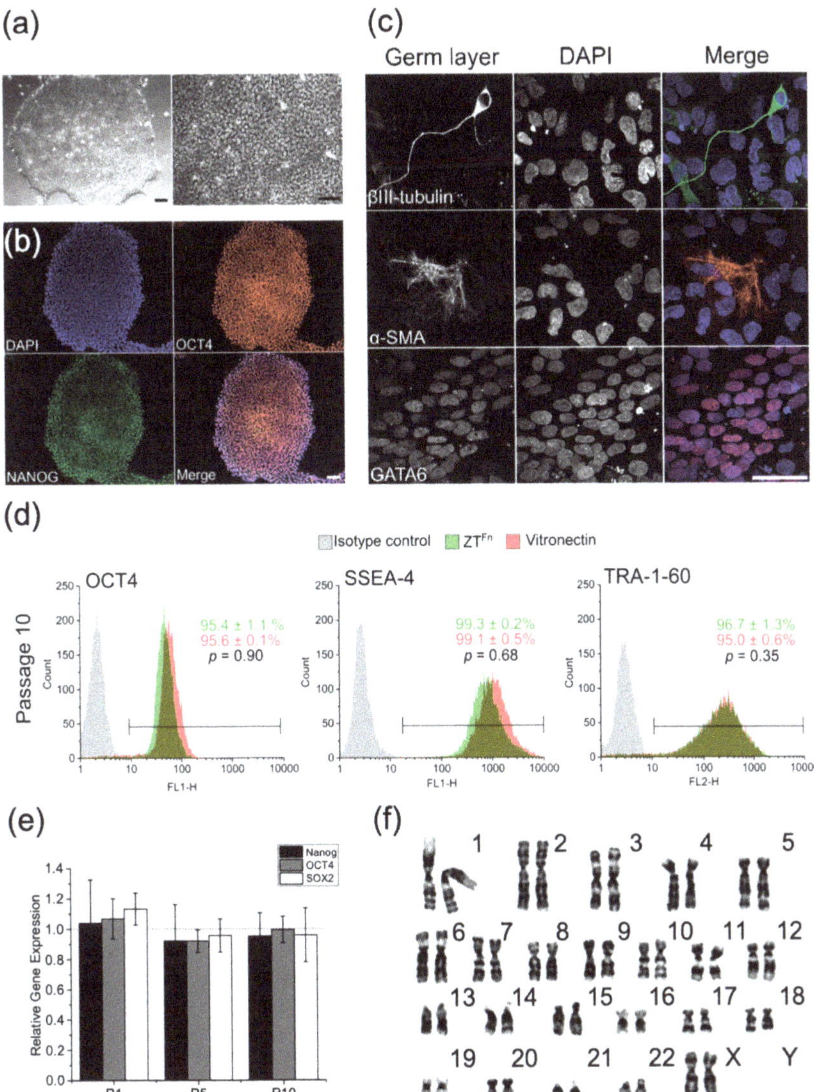

Figure 6. Culturing of human induced pluripotent stem cells (hiPSCs) on the ZT^{Fn} polymer. (**a**) Representative phase contrast micrographs of hiPSCs cultured on ZT^{Fn} for 10 passages. Scale bar = 100 μm. (**b**) Representative epifluorescence micrographs of hiPSCs cultured on ZT^{Fn} for 10 passages. Cells were stained for nuclei (blue), OCT4 (red), and NANOG (green). A merged channel image is shown. Scale bar = 100 μm. (**c**) Confocal micrographs show embryoid body-derived cells stained for markers of the three primary germ layers; β-III tubulin (ectoderm), α-SMA (mesoderm), and GATA-6 (endoderm). Cells were counterstained with DAPI (blue). Scale bar = 50 μm. (**d**) Flow cytometry histograms for pluripotency markers OCT4, SSEA-4, and TRA-1-60 derived from hiPSCs cultured on ZT^{Fn} or vitronectin for 10 passages. The average percentage of positive cells ± SEM is shown (n = 3). (**e**) Quantitative RT-qPCR analysis of *NANOG*, *OCT4*, and *SOX2* expression levels in hiPSCs cultured on ZT^{Fn} for 1, 5, and 10 passages relative to cells cultured on vitronectin. Error bars represent SEM (n = 3). (**f**) Representative karyogram of hiPSCs cultured on ZT^{Fn} for 10 passages (n = 20).

7. Conclusions

The need to sustain and study cells ex vivo continues to drive the development of culturing materials that can support a naturalistic cell behavior, thereby, enabling improved research and therapeutic applications. While a variety of cell substrates are now available, few are up-scalable, stable, and economic. Various available systems, like self-assembling peptides (recently reviewed in [9]), short hydrogelation peptides [57,58], and non-peptide based materials such as poly-L-lactic acid (PLLA) and poly-caprolactone (PCL) [59,60] have been employed in cell growth, but their production requires the use of peptide synthesis, harsh solvents (e.g., DMSO, acetone, and dimethylformamide), or methods such as electrospinning. This results in low material yield, higher production cost, and troubles the incorporation of delicate functional proteins. The functionalization of those systems most commonly uses short peptidic motif sequences, which can result in variable or partial efficiency in integrin targeting. When functionalized with full-length proteins, this is mostly done post-production using click chemistry or, in the case of electrospun fibers, oxygen plasma treatment [59,60], which also increases the overall hydrophilicity of the latter material. These functionalization methods, however, allow little control over both the level of incorporation and the distribution of functional moieties on the material, a particularly important aspect when pursuing co-functionalization.

In this regard, we have developed an up-scalable, stable and economical cell substrate based on the naturally occurring palindromic assembly of two proteins of human origin: Z1Z2 and Tel. The robustness, stability, high protein yield, low production cost, and diverse functionalization possibilities in mild media and under biological conditions, are key advantages of the recombinant ZT system in cell-based applications. The assembled ZT polymer is highly flexible and its building blocks act as robust scaffolds that can be functionalized conveniently through the incorporation of bioactive moieties by gene manipulation. Thereby, the polymer is fully encodable and, predictably, biodegradable. The system has been demonstrated to display efficiently both peptidic motifs and full protein domains, with little effect on its polymerization capability. Both Z_{1212} tandem and Tel components can act as recipients of functional moieties at their termini and, in addition, the Z_{1212} tandem contains four internal sites (the CD loops) that can display independent peptidic motifs. Modified building block variants have proven to be stable upon modification and to present a broad tolerance to exogenous sequences. Thereby, the ZT material offers excellent potential for orthogonal functionalization. It is worth noting that in the biomaterial landscape, cell substrates based on large multi-domain proteins of coarse granularity are very scarce. The larger molecular scale of substrates like ZT translates into an enhanced modularity of high promise towards the creation of complex, plurifunctional biomaterials. The conceivable applications of the ZT polymer are diverse, ranging from serving as a scaffold for cell culturing via the incorporation of cell reactive moieties or as a possible drug display system through the incorporation of therapeutic proteins.

Funding: Work conducting to this review was supported by funding from the Konstanz Research School of Chemical Biology (KoRS-CB), the BBSRC (Biotechnology and Biological Sciences Research Council) and The Wellcome Trust (204401/z/16/z). JRF is supported by an EU Marie Sklodowska-Curie Individual Fellowship (TTNPred, 753054).

Conflicts of Interest: C.J.H., O.M. and P.M. disclose a pending patent on the use of functionalized ZT polymer variants for cell-based applications (WO2019030524-A1).

Substrate Availability: Samples of the ZT material can be obtained from the authors on request.

Abbreviations

AuNPs	Gold nanoparticles
CTPR	Consensus tetratricopeptide repeat protein
ECM	Extracellular matrix
hiPSC	Human induced pluripotent stem cells
hESC	Human embryonic stem cells
Ig	Immunoglobulin

mMSC	Murine mesenchymal stromal cells
NMR	Nuclear magnetic resonance
PAGE	Polyacrylamide gel electrophoresis
PEG	Polyethylene glycol
RGD	Arg-Gln-Asp
SEC	Size-exclusion chromatography
Tel	Telethonin
TEM	Transmission electron microscopy

References

1. Kleinman, H.K.; Martin, G.R. Matrigel: Basement membrane matrix with biological activity. *Semin. Cancer Biol.* **2005**, *15*, 378–386. [CrossRef] [PubMed]
2. Hughes, C.S.; Postovit, L.M.; Lajoie, G.A. Matrigel: A complex protein mixture required for optimal growth of cell culture. *PROTEOMICS* **2010**, *10*, 1886–1890. [CrossRef] [PubMed]
3. Benton, G.; Arnaoutova, I.; George, J.; Kleinman, H.K.; Koblinski, J. Matrigel: From discovery and ECM mimicry to assays and models for cancer research. *Adv. Drug Deliv. Rev.* **2014**, *79–80*, 3–18. [CrossRef] [PubMed]
4. Shafiq, M.; Jung, Y.; Kim, S.H. Insight on stem cell preconditioning and instructive biomaterials to enhance cell adhesion, retention, and engraftment for tissue repair. *Biomaterials* **2016**, *90*, 85–115. [CrossRef] [PubMed]
5. Rodin, S.; Domogatskaya, A.; Ström, S.; Hansson, E.M.; Chien, K.R.; Inzunza, J.; Hovatta, O.; Tryggvason, K. Long-term self-renewal of human pluripotent stem cells on human recombinant laminin-511. *Nat. Biotechnol.* **2010**, *28*, 611–615. [CrossRef] [PubMed]
6. Prowse, A.B.J.; Doran, M.R.; Cooper-White, J.J.; Chong, F.; Munro, T.P.; Fitzpatrick, J.; Chung, T.-L.; Haylock, D.N.; Gray, P.P.; Wolvetang, E.J. Long term culture of human embryonic stem cells on recombinant vitronectin in ascorbate free media. *Biomaterials* **2010**, *31*, 8281–8288. [CrossRef]
7. Kalaskar, D.M.; Downes, J.E.; Murray, P.; Edgar, D.H.; Williams, R.L. Characterization of the interface between adsorbed fibronectin and human embryonic stem cells. *J. R. Soc. Interface* **2013**, *10*. [CrossRef]
8. Ross, A.M.; Nandivada, H.; Ryan, A.L.; Lahann, J. Synthetic substrates for long-term stem cell culture. *Polymer* **2012**, *53*, 2533–2539. [CrossRef]
9. Hellmund, K.S.; Koksch, B. Self-Assembling Peptides as Extracellular Matrix Mimics to Influence Stem Cell's Fate. *Front. Chem.* **2019**, *7*. [CrossRef]
10. Melkoumian, Z.; Weber, J.L.; Weber, D.M.; Fadeev, A.G.; Zhou, Y.; Dolley-Sonneville, P.; Yang, J.; Qiu, L.; Priest, C.A.; Shogbon, C.; et al. Synthetic peptide-acrylate surfaces for long-term self-renewal and cardiomyocyte differentiation of human embryonic stem cells. *Nat. Biotechnol.* **2010**, *28*, 606–610. [CrossRef]
11. Higuchi, A.; Kao, S.-H.; Ling, Q.-D.; Chen, Y.-M.; Li, H.-F.; Alarfaj, A.A.; Munusamy, M.A.; Murugan, K.; Chang, S.-C.; Lee, H.-C.; et al. Long-term xeno-free culture of human pluripotent stem cells on hydrogels with optimal elasticity. *Sci. Rep.* **2015**, *5*, 18136. [CrossRef] [PubMed]
12. Lambshead, J.W.; Meagher, L.; Goodwin, J.; Labonne, T.; Ng, E.; Elefanty, A.; Stanley, E.; O'Brien, C.M.; Laslett, A.L. Long-Term Maintenance of Human Pluripotent Stem Cells on cRGDfK-Presenting Synthetic Surfaces. *Sci. Rep.* **2018**, *8*, 701. [CrossRef] [PubMed]
13. Loo, Y.; Goktas, M.; Tekinay, A.B.; Guler, M.O.; Hauser, C.A.E.; Mitraki, A. Self-Assembled Proteins and Peptides as Scaffolds for Tissue Regeneration. *Adv. Healthc. Mater.* **2015**, *4*, 2557–2586. [CrossRef] [PubMed]
14. Zhang, S. Discovery and design of self-assembling peptides. *Interface Focus* **2017**, *7*, 20170028. [CrossRef] [PubMed]
15. Wu, Y.; Collier, J.H. α-Helical Coiled Coil Peptide Materials for Biomedical Applications. *Wiley Interdiscip. Rev. Nanomed. Nanobiotechnol.* **2017**, *9*. [CrossRef]
16. Seroski, D.T.; Hudalla, G.A. Self-Assembled Peptide and Protein Nanofibers for Biomedical Applications. *Biomed. Appl. Funct. Nanomater. Elsevier* **2018**, 569–598.
17. Beesley, J.L.; Woolfson, D.N. The de novo design of α-helical peptides for supramolecular self-assembly. *Curr. Opin. Biotechnol.* **2019**, *58*, 175–182. [CrossRef]
18. Cai, L.; Heilshorn, S.C. Designing ECM-mimetic materials using protein engineering. *Acta Biomater.* **2014**, *10*, 1751–1760. [CrossRef] [PubMed]

19. Webber, M.J.; Appel, E.A.; Meijer, E.W.; Langer, R. Supramolecular biomaterials. *Nat. Mater.* **2016**, *15*, 13–26. [CrossRef]

20. Wang, Y.; Katyal, P.; Montclare, J.K. Protein-Engineered Functional Materials. *Adv. Healthc. Mater.* **2019**, *8*, e1801374. [CrossRef]

21. Mejias, S.H.; Aires, A.; Couleaud, P.; Cortajarena, A.L. Designed Repeat Proteins as Building Blocks for Nanofabrication. *Adv. Exp. Med. Biol.* **2016**, *940*, 61–81. [PubMed]

22. Cannon, K.A.; Ochoa, J.M.; Yeates, T.O. High-symmetry protein assemblies: Patterns and emerging applications. *Curr. Opin. Struct. Biol.* **2019**, *55*, 77–84. [CrossRef] [PubMed]

23. Hamley, I.W. Protein Assemblies: Nature-Inspired and Designed Nanostructures. *Biomacromolecules* **2019**, *20*, 1829–1848. [CrossRef] [PubMed]

24. Zou, P.; Pinotsis, N.; Lange, S.; Song, Y.-H.; Popov, A.; Mavridis, I.; Mayans, O.M.; Gautel, M.; Wilmanns, M. Palindromic assembly of the giant muscle protein titin in the sarcomeric Z-disk. *Nature* **2006**, *439*, 229–233. [CrossRef] [PubMed]

25. Remaut, H.; Waksman, G. Protein-protein interaction through beta-strand addition. *Trends Biochem. Sci.* **2006**, *31*, 436–444. [CrossRef]

26. Benian, G.M.; Mayans, O. Titin and Obscurin: Giants Holding Hands and Discovery of a New Ig Domain Subset. *J. Mol. Biol.* **2015**, *427*, 707–714. [CrossRef] [PubMed]

27. Zou, P.; Gautel, M.; Geerlof, A.; Wilmanns, M.; Koch, M.H.J.; Svergun, D.I. Solution Scattering Suggests Cross-linking Function of Telethonin in the Complex with Titin. *J. Biol. Chem.* **2003**, *278*, 2636–2644. [CrossRef] [PubMed]

28. Tan, H.; Su, W.; Wang, P.; Zhang, W.; Sattler, M.; Zou, P. Expression and purification of a difficult sarcomeric protein: Telethonin. *Protein Expr. Purif.* **2017**, *140*, 74–80. [CrossRef]

29. Bertz, M.; Wilmanns, M.; Rief, M. The titin-telethonin complex is a directed, superstable molecular bond in the muscle Z-disk. *Proc. Natl. Acad. Sci. USA* **2009**, *106*, 13307–133310. [CrossRef]

30. Bruning, M.; Kreplak, L.; Leopoldseder, S.; Müller, S.A.; Ringler, P.; Duchesne, L.; Fernig, D.G.; Engel, A.; Ucurum-Fotiadis, Z.; Mayans, O. Bipartite design of a self-fibrillating protein copolymer with nanopatterned peptide display capabilities. *Nano Lett.* **2010**, *10*, 4533–4537. [CrossRef]

31. Politou, A.S.; Thomas, D.J.; Pastore, A. The folding and stability of titin immunoglobulin-like modules, with implications for the mechanism of elasticity. *Biophys. J.* **1995**, *69*, 2601–2610. [CrossRef]

32. Marino, M.; Svergun, D.I.; Kreplak, L.; Konarev, P.V.; Maco, B.; Labeit, D.; Mayans, O. Poly-Ig tandems from I-band titin share extended domain arrangements irrespective of the distinct features of their modular constituents. *J. Muscle Res. Cell Motil.* **2005**, *26*, 355–365. [CrossRef] [PubMed]

33. Von Castelmur, E.; Marino, M.; Svergun, D.I.; Kreplak, L.; Ucurum-Fotiadis, Z.; Konarev, P.V.; Urzhumtsev, A.; Labeit, D.; Labeit, S.; Mayans, O. A regular pattern of Ig super-motifs defines segmental flexibility as the elastic mechanism of the titin chain. *Proc. Natl. Acad. Sci. USA* **2008**, *105*, 1186–1191. [CrossRef] [PubMed]

34. Marino, M.; Zou, P.; Svergun, D.; Garcia, P.; Edlich, C.; Simon, B.; Wilmanns, M.; Muhle-Goll, C.; Mayans, O. The Ig doublet Z1Z2: A model system for the hybrid analysis of conformational dynamics in Ig tandems from titin. *Struct. Lond. Engl. 1993* **2006**, *14*, 1437–1447. [CrossRef]

35. Lee, E.H.; Hsin, J.; Mayans, O.; Schulten, K. Secondary and tertiary structure elasticity of titin Z1Z2 and a titin chain model. *Biophys. J.* **2007**, *93*, 1719–1735. [CrossRef]

36. Lee, E.H.; Hsin, J.; von Castelmur, E.; Mayans, O.; Schulten, K. Tertiary and secondary structure elasticity of a six-Ig titin chain. *Biophys. J.* **2010**, *98*, 1085–1095. [CrossRef] [PubMed]

37. Zacharchenko, T.; von Castelmur, E.; Rigden, D.J.; Mayans, O. Structural advances on titin: Towards an atomic understanding of multi-domain functions in myofilament mechanics and scaffolding. *Biochem. Soc. Trans.* **2015**, *43*, 850–855. [CrossRef]

38. Iadanza, M.G.; Jackson, M.P.; Hewitt, E.W.; Ranson, N.A.; Radford, S.E. A new era for understanding amyloid structures and disease. *Nat. Rev. Mol. Cell Biol.* **2018**, *19*, 755–773. [CrossRef] [PubMed]

39. Bruning, M.; Barsukov, I.; Franke, B.; Barbieri, S.; Volk, M.; Leopoldseder, S.; Ucurum, Z.; Mayans, O. The intracellular Ig fold: A robust protein scaffold for the engineering of molecular recognition. *Protein Eng. Des. Sel. PEDS* **2012**, *25*, 205–212. [CrossRef]

40. Hill, C.J.; Fleming, J.R.; Mousavinejad, M.; Nicholson, R.; Tzokov, S.B.; Bullough, P.A.; Bogomolovas, J.; Morgan, M.R.; Mayans, O.; Murray, P. Self-Assembling Proteins as High-Performance Substrates for Embryonic Stem Cell Self-Renewal. *Adv. Mater. Deerfield Beach Fla* **2019**, *31*, e1807521. [CrossRef]

41. Halaby, D.M.; Poupon, A.; Mornon, J. The immunoglobulin fold family: Sequence analysis and 3D structure comparisons. *Protein Eng.* **1999**, *12*, 563–571. [CrossRef] [PubMed]
42. Ruoslahti, E. RGD and other recognition sequences for integrins. *Annu. Rev. Cell Dev. Biol.* **1996**, *12*, 697–715. [CrossRef] [PubMed]
43. Hersel, U.; Dahmen, C.; Kessler, H. RGD modified polymers: Biomaterials for stimulated cell adhesion and beyond. *Biomaterials* **2003**, *24*, 4385–4415. [CrossRef]
44. Bellis, S.L. Advantages of RGD peptides for directing cell association with biomaterials. *Biomaterials* **2011**, *32*, 4205–4210. [CrossRef] [PubMed]
45. Kapp, T.G.; Rechenmacher, F.; Neubauer, S.; Maltsev, O.V.; Cavalcanti-Adam, E.A.; Zarka, R.; Reuning, U.; Notni, J.; Wester, H.-J.; Mas-Moruno, C.; et al. A Comprehensive Evaluation of the Activity and Selectivity Profile of Ligands for RGD-binding Integrins. *Sci. Rep.* **2017**, *7*, 39805. [CrossRef] [PubMed]
46. Iozzo, R.V. Matrix proteoglycans: From molecular design to cellular function. *Annu. Rev. Biochem.* **1998**, *67*, 609–652. [CrossRef] [PubMed]
47. Weber, I.T.; Harrison, R.W.; Iozzo, R.V. Model structure of decorin and implications for collagen fibrillogenesis. *J. Biol. Chem.* **1996**, *271*, 31767–31770. [CrossRef] [PubMed]
48. Salinas, C.N.; Anseth, K.S. Decorin moieties tethered into PEG networks induce chondrogenesis of human mesenchymal stem cells. *J. Biomed. Mater. Res. A* **2009**, *90*, 456–464. [CrossRef]
49. Potts, J.R.; Campbell, I.D. Structure and function of fibronectin modules. *Matrix Biol. J. Int. Soc. Matrix Biol.* **1996**, *15*, 313–320. [CrossRef]
50. Hill, C.J.; Murray, P.; Mayans, O. Fusion Proteins Assembling into Scaffolds and Promoting Stem Cell Renewal 2019. Unpublished work.
51. Lam, J.; Segura, T. The modulation of MSC integrin expression by RGD presentation. *Biomaterials* **2013**, *34*, 3938–3947. [CrossRef]
52. Barczyk, M.; Carracedo, S.; Gullberg, D. Integrins. *Cell Tissue Res.* **2010**, *339*, 269–280. [CrossRef] [PubMed]
53. Pytela, R.; Pierschbacher, M.D.; Ruoslahti, E. A 125/115-kDa cell surface receptor specific for vitronectin interacts with the arginine-glycine-aspartic acid adhesion sequence derived from fibronectin. *Proc. Natl. Acad. Sci. USA* **1985**, *82*, 5766–5770. [CrossRef] [PubMed]
54. Aota, S.; Nomizu, M.; Yamada, K.M. The short amino acid sequence Pro-His-Ser-Arg-Asn in human fibronectin enhances cell-adhesive function. *J. Biol. Chem.* **1994**, *269*, 24756–24761. [PubMed]
55. Mas-Moruno, C.; Fraioli, R.; Rechenmacher, F.; Neubauer, S.; Kapp, T.G.; Kessler, H. αvβ3- or α5β1-Integrin-Selective Peptidomimetics for Surface Coating. *Angew. Chem. Int. Ed.* **2016**, *55*, 7048–7067. [CrossRef] [PubMed]
56. Humphries, J.D.; Byron, A.; Humphries, M.J. Integrin ligands at a glance. *J. Cell Sci.* **2006**, *119*, 3901–3903. [CrossRef] [PubMed]
57. Cheng, G.; Castelletto, V.; Jones, R.R.; Connon, C.J.; Hamley, I.W. Hydrogelation of self-assembling RGD-based peptides. *Soft Matter* **2011**, *7*, 1326–1333. [CrossRef]
58. Loo, Y.; Lakshmanan, A.; Ni, M.; Toh, L.L.; Wang, S.; Hauser, C.A.E. Peptide Bioink: Self-Assembling Nanofibrous Scaffolds for Three-Dimensional Organotypic Cultures. *Nano Lett.* **2015**, *15*, 6919–6925. [CrossRef] [PubMed]
59. Schaub, N.J.; Le Beux, C.; Miao, J.; Linhardt, R.J.; Alauzun, J.G.; Laurencin, D.; Gilbert, R.J. The Effect of Surface Modification of Aligned Poly-L-Lactic Acid Electrospun Fibers on Fiber Degradation and Neurite Extension. *PLoS ONE* **2015**, *10*, e0136780. [CrossRef]
60. Klinkhammer, K.; Bockelmann, J.; Simitzis, C.; Brook, G.A.; Grafahrend, D.; Groll, J.; Möller, M.; Mey, J.; Klee, D. Functionalization of electrospun fibers of poly(epsilon-caprolactone) with star shaped NCO-poly(ethylene glycol)-stat-poly(propylene glycol) for neuronal cell guidance. *J. Mater. Sci. Mater. Med.* **2010**, *21*, 2637–2651. [CrossRef]

 International Journal of
Molecular Sciences

Review

Strategies to Obtain Designer Polymers Based on Cyanobacterial Extracellular Polymeric Substances (EPS)

Sara B. Pereira [1,2,*,†], Aureliana Sousa [1,3,†], Marina Santos [1,2,4], Marco Araújo [1,3], Filipa Serôdio [1,3], Pedro Granja [1,3,5] and Paula Tamagnini [1,2,6]

[1] i3S - Instituto de Investigação e Inovação em Saúde, Universidade do Porto, Rua Alfredo Allen, 208, 4200-135 Porto, Portugal; filipa@ineb.up.pt (A.S.); marina.santos@ibmc.up.pt (M.S.); marco.araujo@i3s.up.pt (M.A.); filipa.serodio@ineb.up.pt (F.S.); pgranja@i3s.up.pt (P.G.); pmtamagn@ibmc.up.pt (P.T.)

[2] IBMC - Instituto de Biologia Celular e Molecular, Universidade do Porto, Rua Alfredo Allen, 208, 4200-135 Porto, Portugal

[3] INEB - Instituto de Engenharia Biomédica, Universidade do Porto, Rua Alfredo Allen, 208, 4200-135 Porto, Portugal

[4] ICBAS - Instituto de Ciências Biomédicas Abel Salazar, Rua de Jorge Viterbo Ferreira 228, 4050-313 Porto, Portugal

[5] FEUP - Faculdade de Engenharia, Departamento de Engenharia Metalúrgica e Materiais, Universidade do Porto, Rua Dr. Roberto Frias, 4200-465 Porto, Portugal

[6] FCUP - Faculdade de Ciências, Departamento de Biologia, Universidade do Porto, Rua do Campo Alegre, Edifício FC4, 4169-007 Porto, Portugal

* Correspondence: sarap@ibmc.up.pt; Tel.: +351-220-408-800

† These authors contributed equally to this work.

Received: 1 October 2019; Accepted: 12 November 2019; Published: 14 November 2019

Abstract: Biopolymers derived from polysaccharides are a sustainable and environmentally friendly alternative to the synthetic counterparts available in the market. Due to their distinctive properties, the cyanobacterial extracellular polymeric substances (EPS), mainly composed of heteropolysaccharides, emerge as a valid alternative to address several biotechnological and biomedical challenges. Nevertheless, biotechnological/biomedical applications based on cyanobacterial EPS have only recently started to emerge. For the successful exploitation of cyanobacterial EPS, it is important to strategically design the polymers, either by genetic engineering of the producing strains or by chemical modification of the polymers. This requires a better understanding of the EPS biosynthetic pathways and their relationship with central metabolism, as well as to exploit the available polymer functionalization chemistries. Considering all this, we provide an overview of the characteristics and biological activities of cyanobacterial EPS, discuss the challenges and opportunities to improve the amount and/or characteristics of the polymers, and report the most relevant advances on the use of cyanobacterial EPS as scaffolds, coatings, and vehicles for drug delivery.

Keywords: EPS-based biomaterials; cyanobacteria; designer biopolymers; extracellular polymeric substances (EPS), metabolic engineering; polymer functionalization

---

1. Introduction

Biopolymers are macromolecules produced by different organisms or derived from natural resources [1]. Owing to their biocompatibility, non-toxicity, flexibility, functionality, biodegradability, and possibility to be recycled by biological processes, they constitute a sustainable alternative to petrochemical-derived polymers [1–3]. Polysaccharides are a highly abundant and diverse group of biopolymers that can be found in all domains of life [4]. In fact, the most abundant biopolymers, cellulose and chitin, are polysaccharides [5,6]. In addition, polysaccharides display an enormous variation of physicochemical properties, such as solubility, viscosity, gelling capacity, chain length (degree of polymerization), linkage types, and patterns, which confers them the versatility to be used in a vast range of applications. Therefore, it is not surprising that they have been used for a long time in important commercial areas such as food, pharmaceuticals, biomedicine, electronics, and bioremediation [7].

Nowadays, the market of polysaccharides is dominated by the polymers isolated from plants (e.g., cellulose, starch, and pectins), algae (agar, alginate, carrageenan) and animals (heparin, chondroitin sulfate, hyaluronic acid), whereas those produced by bacteria still represent a small fraction of the global market [8]. However, the interest in microbial polysaccharides, particularly those of bacterial origin, is rapidly growing since they usually have shorter production times and are easier to isolate. In addition, it is possible to obtain these polymers while avoiding the use of environmentally damaging chemicals, harvesting from oceans or competition for valuable land and animal-related ethical issues [9,10]. Bacterial polysaccharide production can also occur under controlled conditions developing polymers with consistent features, whereas plant and algal polysaccharides are easily affected by climatological and geological environmental conditions [1]. Furthermore, bacterial synthesis offers an attractive alternative for the sustainable production of tailored biopolymers, reducing downstream processing [10]. Despite these advantages, the implementation of bacterial polymers in the market is hindered mainly by their high production costs [1]. Thus, the potential of these polymers is mainly in high-value market niches, such as cosmetics, pharmaceuticals, and biomedicine, in which other polymers fail to comply with the required degree of purity or lack specific functional properties. In such applications, product quality surpasses cost production and product yield issues [11].

Among bacteria, cyanobacteria emerge as excellent candidates for the production of polysaccharide polymers since most of the strains produce extracellular polymeric substances (EPS), mainly composed by heteropolysaccharides, with a distinctive set of industrially-desirable features including (i) strong anionic nature, (ii) presence of sulfate groups, (iii) high variety of possible structural conformations, and (iv) amphiphilic behavior [12,13]. In addition, the use of cyanobacteria as cell factories eliminates the need for carbon feedstocks, since their photoautotrophic metabolism allows a low-cost production while contributing to carbon dioxide sequestration. For this purpose, a deeper knowledge on the cyanobacterial EPS biosynthetic pathways is required, to both enhance productivity and engineer structural and compositional variants tailored for a given application.

In the last years, significant advances were made on the characterization of cyanobacterial EPS [13–16] and the validation of their biotechnological and biomedical potential as metal chelators [15, 17–21], flocculating, emulsifying or rheology modifiers [22–25] and/or agents with valuable biological activities (e.g., antiviral, antimicrobial, anticoagulant, antitumor) [26–34]. However, the development of biomaterials based on these polymers has only recently started to be explored [35–39]. In this context, the repertoire of cyanobacterial EPS-based biomaterials can be significantly expanded through the development of strategies to obtain designer biopolymers with specialized features, by either metabolic engineering and/or chemical functionalization. Therefore, the main aim of this review is to provide the state-of-the-art on the characteristics and biological properties of cyanobacterial EPS, discuss opportunities to improve the production/characteristics of polymers by metabolic engineering, list the strategies for their extraction, purification, and functionalization, and describe the most relevant advances on the production of cyanobacterial EPS-based biomaterials.

2. Cyanobacterial Extracellular Polymeric Substances (EPS)

2.1. Polymer Characteristics

The cyanobacterial EPS can remain attached to the cell surface as sheaths, capsules, or slimes, or be released into the surrounding environment, being designated as released polysaccharides (RPS) [12]. These polymers are heteropolysaccharides, being usually composed by a higher number of different monosaccharides (up to 13) than those produced by other bacteria, which typically contain four or fewer monomers [13]. Sugars commonly found in cyanobacterial EPS include the hexoses glucose, galactose, mannose, and fructose; the pentoses ribose, xylose, and arabinose; the deoxyhexoses fucose and rhamnose; the acidic hexoses glucuronic and galacturonic acids, the amino sugars glucosamine, galactosamine, N-acetyl galactosamine, and N-acetyl glucosamine [13].

Glucose is frequently the most abundant monosaccharide in cyanobacterial EPS. However, rhamnose, xylose, arabinose, fucose, mannose, and uronic acids have been found in higher amounts than glucose in some cyanobacterial polymers [40]. For example, uronic acids were the only constituents identified in the RPS of *Microcystis wesenbergii* [41], while another uncharacteristic case is that of *Cyanothece* sp. 113, which produces an extracellular polysaccharide constituted entirely by D-glucose [42]. Despite these examples, most cyanobacterial EPS have a strain-specific heterogeneous composition, which contributes to the astonishing diversity of cyanobacterial polymers. The high diversity of monosaccharidic building blocks, and consequent variety of linkages, is considered the main reason for the increased complexity and broad range of possible conformations of cyanobacterial EPS, setting them apart from other bacterial polymers [12,13]. For example, the EPS produced by *Mastigocladus laminosus* and *Cyanospira capsulata* contain repeating units of 15 monosaccharides [12]. Branching can also occur on different positions of a monosaccharide resulting in even higher structural diversity [43]. As a result, the cyanobacterial EPS usually possess a high molecular mass (in the order of MDa [14,34]) which has a direct influence on the rheological properties of the polymers [12]. The complexity of the EPS produced by cyanobacteria also make their structure elucidation challenging, and thus it is not surprising that cellulose is probably the best characterized polysaccharide in cyanobacteria [44].

Many cyanobacterial EPS also possess two different uronic acids (from 2% up to 80% of the total EPS dry weight, commonly between 15% and 30%), which is a rare feature in microbial EPS. In addition, they usually contain sulfate groups, which are usually present in the EPS produced by archaea and eukaryotic EPS but absent in those produced by bacteria. These last two features contribute to the overall anionic charge of the cyanobacterial polymers, making them suitable for a vast array of applications [12,45]. Importantly, these features are crucial for the functionalization of the polymers and contribute to their capacity to retain water and form hydrated gels [43]. The cyanobacterial EPS are also amphiphilic molecules, combining an hydrophilic fraction composed of sulfated sugars, uronic acids and ketal-linked pyruvyl groups and hydrophobic groups including ester-linked acetyl groups, deoxysugars (e.g., rhamnose and fucose) and peptidic fraction [12,13]. The presence of these hydrophobic groups strongly contributes to the emulsifying properties of polysaccharides [13].

Overall, the above-mentioned characteristics of these highly complex polymers make them very attractive for the biotechnological and biomedical fields.

2.2. Relevant Biological Activities

Initially, the research on cyanobacterial EPS was mainly focused on the potential of these polymers as bioremediation agents for the treatment of industrial and domestic wastewaters, namely for the removal of ammonia, phosphates, and heavy metals [17–20,46–48]. However, as knowledge accumulated, their putative antiviral, antimicrobial, antioxidant, anticoagulant, immunomodulatory, and antitumor activities started to be unveiled [26,29,30,32–34,49–58], opening the way for the use of cyanobacterial EPS in biomedical applications.

Due to the limited structural information available for cyanobacterial EPS, the relationship between their structures and biological activities is far from being understood. However, the available data suggests that the negative charge and presence of sulfate groups contributes significantly to the antiviral activity displayed by several polymers [29,30,49–51,59]. These effects are likely due to inhibition of fusion of the enveloped virus with its target membrane, either by impairing the virus–cell attachment or the direct interaction of the negative charges of the polymer with positive charges on the virus surface [60,61]. The antiviral activity of the polymers seems to be mainly dependent on the number of negative charges and the molecular weight [60]. In the case of the sulphated polymer calcium spirulan, isolated from *Arthrospira platensis*, it was suggested that the presence of sulfate groups provides an additional contribution to the antiviral activity of these polymers by chelating calcium ions, which helps to retain the molecular conformation of the polymer [51].

The antimicrobial activity of cyanobacterial products is also well documented in the literature (reviewed in [28]). However, many of the available data were obtained using crude extracts [62,63], and thus, it is not always easy to uncouple the effects of the EPS from those resulting from the other molecules. Despite these constraints, it was demonstrated that the EPS produced by *Synechocystis* sp. R10 and *Gloecapsa* sp. Gacheva 2007/R-06/1 display antimicrobial activity against a broad spectrum of the most common food-borne pathogens [31]. Extracts of EPS released by the cyanobacterium *Arthrospira platensis* also showed antimicrobial activity against both Gram-positive and Gram-negative bacteria. Importantly, different EPS extracts showed different activities, indicating the presence of different components that differ in their solubility in the solvents employed [52].

A strong correlation between the sulfate content of cyanobacterial polymers and its antioxidative and anticoagulant activities was also found [26,32,33,53], and the immunomodulatory effects of specific cyanobacterial EPS were demonstrated [54]. The presence of sulfate has also been associated to the antitumor activity displayed by some EPS [34,55], although further studies are required to unveil the exact contribution of the sulfate groups. The mechanism of selective cytotoxicity displayed by different EPS with antitumor properties is also being evaluated. Studies performed with EPS isolated from *Aphanothece halophytica*, *Nostoc sphaeroides*, *Aphanizomenon flosaquae*, and *Synechocystis* $\Delta sigF$ revealed that the antitumor effect of these polymers is due to the induction of apoptosis in the tumor cells [27,34,57,58].

The vast range of biological activities displayed by cyanobacterial EPS opens a new set of possibilities for its use. However, for this process to be viable, it is necessary to complement these investigations with efforts aiming at optimizing polymer yield and tailoring its composition for specific applications.

3. Strategies to Optimize Cyanobacterial EPS Production and/or Polymer Characteristics

Due to their minimal nutritional requirements, cyanobacteria constitute a sustainable platform for polymer production. Moreover, depending on the environmental conditions (e.g., favorable and regular conditions), their photosynthetic metabolism allows large-scale cultivation outdoors, either in closed systems or open ponds, minimizing the costs of energy supply compared to the cultivation of e.g., heterotrophic bacteria [64]. Nevertheless, it is important to take into consideration that some cyanobacterial strains can produce toxins, and although these strains are not used for EPS production, it is essential to monitor possible contamination of cultures and/or polymers with these substances, particularly in open systems.

Despite the advantages of using cyanobacteria for EPS production, to achieve economic viability it is necessary to optimize the production process, by (i) evaluating the best cultivation system and/or photobioreactor geometry (ii) determining the most favorable growth conditions including nutrients (carbon, macroelements, microelements), temperature, light and gases exchange, (iii) establishing of a zero-waste value chain by re-utilizing waste biomass, and (iv) optimizing downstream processing including extraction and purification of the EPS. These parameters may vary significantly depending on the strain, as already well established for the effect of the growth conditions on EPS production [12,13,15,16,65], and will not be discussed here. It is however important to emphasize that, depending on the strain, changes in the cultivation/growth conditions can affect both the amount as well as the composition of the EPS. Metabolic engineering approaches also provide an opportunity to optimize the amount of EPS produced and/or the polymers' characteristics in order to meet industrial demands [11,66]. However, the limited information available on the cyanobacterial EPS biosynthetic process has limited the use of this approach, but the information available in the literature can provide important clues guiding future actions.

3.1. Metabolic Engineering of EPS-Producing Strains

The connection between central metabolic pathways and EPS biosynthesis has been elucidated for several bacteria, opening the way for the successful optimization of EPS-producing strains such as the xanthan-producing *Xhantomonas campestris* [1,66]. More recently, the mechanisms of EPS production by cyanobacteria started to be unveiled, mainly using the model strain *Synechocystis* sp. PCC 6803 (hereafter *Synechocystis*) [67–71]. Nevertheless, more studies are necessary to fully understand this process in cyanobacteria.

Studies performed in several bacteria point out that, regardless of the variety of surface polysaccharides produced, their biosynthetic pathways are relatively conserved [72]. Generally, the EPS biosynthetic pathway starts with the activation of monosaccharides and its conversion into sugar nucleotides; then, the monosaccharides are sequentially transferred from the sugar nucleotide donors to carrier molecules and assembled as repeating units. Finally, the EPS are exported to the exterior of the cell [1,72]. These steps require the participation of three groups of proteins, namely (1) enzymes involved in the biosynthesis of the sugars nucleotides, (2) glycosyltransferases to transfer the sugars to specific acceptors, and (3) proteins involved in EPS assembly, polymerization, and export [1,73,74] (Figure 1).

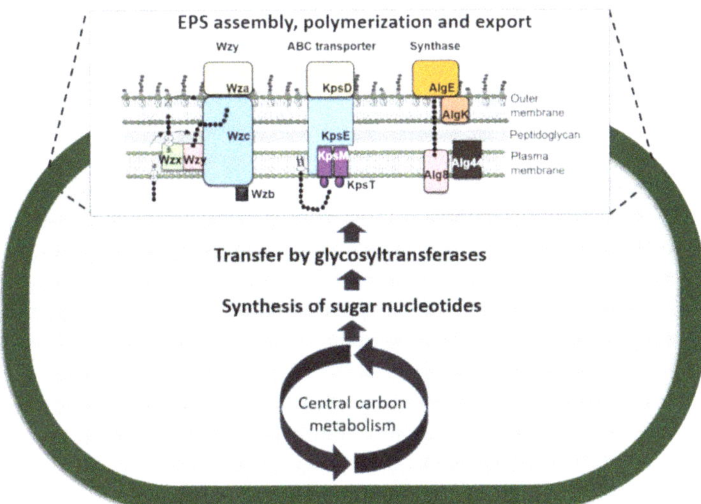

Figure 1. Sequence and compartmentalization of the events leading to the production of bacterial extracellular polymeric substances (EPS). EPS assembly, polymerization, and export usually follows one of three main mechanisms: the Wzy-, ABC transporter- or Synthase-dependent pathways. Adapted from [69].

All steps of the biosynthetic process offer opportunities for optimizing of the amount of EPS produced and/or its quality through genetic manipulation [11]. Here, we discuss the opportunities to improve cyanobacterial EPS production/characteristics by targeting carbon availability, synthesis of sugar nucleotide precursors, assembly of the repeating unit, and polymerization and export of the polymer.

3.1.1. Carbon Availability

The production of polysaccharides is a carbon-intensive and energy-demanding process that competes with cell's growth for available carbon resources. Thus, one of the strategies to improve EPS production consists of increasing the carbon pool of the cells, either by boosting the photosynthetic efficiency and/or the inorganic carbon intake. Previously, it was shown that the overexpression of the endogenous *Synechocystis* bicarbonate transporter BicA led to an increase in EPS production [75], and that high CO_2 pressure boosts the generation of these polymers in *Synechococcus* sp. PCC 8806 [76]. Carbon availability can also be increased by eliminating the carbon sinks and competing pathways, such as the production of glycogen, sucrose, and compatible solutes (e.g., glucosylglycerol). The branching points between *Synechocystis'* primary metabolism and sugar nucleotide, glycogen, sucrose, and glucosylglycerol pathways is depicted in Figure 2. Glycogen is a glucose storage polymer that, in cyanobacteria, can accumulate to levels of more than 50% of the cellular dry weight, depending on the growth conditions [77]. A *Synechocystis* mutant (Δ*glgC*) unable to produce glycogen possesses a higher energy charge and produces more organic acids [78]. The overexpression of the glycogen debranching enzyme GlgP, also results in massive decline of the glycogen content [79], compensating the carbon drain in an ethanol-producing *Synechocystis* mutant [79]. Although these studies unequivocally demonstrate that glycogen depletion increases the availability of carbon, it remains to be shown if this carbon surplus can be efficiently redirected towards EPS production. Regarding sucrose metabolism, the overexpression of Ugp (responsible for converting uridine triphosphate (UTP) and glucose-1-phosphate into uridine diphosphate (UDP)-glucose that serves as a substrate for sucrose and EPS synthesis) inhibited sucrose accumulation in *Synechocystis* under salt stress [80], raising the hypothesis that this effect may be due to the shift of carbon flux towards the synthesis of the exopolysaccharides [81]. A relationship between the glucosylglycerol metabolism and EPS synthesis in *Synechocystis* was also found. In this case, a mutant in a glucosylhydrolase (GghA) released higher amounts of polysaccharides (RPS) to the medium, suggesting a function of glucosylglycerol degradation via GghA in the synthesis and/or attachment of EPS to *Synechocystis* cells [82].

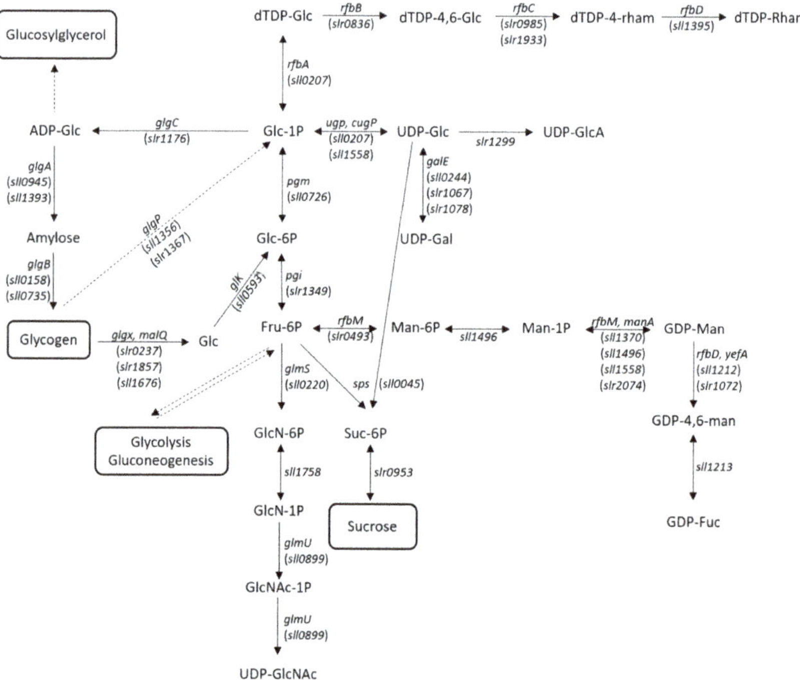

Figure 2. Branching points between the central carbon metabolism and the sugar nucleotide, glycogen, sucrose, and glucosylglycerol pathways in *Synechocystis* sp. PCC 6803 (based on Kegg database (https://www.genome.jp/kegg/) and [77,80,83,84]). Dotted lines indicate that the intermediary reactions

are not represented. The locus tag of the genes encoding the enzymes are indicated in parenthesis. ADP-Glc: ADP-glucose; dTDP-4,6-Glc: dTDP-4-oxo-6-deoxy-D-glucose; dTDP-4-rham: dTDP-4-dehydro-beta-L-rhamnose; dTDP-Glc: dTDP-D-glucose; dTDP-rham: dTDP-L-rhamnose; Fru-6P: D-Fructose 6-phosphate; GDP-4,6-man: GDP-4-dehydro-6-deoxy-D-mannose; GDP-Fuc: GDP-L-fucose; GDP-Man: GDP-D-mannose; Glc: glucose; Glc-1P: D-glucose 1-phosphate; Glc-6P: D-glucose 6-phosphate; GlcN-1P: D-Glucosamine 1-phosphate; GlcN-6P: D-Glucosamine 6-phosphate; GlcNAc-1P: N-Acetyl-D-glucosamine 1-phosphate; Man-1P: D-Mannose 1-phosphate; Man-6P: D-Mannose 6-phosphate; Suc-6P: Sucrose 6-phosphate; UDP-Gal: UDP-D-galactose; UDP-Glc: UDP-D-glucose; UDP-GlcA: UDP-D-glucuronate; UDP-GlcNAc: UDP-N-acetyl-D-glucosamine.

3.1.2. Synthesis of Sugar Nucleotide Precursors

A common bottleneck in microbial EPS production is the insufficient levels of sugar nucleotides [66,74]. This aspect is particularly relevant in Gram-negative bacteria, as these precursors are also required for the production of other surface polysaccharides, including the O-antigen of the lipopolysaccharides (LPS) and the S-layer glycans [85,86]. Thus, another strategy to increase cyanoabacterial EPS production consists of increasing the levels of sugar nucleotide precursors. However, the success of this approach is still controversial [74], since it is necessary to balance the carbon supply for sugar nucleotide synthesis with glycolysis [66,74]. Higher levels of sugar nucleotides can be achieved by overexpressing enzymes such as Ugp involved in the branching-point between the cell's primary metabolism and the sugar nucleotide pathway [10,66,74,87], as previously suggested (Figure 2) [80]. It is also necessary to consider the energetic requirements of sugar nucleotide synthesis. Availability of high-energy compounds such as adenosine triphosphate (ATP) and UTP may limit sugar nucleotide production, and therefore, strategies to increase the levels of cellular energy may also be advantageous for EPS production [66]. Finally, increasing or decreasing the synthesis of a certain type of nucleotide sugar precursor may have an impact on the EPS monosaccharidic composition [74]. Targeted modifications to obtain improved EPS for different applications are the increase in uronic acids (e.g., by targeting UDP-glucose dehydrogenase) and amino sugars (e.g., through modification of UDP-N-acetylglucosamine pyrophosphorylase). Enrichment in rare sugars such as rhamnose and fucose can also be advantageous to confer unique physical and bioactive properties to the polymers [8]. Recently, *Synechocystis'* mutants in the tyrosine kinase Sll0923 (Wzc homologue) and/or the low molecular weight tyrosine phosphatase Slr0328 (Wzb homologue) produced EPS enriched in rhamnose [70]. Similar results had been obtained for a mutant in the ATP-binding component (Sll0982; KpsT homologue) of an EPS-related ABC transporter [68], raising the hypothesis that rhamnose metabolism is closely associated with the last steps of EPS production. This is further supported by the presence of *slr0985*, encoding a dTDP-4-dehydrorhamnose 3,5-epimerase, in close proximity to *wzc* and *kpsT* [70].

3.1.3. Assembly of the Repeating Unit

Genetic engineering of glycosyltransferases offers a great opportunity for the optimization of the polymers' composition and structure [74]. Overexpression of a native glycosyltransferase may increase the incorporation of the substrate sugar, provided that sufficient amounts of the sugar nucleotide are available. Alternatively, new monosaccharides may be introduced into the polymer by heterologously expressing the corresponding glycosyltransferase genes [66]. New insights into the mechanism and structure of these enzymes will enable approaches to broaden the substrate specificity and/or to swap substrate and acceptor domains from different glycosyltransferases [66,88]. However, further knowledge on this class of enzymes is necessary, as most of the cyanobacterial glycosyltransferases identified have not been characterized biochemically, making it difficult to understand their exact role in the synthesis of EPS [15]. The enzymes responsible for methylation, acetylation and pyruvylation of the EPS can also be targeted to modulate the rheological behavior of the polymers [66]. Interestingly, a *Synechocystis* mutant in a putative methyltransferase (Slr1610) displayed differences in both the molecular weight and monosaccharidic composition of its EPS compared to the wildtype [68]. Despite the significant contribution of the sulfate groups for the biological activities of the polymers, genetic engineering strategies aiming to tailor the sulfate levels in cyanobacterial EPS remain unexplored. This could be achieved by targeting the sulfotransferases responsible for the transfer of sulfate to the polymers.

3.1.4. Polymerization and Export of the Polymer

A clear understanding of the last steps of EPS production and the structure/function of the proteins that participate in this process is essential to enable the rational design of engineering strategies (e.g., enzyme engineering, random mutagenesis and/or site-directed evolution) aiming at improving EPS production and/or tailoring the polymer length [10,88,89]. This last aspect is important to determine the rheological properties of the polymers as well as its potential for the production of biomaterials [66]. Therefore, targeted modification of the molecular weight by engineering the proteins involved in the polymerization, export, or degradation of the polymer (e.g., synthases, polymerases, glucosidases) represents a possibility to obtain new polymer variants [88], as successfully shown for xanthan gum and bacterial alginate [90,91].

Although the knowledge on the last steps of EPS production in cyanobacteria is limited, these mechanisms seem to be relatively conserved throughout bacteria, with the polymerization and export of the polymers usually following one of three main mechanisms: the Wzy-, ABC transporter-, or synthase-dependent pathways [88]. However, a phylum-wide analysis of cyanobacterial genomes reveled that most strains harbor gene-encoding proteins related to the three pathways but often not the complete set defining a single pathway, implying a more complex scenario than that observed for other bacteria [69]. This complexity raises the hypothesis of functional redundancy, either owing to the existence of multiple copies for some of the EPS-related genes/proteins and/or a crosstalk between the components of the different assembly and export pathways [69,70]. In agreement, mutational analyses showed that proteins related to both the Wzy- and the ABC-dependent pathways operate in *Synechocystis'* EPS production, although their exact roles have only recently started to be elucidated [67,68,70]. Further knowledge is required to identify the bottlenecks in polymer export and pinpoint the best candidates for chain length regulation in cyanobacteria. Despite that, it was recently shown that the truncation of the C-terminal region of the *Synechocystis'* polysaccharide copolymerase Wzc leads to an increase of the EPS attached to the cell [70] and that the deletion of a monooxigenase involved in polysaccharide degradation and recycling results in increased levels of RPS [92]. More studies are necessary to determine if these or similar modifications affect the length of the polymers obtained.

4. Isolation, Purification, and Functionalization of Cyanobacterial EPS

The isolation and purification of the polymers must be cost effective, scalable, and easy to perform. It is also important to take into consideration that the methods selected influence the polymers' yield and quality [15] and, thus, it may be necessary to adapt the protocols to the characteristics of the polymers and their final application [93]. One of the main aspects to consider is whether the EPS are attached to the cells or released to the culture medium (RPS). In the case of the EPS attached to the cells, detachment can be achieved using formaldehyde, glutaraldehyde, ethylenediaminetetraacetic acid (EDTA), sodium hydroxide, sonication, heating, cell washing with water, complexation, or ionic resins [15,16]. To select one of these methods, it is important to not only evaluate the yield, but also the levels of contamination of the polysaccharides with other cellular components. In addition, RPS are much easier to recover, being usually separated from cells by filtration and/or centrifugation. Once isolated, polymers are usually precipitated using ice cold absolute alcohols such as methanol, ethanol, or isopropanol and recovered [16,93]. The polarity of the alcohol and the low temperatures used have an impact on the yield of the polysaccharides and on the co-precipitation of impurities [16]. Despite the efficiency of selective alcohol precipitation, the costs and requirement of large amounts of precipitating agents led to the search of alternative techniques more suitable at the industrial scale, such as tangential ultrafiltration [16,94]. However, this methodology may need to be improved to minimize the problems of high viscosity of polymer solutions resulting in membrane clogging [16]. Tangential ultrafiltration can also be used to obtain a concentrated polymer solution before precipitation or spray-drying of the polymers, thus increasing the efficiency of these processes.

After isolation of the EPS, contaminants such as inorganic salts, heavy metals, proteins, polyphenols, endotoxins, nucleic acids, or cell debris may still be present in the polymer solution. However, it is necessary to have polysaccharides with high purity levels to accurately determine their structure and composition and to obtain reproducible results for therapeutic applications [95]. Inorganic salts, monosaccharides, oligosaccharides and low molecular weight non-polar substances can be removed by dialysis. The choice of device, the molecular weight cut-off, and duration of the dialysis is very important to determine the success of this method. However, at an industrial scale, dialysis may not be a viable option. An alternative way to remove inorganic salts is through ion exchange resins, normally in the form of beads [96]. Removal of peptides and proteins can be achieved using different methods, including protease (e.g., pronase) treatment or the Sevag method (usually less efficient) [96,97]. Trichlorotrifluoroethane and trichloroacetic acids can also be used to remove proteins from the polysaccharide's solution. However, it is necessary to consider that the first is highly volatile and, thus, has to be employed at 4 °C limiting its use, while the trichloroacetic acid is widely used but its acidity can damage the polymer structure [96,97]. The levels of polyphenol contaminants are usually reduced with charcoal washes and centrifugations, hydrogen peroxide method or functionalized resins with imidazole and pyridine [95,98]. The selection of the best purification methods depends on the characteristics of the polymers, the methods used for their isolation, and the envisaged application.

The presence of endotoxins is one of the major issues to be addressed before any biomaterial is consider safe to be used. Endotoxins are mainly due to the presence of LPS, with lipid A being responsible for most of the biological activity of these contaminants [99]. Endotoxins can significantly affect the biological effects of the polymers by eliciting a wide range of cellular responses that compromise cell viability [100,101]. Therefore, limits are imposed by regulatory entities ([102], pp. 171–175, 520–523) As an example, the food and drug administration (FDA) adopted the US Pharmacopoeia endotoxin reference standard, limiting the amount of endotoxins in eluates from medical devices to 0.5 Eurotoxin Units (EU)/mL [103]. Endotoxins are highly heat-stable and not easily destroyed by standard autoclave programs [104]. However, they can be removed by other techniques including ultrafiltration, two-phase extraction, and adsorption [99], although the efficiency of these methods depends on the characteristics of the polymer.

Depending on the application, it may be necessary to isolate fractions of the polymers with specific molecular weights. Fractionation is usually achieved by ultracentrifugation, with the added advantage of simultaneous elimination of contaminants [105]. Filtration and ultrafiltration are also popular alternatives, however, depending on the material of the filter membrane, the polysaccharides can be retained in the filter, decreasing the yield of the purification [106]. Other methods include affinity chromatography, gel chromatography, anion exchange chromatography, cellulose column chromatography, quaternary ammonium salt precipitation, graded precipitation methods, and preparative zone electrophoresis (reviewed in [96]).

The development of polysaccharide-based biomaterials often requires the chemical functionalization of the polymers. In this context, the characteristics of cyanobacterial EPS, offer a vast range of opportunities for targeted modifications (Figure 3). Successful examples of these functionalization reactions have already been described for other bacterial EPS [107–114]. The hydroxyl groups present in hexoses, pentoses, deoxyhexoses, uronic acids, and aminosugars can act as nucleophiles in base-catalyzed esterification reactions in the presence of anhydrides, esters, or carboxylic acids (Figure 3A–C). This strategy has been successfully used to fabricate photocrosslinkable hydrogels based on dextran and hyaluronic acid [108,110,114]. Another approach consists of the oxidation of diols in the presence of sodium periodate to generate reactive aldehydes, which can further react with primary amines in reductive amination reactions (Figure 3D) to produce hydrogels [107,112]. Hydroxyl groups can also undergo free radical polymerization reactions to generate graft copolymers for drug delivery (Figure 3E), as previously demonstrated for xanthan gum [109]. The carboxylic groups present in uronic acid residues allow the polymers' functionalization through esterification or carbodiimide reactions, with the latter being particularly interesting for bioconjugation (Figure 3F) [111,113]. On the other hand, free amino groups from glucosamine residues can react with anhydrides and carboxylic acids to form amides (Figure 3G,H), or with aldehydes to form Schiff bases, which can be further reduced to imines. Overall, these chemical modifications are valuable strategies to obtain designer polymers with improved properties suitable for the development of novel biomaterials.

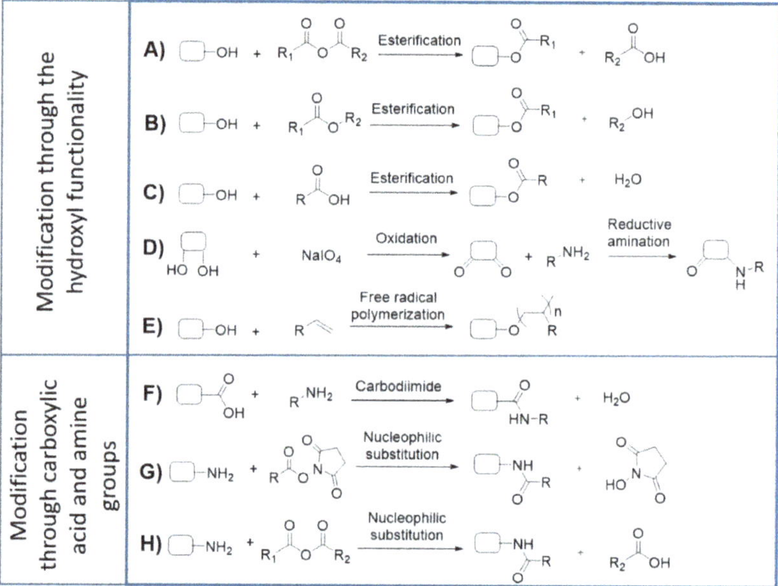

Figure 3. Representative strategies for functionalization of EPS by chemical modification. EPS can undergo esterification with anhydrides (**A**), esters (**B**), carboxylic acids (**C**), periodate-mediated oxidation followed by reductive amination (**D**), free radical polymerization with vinyl moiety (**E**), carbodiimide coupling (**F**) and nucleophilic substitution with esters and anhydrides (**G,H**), depending on the target functional group.

For their use in biomedical applications, the polymers and/or derived biomaterials have to be biocompatible, i.e., be able to "perform with an appropriate host response in a specific application" [115]. Biocompatibility is usually evaluated in vitro by accessing the effects that biopolymers or biomaterials have on living cells [116]. Several guidelines are described in international standard protocols, with the material's toxicity (defined as cytotoxicity) being the most common and widely used parameter evaluated (ISO 10993-5) [117]. Depending on the application, biotolerability, i.e., "the ability to reside in the body for long periods of time with only low degrees of inflammatory reaction" is an important issue to consider. This property is particularly important for non-degrading or slow-degrading implant materials [115]. Other important biosafety tests include the evaluation of the mutagenic and carcinogenic potential [118,119].

5. Development and Possible Applications of Cyanobacterial EPS-Based Biomaterials

Over the past few years, the development of biomaterials for therapeutic applications has become a rapidly expanding multidisciplinary field of research, with an increasing interest in uncovering novel polysaccharide-based scaffolds, coatings, and drug carriers [120]. Despite the potential of the cyanobacterial EPS and the vast range of opportunities to further improve the characteristics of the polymers by genetic engineering and/or chemical modification, the number of studies reporting their use as biomaterial is still very limited. Nevertheless, the available data represent an important step to validate the potential of cyanobacterial EPS.

The RPS produced by the cyanobacterium *Trichormus variabilis* VRUC 168 were combined with diacrylated polyethylene glycol to produce photopolymerizable hybrid hydrogels [35]. These gels were stable over time and resistant to dehydration and spontaneous hydrolysis, being successfully used as matrices for the active form of the enzyme thiosulfate:cyanide sulfur transferase, as well as for 3D culture system of human mesenchymal stem cells (hMSCs). In another study, the RPS produced by *Nostoc commune* were combined with glycerol to prepare biopolymeric films suitable for the development of new materials, including coatings and membranes [38]. Importantly, the simple and effective methodology developed allows control of the films' thickness and mechanical properties, thus expanding the repertoire of applications in the food and biomedical industries. The polymer produced by the strong RPS producer *Cyanothece* sp. CCY 0110 [14] was also shown to be a promising vehicle for topical administration of therapeutic macromolecules. This polymer was able to spontaneously assemble with functional proteins into a new phase with gel-like behavior, and the proteins were released progressively and structurally intact near physiological conditions, primarily through the swelling of the polymer–protein matrix. The release kinetics could be modulated by the addition of divalent cations, such as calcium [37]. The same polymer combined with arabic gum was also used to generate microparticles capable of encapsulating vitamin B12 [36]. More recently, the RPS isolated from this *Cyanothece* strain was used to produce an anti-adhesive coating, obtained by spin coating (for details see [39]). This coating efficiently prevents the adhesion of relevant etiological agents, even in the presence of plasma proteins, being an important step towards the establishment of a new technological platform capable of preventing medical device-associated infections [39].

6. Conclusions and Future Perspectives

Owing to their characteristics and biological activities, the EPS produced by cyanobacteria are a promising platform for biotechnological and biomedical applications, including the development of novel biomaterials for therapeutic applications. However, their successful exploitation largely depends on combined efforts to optimize the amount of EPS produced and tailor their characteristics. The recent advances in the knowledge of cyanobacterial EPS biosynthetic pathways pave the way for the generation of genetically modified strains. However, there are still challenges to address, including (i) a better understanding of the relationship between central metabolism and the synthesis of sugar nucleotides, (ii) the identification and characterizing of other key components of the EPS production machinery, and (iii) elucidation of the regulatory networks of the EPS production process. Further studies, taking into account high throughput data obtained from systems biology approaches and structural information of both proteins and polymers, will be crucial to address these issues. Moving beyond cellular processes, the chemical functionalization of the polymers can also significantly increase the repertoire of cyanobacterial EPS suitable for targeted applications. The implementation of this strategy is currently limited by the lack of knowledge on the structure of cyanobacterial polymers. However, the advent of new technologies and approaches will help to overtake this bottleneck. The results obtained in the (yet limited number of) studies reporting the use of cyanobacterial EPS-based biotechnology validate their potential, encouraging future endeavors.

Author Contributions: S.B.P.; A.S.; M.S.; M.A.; and F.S. performed the literature survey and prepared the original draft. S.B.P.; P.G.; and P.T. reviewed and edited the manuscript. All of the authors contributed to the final version of the manuscript.

Funding: This work was financed by FEDER—Fundo Europeu de Desenvolvimento Regional funds through the COMPETE 2020 - Operacional Programme for Competitiveness and Internationalisation (POCI), Portugal 2020, and by Portuguese funds through FCT—Fundação para a Ciência e a Tecnologia/Ministério da Ciência, Tecnologia e

Ensino Superior in the framework of the project POCI-01-0145-FEDER-028779, contract DL57/2016/CP1327/CT0007 and fellowship SFRH/BD/119920/2016.

Conflicts of Interest: The authors declare no conflict of interest.

Abbreviations

ATP	Adenosine triphosphate
EDTA	Ethylenediaminetetraacetic acid
EPS	Extracellular polymeric substances
FDA	Food and drug administration
hMSCs	Human mesenchymal stem cells
ISO	International organization for standardization
LPS	Lipopolysaccharides
RPS	Released polysaccharides
UDP	Uridine diphosphate
UTP	Uridine triphosphate

References

1. Ates, O. Systems Biology of Microbial Exopolysaccharides Production. *Front. Bioeng. Biotechnol.* **2015**, *3*, 200. [CrossRef]
2. Yadav, P.; Yadav, H.; Shah, V.G.; Shah, G.; Dhaka, G. Biomedical Biopolymers, their Origin and Evolution in Biomedical Sciences: A Systematic Review. *J. Clin. Diagn. Res.* **2015**, *9*, ZE21–ZE25. [CrossRef] [PubMed]
3. Bilal, M.; Iqbal, H.M.N. Naturally-derived biopolymers: Potential platforms for enzyme immobilization. *Int. J. Biol. Macromol.* **2019**, *130*, 462–482. [CrossRef] [PubMed]
4. Schmid, J. Recent insights in microbial exopolysaccharide biosynthesis and engineering strategies. *Curr. Opin. Biotechnol.* **2018**, *53*, 130–136. [CrossRef] [PubMed]
5. Klemm, D.; Heublein, B.; Fink, H.-P.; Bohn, A. Cellulose: Fascinating Biopolymer and Sustainable Raw Material. *Angew. Chem. Int. Ed.* **2005**, *44*, 3358–3393. [CrossRef] [PubMed]
6. Chen, X.; Gao, Y.; Wang, L.; Chen, H.; Yan, N. Effect of Treatment Methods on Chitin Structure and Its Transformation into Nitrogen-Containing Chemicals. *Chempluschem* **2015**, *80*, 1565–1572. [CrossRef]
7. Dassanayake, R.; Acharya, S.; Abidi, N. Biopolymer-Based Materials from Polysaccharides: Properties, Processing, Characterization and Sorption Applications. In *Advanced Sorption Process Applications*; Edebali, S., Ed.; IntechOpen: London, UK, 2018; pp. 1–24.
8. Roca, C.; Alves, V.D.; Freitas, F.; Reis, M.A. Polysaccharides enriched in rare sugars: Bacterial sources, production and applications. *Front. Microbiol.* **2015**, *6*, 288. [CrossRef] [PubMed]
9. Freitas, F.; Torres, C.A.V.; Reis, M.A.M. Engineering aspects of microbial exopolysaccharide production. *Bioresour. Technol.* **2017**, *245*, 1674–1683. [CrossRef]
10. Anderson, L.A.; Islam, M.A.; Prather, K.L.J. Synthetic biology strategies for improving microbial synthesis of "green" biopolymers. *J. Biol. Chem.* **2018**, *293*, 5053–5061. [CrossRef]
11. Freitas, F.; Alves, V.D.; Reis, M.A.M. Advances in bacterial exopolysaccharides: From production to biotechnological applications. *Trends Biotechnol.* **2011**, *29*, 388–398. [CrossRef]
12. Pereira, S.; Zille, A.; Micheletti, E.; Moradas-Ferreira, P.; De Philippis, R.; Tamagnini, P. Complexity of cyanobacterial exopolysaccharides: Composition, structures, inducing factors and putative genes involved in their biosynthesis and assembly. *FEMS Microbiol. Rev.* **2009**, *33*, 917–941. [CrossRef] [PubMed]
13. Rossi, F.; De Philippis, R. Exocellular Polysaccharides in Microalgae and Cyanobacteria: Chemical Features, Role and Enzymes and Genes Involved in Their Biosynthesis. In *The Physiology of Microalgae*; Borowitzka, M.A., Beardall, J., Raven, J.A., Eds.; Springer International Publishing: Cham, Switzerland, 2016; pp. 565–590.
14. Mota, R.; Guimaraes, R.; Buttel, Z.; Rossi, F.; Colica, G.; Silva, C.J.; Santos, C.; Gales, L.; Zille, A.; De Philippis, R.; et al. Production and characterization of extracellular carbohydrate polymer from *Cyanothece* sp. CCY 0110. *Carbohydr. Polym.* **2013**, *92*, 1408–1415. [CrossRef] [PubMed]
15. Bhunia, B.; Prasad Uday, U.S.; Oinam, G.; Mondal, A.; Bandyopadhyay, T.K.; Tiwari, O.N. Characterization, genetic regulation and production of cyanobacterial exopolysaccharides and its applicability for heavy metal removal. *Carbohydr. Polym.* **2018**, *179*, 228–243. [CrossRef]

16. Delattre, C.; Pierre, G.; Laroche, C.; Michaud, P. Production, extraction and characterization of microalgal and cyanobacterial exopolysaccharides. *Biotechnol. Adv.* **2016**, *34*, 1159–1179. [CrossRef] [PubMed]

17. Pereira, S.; Micheletti, E.; Zille, A.; Santos, A.; Moradas-Ferreira, P.; Tamagnini, P.; De Philippis, R. Using extracellular polymeric substances (EPS)-producing cyanobacteria for the bioremediation of heavy metals: Do cations compete for the EPS functional groups and also accumulate inside the cell? *Microbiology* **2011**, *157*, 451–458. [CrossRef] [PubMed]

18. Mota, R.; Rossi, F.; Andrenelli, L.; Pereira, S.B.; De Philippis, R.; Tamagnini, P. Released polysaccharides (RPS) from *Cyanothece* sp. CCY 0110 as biosorbent for heavy metals bioremediation: Interactions between metals and RPS binding sites. *Appl. Microbiol. Biotechnol.* **2016**, *100*, 7765–7775. [CrossRef] [PubMed]

19. De Philippis, R.; Colica, G.; Micheletti, E. Exopolysaccharide-producing cyanobacteria in heavy metal removal from water: Molecular basis and practical applicability of the biosorption process. *Appl. Microbiol. Biotechnol.* **2011**, *92*, 697–708. [CrossRef]

20. Gupta, P.; Diwan, B. Bacterial Exopolysaccharide mediated heavy metal removal: A Review on biosynthesis, mechanism and remediation strategies. *Biotechnol. Rep.* **2017**, *13*, 58–71. [CrossRef]

21. Ozturk, S.; Aslim, B.; Suludere, Z.; Tan, S. Metal removal of cyanobacterial exopolysaccharides by uronic acid content and monosaccharide composition. *Carbohydr. Polym.* **2014**, *101*, 265–271. [CrossRef]

22. Han, P.-p.; Sun, Y.; Wu, X.-y.; Yuan, Y.-j.; Dai, Y.-j.; Jia, S.-r. Emulsifying, Flocculating, and Physicochemical Properties of Exopolysaccharide Produced by Cyanobacterium *Nostoc flagelliforme*. *Appl. Biochem. Biotechnol.* **2014**, *172*, 36–49. [CrossRef]

23. Jindal, N.; Singh, D.P.; Khattar, J.I.S. Kinetics and physico-chemical characterization of exopolysaccharides produced by the cyanobacterium *Oscillatoria formosa*. *World J. Microbiol. Biotechnol.* **2011**, *27*, 2139–2146. [CrossRef]

24. Paniagua-Michel, J.d.J.; Olmos-Soto, J.; Morales-Guerrero, E.R. Chapter Eleven-Algal and Microbial Exopolysaccharides: New Insights as Biosurfactants and Bioemulsifiers. In *Advances in Food and Nutrition Research*, 1st ed.; Kim, S.-K., Ed.; Academic Press: Amsterdam, The Netherlands, 2014; pp. 221–257.

25. Bhatnagar, M.; Bhatnagar, A. Diversity of Polysaccharides in Cyanobacteria. In *Microbial Diversity in Ecosystem Sustainability and Biotechnological Applications: Volume 1. Microbial Diversity in Normal & Extreme Environments*; Satyanarayana, T., Johri, B.N., Das, S.K., Eds.; Springer Singapore: Singapore, 2019; pp. 447–496.

26. Hussein, M.H.; Abou-ElWafa, G.S.; Shaaban-Dessuuki, S.A.; Hassan, N.I. Characterization and Antioxidant Activity of Exopolysaccharide Secreted by *Nostoc carneum*. *Int. J. Pharmacol.* **2015**, *11*, 432–439. [CrossRef]

27. Li, H.; Su, L.; Chen, S.; Zhao, L.; Wang, H.; Ding, F.; Chen, H.; Shi, R.; Wang, Y.; Huang, Z. Physicochemical Characterization and Functional Analysis of the Polysaccharide from the Edible Microalga *Nostoc sphaeroides*. *Molecules* **2018**, *23*, 508. [CrossRef] [PubMed]

28. Dewi, I.C.; Falaise, C.; Hellio, C.; Bourgougnon, N.; Mouget, J.-L. Chapter 12 - Anticancer, Antiviral, Antibacterial, and Antifungal Properties in Microalgae. In *Microalgae in Health and Disease Prevention*; Levine, I.A., Fleurence, J., Eds.; Academic Press: Amsterdam, The Netherlands, 2018; pp. 235–261.

29. Kanekiyo, K.; Hayashi, K.; Takenaka, H.; Lee, J.B.; Hayashi, T. Anti-herpes simplex virus target of an acidic polysaccharide, nostoflan, from the edible blue-green alga *Nostoc flagelliforme*. *Biol. Pharm. Bull.* **2007**, *30*, 1573–1575. [CrossRef] [PubMed]

30. Ahmadi, A.; Moghadamtousi, S.Z.; Abubakar, S.; Zandi, K. Antiviral Potential of Algae Polysaccharides Isolated from Marine Sources: A Review. *Biomed. Res. Int.* **2015**, 825203. [CrossRef] [PubMed]

31. Najdenski, H.M.; Gigova, L.G.; Iliev, I.I.; Pilarski, P.S.; Lukavsky, J.; Tsvetkova, I.V.; Ninova, M.S.; Kussovski, V.K. Antibacterial and antifungal activities of selected microalgae and cyanobacteria. *Int. J. Food Sci. Technol.* **2013**, *48*, 1533–1540. [CrossRef]

32. Parwani, L.; Bhatnagar, M.; Bhatnagar, A.; Sharma, V. Antioxidant and iron-chelating activities of cyanobacterial exopolymers with potential for wound healing. *J. Appl. Phycol.* **2014**, *26*, 1473–1482. [CrossRef]

33. Majdoub, H.; Ben Mansour, M.; Chaubet, F.; Roudesli, M.S.; Maaroufi, R.M. Anticoagulant activity of a sulfated polysaccharide from the green alga *Arthrospira platensis*. *Biochim. Biophys. Acta* **2009**, *1790*, 1377–1381. [CrossRef]

34. Flores, C.; Lima, R.T.; Adessi, A.; Sousa, A.; Pereira, S.B.; Granja, P.L.; De Philippis, R.; Soares, P.; Tamagnini, P. Characterization and antitumor activity of the extracellular carbohydrate polymer from the cyanobacterium *Synechocystis* ΔsigF mutant. *Int. J. Biol. Macromol.* **2019**, *136*, 1219–1227. [CrossRef]

35. Bellini, E.; Ciocci, M.; Savio, S.; Antonaroli, S.; Seliktar, D.; Melino, S.; Congestri, R. *Trichormus variabilis* (Cyanobacteria) Biomass: From the Nutraceutical Products to Novel EPS-Cell/Protein Carrier Systems. *Mar. Drugs* **2018**, *16*, 298. [CrossRef]

36. Estevinho, B.N.; Mota, R.; Leite, J.P.; Tamagnini, P.; Gales, L.; Rocha, F. Application of a cyanobacterial extracellular polymeric substance in the microencapsulation of vitamin B12. *Powder Technol.* **2019**, *343*, 644–651. [CrossRef]

37. Leite, J.P.; Mota, R.; Durão, J.; Neves, S.C.; Barrias, C.C.; Tamagnini, P.; Gales, L. Cyanobacterium-Derived Extracellular Carbohydrate Polymer for the Controlled Delivery of Functional Proteins. *Macromol. Biosci.* **2017**, *17*, 1600206. [CrossRef] [PubMed]

38. Rodriguez, S.; Torres, F.G.; López, D. Preparation and Characterization of Polysaccharide Films from the Cyanobacteria *Nostoc commune. Polym. Renew. Resour.* **2017**, *8*, 133–150. [CrossRef]

39. Costa, B.; Mota, R.; Parreira, P.; Tamagnini, P. ; L.; Martins, M.C.; Costa, F. Broad-Spectrum Anti-Adhesive Coating Based on an Extracellular Polymer from a Marine Cyanobacterium. *Mar. Drugs* **2019**, *17*, 243. [CrossRef] [PubMed]

40. Li, P.; Harding, S.E.; Liu, Z. Cyanobacterial exopolysaccharides: Their nature and potential biotechnological applications. *Biotechnol. Genet. Eng. Rev.* **2001**, *18*, 375–404. [CrossRef]

41. Forni, C.; Telo, F.R.; Caiola, M.G. Comparative analysis of the polysaccharides produced by different species of *Microcystis* (Chroococcales, Cyanophyta). *Phycologia* **1997**, *36*, 181–185. [CrossRef]

42. Chi, Z.; Su, C.D.; Lu, W.D. A new exopolysaccharide produced by marine *Cyanothece* sp 113. *Bioresour. Technol.* **2007**, *98*, 1329–1332. [CrossRef]

43. Kehr, J.-C.; Dittmann, E. Biosynthesis and Function of Extracellular Glycans in Cyanobacteria. *Life* **2015**, *5*, 164–180. [CrossRef]

44. Nobles, D.R.; Romanovicz, D.K.; Brown, R.M. Cellulose in Cyanobacteria. Origin of Vascular Plant Cellulose Synthase? *Plant. Physiol.* **2001**, *127*, 529–542. [CrossRef]

45. Colica, G.; De Philippis, R. Exopolysaccharides from cyanobacteria and their possible industrial applications. In *Cyanobacteria*; Sharma, N.K., Rai, A.K., Stal, L.J., Eds.; John Wiley & Sons, Ltd.: New York, NY, USA, 2013; pp. 197–207.

46. Li, H.; Zhao, Q.Y.; Huang, H. Current states and challenges of salt-affected soil remediation by cyanobacteria. *Sci. Total Environ.* **2019**, *669*, 258–272. [CrossRef]

47. Anahas, A.M.P.; Muralitharan, G. Characterization of heterocystous cyanobacterial strains for biodiesel production based on fatty acid content analysis and hydrocarbon production. *Energ. Convers. Manage.* **2018**, *157*, 423–437. [CrossRef]

48. De Philippis, R.; Micheletti, E. Heavy Metal Removal with Exopolysaccharide-Producing Cyanobacteria. In *Heavy Metals in the Environment*; Wang, L.K., Chen, J.P., Hung, Y.T., Shammas, N.K., Eds.; CRC Press: Boca Raton, FL, USA, 2009; pp. 89–122.

49. Radonić, A.; Thulke, S.; Achenbach, J.; Kurth, A.; Vreemann, A.; König, T.; Walter, C.; Possinger, K.; Nitsche, A. Anionic Polysaccharides From Phototrophic Microorganisms Exhibit Antiviral Activities to Vaccinia Virus. *J. Antivir. Antiretrovir.* **2011**, *2*, 51–55. [CrossRef]

50. Hayashi, K.; Hayashi, T.; Kojima, I. A Natural Sulfated Polysaccharide, Calcium Spirulan, Isolated from *Spirulina platensis*: *In Vitro* and *ex Vivo* Evaluation of Anti-Herpes Simplex Virus and Anti-Human Immunodeficiency Virus Activities. *AIDS Res. Hum. Retrovir.* **1996**, *12*, 1463–1471. [CrossRef] [PubMed]

51. Hayashi, T.; Hayashi, K.; Maeda, M.; Kojima, I. Calcium spirulan, an inhibitor of enveloped virus replication, from a blue-green alga *Spirulina platensis. J. Nat. Prod.* **1996**, *59*, 83–87. [CrossRef] [PubMed]

52. Challouf, R.; Trabelsi, L.; Ben Dhieb, R.; El Abed, O.; Yahia, A.; Ghozzi, K.; Ben Ammar, J.; Omran, H.; Ben Ouada, H. Evaluation of cytotoxicity and biological activities in extracellular polysaccharides released by cyanobacterium *Arthrospira platensis. Braz. Arch. Biol. Technol.* **2011**, *54*, 831–838. [CrossRef]

53. Volk, R.-B.; Venzke, K.; Blaschek, W.; Alban, S. Complement Modulating and Anticoagulant Effects of a Sulfated Exopolysaccharide Released by the Cyanobacterium *Synechocystis aquatilis. Planta Med.* **2006**, *72*, 1424–1427. [CrossRef]

54. Gudmundsdottir, A.B.; Omarsdottir, S.; Brynjolfsdottir, A.; Paulsen, B.S.; Olafsdottir, E.S.; Freysdottir, J. Exopolysaccharides from *Cyanobacterium aponinum* from the Blue Lagoon in Iceland increase IL-10 secretion by human dendritic cells and their ability to reduce the IL-17(+)ROR gamma t(+)/IL-10(+)FoxP3(+) ratio in CD4(+) T cells. *Immunol. Lett.* **2015**, *163*, 157–162. [CrossRef]

55. Mishima, T.; Murata, J.; Toyoshima, M.; Fujii, H.; Nakajima, M.; Hayashi, T.; Kato, T.; Saiki, I. Inhibition of tumor invasion and metastasis by calcium spirulan (Ca-SP), a novel sulfated polysaccharide derived from a blue-green alga, *Spirulina platensis. Clin. Exp. Metastasis.* **1998**, *16*, 541–550. [CrossRef]

56. Gigova, L.; Toshkova, R.; Gardeva, E.; Gacheva, G.J.; Ivanova, N.; Yossifova, L.; Petkov, G. Growth inhibitory activity of selected microalgae and cyanobacteria towards human cervical carcinoma cells (HeLa). *J. Pharm. Res.* **2011**, *4*, 4702–4707.

57. Xue, X.; Lv, Y.; Liu, Q.; Zhang, X.; Zhao, Y.; Zhang, L.; Xu, S. Extracellular polymeric substance from *Aphanizomenon flos-aquae* induces apoptosis via the mitochondrial pathway in A431 human epidermoid carcinoma cells. *Exp. Ther. Med.* **2015**, *10*, 927–932. [CrossRef]

58. Ou, Y.; Xu, S.; Zhu, D.; Yang, X. Molecular mechanisms of exopolysaccharide from *Aphanothece halaphytica* (EPSAH) induced apoptosis in HeLa cells. *PLoS ONE* **2014**, *9*, e87223. [CrossRef] [PubMed]

59. Kanekiyo, K.; Lee, J.B.; Hayashi, K.; Takenaka, H.; Hayakawa, Y.; Endo, S.; Hayashi, T. Isolation of an antiviral polysaccharide, nostoflan, from a terrestrial cyanobacterium, *Nostoc flagelliforme. J. Nat. Prod.* **2005**, *68*, 1037–1041. [CrossRef]

60. Lüscher-Mattii, M. Polyanions—A Lost Chance in the Fight against HIV and other Virus Diseases? *Antivir. Chem. Chemother.* **2000**, *11*, 249–259. [CrossRef] [PubMed]

61. Witvrouw, M.; DeClercq, E. Sulfated polysaccharides extracted from sea algae as potential antiviral drugs. *Gen. Pharmacol.* **1997**, *29*, 497–511. [CrossRef]

62. Vijayakumar, S.; Menakha, M. Pharmaceutical applications of cyanobacteria-A review. *J. Acute Med.* **2015**, *5*, 15–23. [CrossRef]

63. Nunnery, J.K.; Mevers, E.; Gerwick, W.H. Biologically active secondary metabolites from marine cyanobacteria. *Curr. Opin. Biotechnol.* **2010**, *21*, 787–793. [CrossRef]

64. Pathak, J.; Rajneesh; Maurya, P.K.; Singh, S.P.; Häder, D.-P.; Sinha, R.P. Cyanobacterial Farming for Environment Friendly Sustainable Agriculture Practices: Innovations and Perspectives. *Front. Environ. Sci.* **2018**, *6*, 7. [CrossRef]

65. De Philippis, R.; Vincenzini, M. Exocellular polysaccharides from cyanobacteria and their possible applications. *FEMS Microbiol. Rev.* **1998**, *22*, 151–175. [CrossRef]

66. Ruffing, A.; Chen, R.R. Metabolic Engineering of Microorganisms for Oligosaccharide and Polysaccharide Production. In *Microbial Production of Biopolymers and Polymer Precursors: Applications and Perspectives*; Rehm, B.H.A., Ed.; Caister Academic Press: Norfolk, UK, 2009; pp. 197–228.

67. Jittawuttipoka, T.; Planchon, M.; Spalla, O.; Benzerara, K.; Guyot, F.; Cassier-Chauvat, C.; Chauvat, F. Multidisciplinary Evidences that *Synechocystis* PCC6803 Exopolysaccharides Operate in Cell Sedimentation and Protection against Salt and Metal Stresses. *PLoS ONE* **2013**, *8*, e55564. [CrossRef]

68. Fisher, M.L.; Allen, R.; Luo, Y.; Curtiss, R.; III. Export of Extracellular Polysaccharides Modulates Adherence of the Cyanobacterium *Synechocystis. PLoS ONE* **2013**, *8*, e74514. [CrossRef]

69. Pereira, S.B.; Mota, R.; Vieira, C.P.; Vieira, J.; Tamagnini, P. Phylum-wide analysis of genes/proteins related to the last steps of assembly and export of extracellular polymeric substances (EPS) in cyanobacteria. *Sci. Rep.* **2015**, *5*, 14835. [CrossRef] [PubMed]

70. Pereira, S.B.; Santos, M.; Leite, J.P.; Flores, C.; Eisfeld, C.; Büttel, Z.; Mota, R.; Rossi, F.; De Philippis, R.; Gales, L.; et al. The role of the tyrosine kinase Wzc (Sll0923) and the phosphatase Wzb (Slr0328) in the production of extracellular polymeric substances (EPS) by *Synechocystis* PCC 6803. *Microbiologyopen* **2019**, *8*, e753. [CrossRef] [PubMed]

71. Allen, R.; Rittmann, B.E.; Curtiss, R. Axenic Biofilm Formation and Aggregation by *Synechocystis* PCC 6803 is Induced by Changes in Nutrient Concentration, and Requires Cell Surface Structures. *Appl. Environ. Microbiol.* **2019**, *85*, e02192-18. [CrossRef] [PubMed]

72. Whitfield, C.; Larue, K. Stop and go: Regulation of chain length in the biosynthesis of bacterial polysaccharides. *Nat. Struct. Mol. Biol.* **2008**, *15*, 121–123. [CrossRef] [PubMed]

73. Reeves, P.R.; Hobbs, M.; Valvano, M.A.; Skurnik, M.; Whitfield, C.; Coplin, D.; Kido, N.; Klena, J.; Maskell, D.; Raetz, C.R.; et al. Bacterial polysaccharide synthesis and gene nomenclature. *Trends Microbiol.* **1996**, *4*, 495–503. [CrossRef]

74. Wang, J.; Salem, D.R.; Sani, R.K. Extremophilic exopolysaccharides: A review and new perspectives on engineering strategies and applications. *Carbohydr. Polym.* **2019**, *205*, 8–26. [CrossRef]

75. Kamennaya, N.A.; Eun Ahn, S.; Park, H.; Bartal, R.; Sasaki, K.A.; Holman, H.-Y.; Jansson, C. Installing extra bicarbonate transporters in the cyanobacterium *Synechocystis* sp. PCC6803 enhances biomass production. *Metab. Eng.* **2015**, *29*, 76–85. [CrossRef]

76. Kamennaya, N.A.; Zemla, M.; Mahoney, L.; Chen, L.; Holman, E.; Holman, H.-Y.; Auer, M.; Ajo-Franklin, C.M.; Jansson, C. High pCO2-induced exopolysaccharide-rich ballasted aggregates of planktonic cyanobacteria could explain Paleoproterozoic carbon burial. *Nat. Commun.* **2018**, *9*, 2116. [CrossRef]

77. Zilliges, Y. Glycogen: A Dynamic Cellular Sink and Reservoir for Carbon. In *The Cell Biology of Cyanobacteria*; Flores, E., Herrero, A., Eds.; Caister Academic Press: Norfolk, UK, 2014; pp. 189–210.

78. Cano, M.; Holland, S.C.; Artier, J.; Burnap, R.L.; Ghirardi, M.; Morgan, J.A.; Yu, J. Glycogen Synthesis and Metabolite Overflow Contribute to Energy Balancing in Cyanobacteria. *Cell Rep.* **2018**, *23*, 667–672. [CrossRef]

79. Pade, N.; Mikkat, S.; Hagemann, M. Ethanol, glycogen and glucosylglycerol represent competing carbon pools in ethanol-producing cells of *Synechocystis* sp. PCC 6803 under high-salt conditions. *Microbiology* **2017**, *163*, 300–307. [CrossRef]

80. Du, W.; Liang, F.; Duan, Y.; Tan, X.; Lu, X. Exploring the photosynthetic production capacity of sucrose by cyanobacteria. *Metab. Eng.* **2013**, *19*, 17–25. [CrossRef] [PubMed]

81. Ozturk, S.; Aslim, B. Modification of exopolysaccharide composition and production by three cyanobacterial isolates under salt stress. *Environ. Sci. Pollut. Res. Int.* **2010**, *17*, 595–602. [CrossRef] [PubMed]

82. Kirsch, F.; Pade, N.; Klähn, S.; Hess, W.R.; Hagemann, M. The glucosylglycerol-degrading enzyme GghA is involved in acclimation to fluctuating salinities by the cyanobacterium *Synechocystis* sp. strain PCC 6803. *Microbiology* **2017**, *163*, 1319–1328. [CrossRef] [PubMed]

83. Osanai, T.; Azuma, M.; Tanaka, K. Sugar catabolism regulated by light- and nitrogen-status in the cyanobacterium *Synechocystis* sp. PCC 6803. *Photochem. Photobiol. Sci.* **2007**, *6*, 508–514. [CrossRef] [PubMed]

84. Maeda, K.; Narikawa, R.; Ikeuchi, M. CugP Is a Novel Ubiquitous Non-GalU-Type Bacterial UDP-Glucose Pyrophosphorylase Found in Cyanobacteria. *J. Bacteriol.* **2014**, *196*, 2348–2354. [CrossRef] [PubMed]

85. Whitfield, C.; Trent, M.S. Biosynthesis and Export of Bacterial Lipopolysaccharides. *Annu. Rev. Biochem.* **2014**, *83*, 99–128. [CrossRef]

86. Ristl, R.; Steiner, K.; Zarschler, K.; Zayni, S.; Messner, P.; Schaffer, C. The S-Layer Glycome-Adding to the Sugar Coat of Bacteria. *Int. J. Microbiol.* **2011**, *2011*, 127870. [CrossRef]

87. Angermayr, S.A.; Gorchs Rovira, A.; Hellingwerf, K.J. Metabolic engineering of cyanobacteria for the synthesis of commodity products. *Trends Biotechnol.* **2015**, *33*, 352–361. [CrossRef]

88. Schmid, J.; Sieber, V.; Rehm, B. Bacterial exopolysaccharides: Biosynthesis pathways and engineering strategies. *Front. Microbiol.* **2015**, *6*, 496. [CrossRef]

89. Rehm, B.H. Bacterial polymers: Biosynthesis, modifications and applications. *Nat. Rev. Microbiol.* **2010**, *8*, 578–592. [CrossRef]

90. Galvan, E.M.; Ielmini, M.V.; Patel, Y.N.; Bianco, M.I.; Franceschini, E.A.; Schneider, J.C.; Ielpi, L. Xanthan chain length is modulated by increasing the availability of the polysaccharide copolymerase protein GumC and the outer membrane polysaccharide export protein GumB. *Glycobiology* **2013**, *23*, 259–272. [CrossRef] [PubMed]

91. Diaz-Barrera, A.; Soto, E.; Altamirano, C. Alginate production and *alg8* gene expression by *Azotobacter vinelandii* in continuous cultures. *J. Ind. Microbiol. Biotechnol.* **2012**, *39*, 613–621. [CrossRef] [PubMed]

92. Miranda, H.; Immerzeel, P.; Gerber, L.; Hörnaeus, K.; Lind, S.B.; Pattanaik, B.; Lindberg, P.; Mamedov, F.; Lindblad, P. Sll1783, a monooxygenase associated with polysaccharide processing in the unicellular cyanobacterium *Synechocystis* PCC 6803. *Physiol. Plant.* **2017**, *161*, 182–195. [CrossRef] [PubMed]

93. Flores, C.; Tamagnini, P. Looking Outwards: Isolation of Cyanobacterial Released Carbohydrate Polymers and Proteins. *JoVE* **2019**, *147*, e59590. [CrossRef]

94. Patel, A.K.; Laroche, C.; Marcati, A.; Ursu, A.V.; Jubeau, S.; Marchal, L.; Petit, E.; Djelveh, G.; Michaud, P. Separation and fractionation of exopolysaccharides from *Porphyridium cruentum*. *Bioresour. Technol.* **2013**, *145*, 345–350. [CrossRef]

95. Ghisalberti, E.L. Detection and Isolation of Bioactive Natural Products. In *Bioactive Natural Products: Detection, Isolation, and Structural Determination*, 2nd ed.; Colegate, S.M., Molyneux, R.J., Eds.; CRC Press: Boca Raton, FL, USA, 2007; pp. 11–76.

96. Shi, L. Bioactivities, isolation and purification methods of polysaccharides from natural products: A review. *Int. J. Biol. Macromol.* **2016**, *92*, 37–48. [CrossRef]

97. Chaplin, M.F.; Kennedy, J.F. *Carbohydrate Analysis: A Practical Approach*, 2nd ed.; Oxford University Press: Oxford, UK, 1994.

98. Silva, M.; Castellanos, L.; Ottens, M. Capture and Purification of Polyphenols Using Functionalized Hydrophobic Resins. *Ind. Eng. Chem. Res.* **2018**, *57*, 5359–5369. [CrossRef]

99. Petsch, D.; Anspach, F.B. Endotoxin removal from protein solutions. *J. Biotechnol.* **2000**, *76*, 97–119. [CrossRef]

100. Gao, B.; Tsan, M.-F. Endotoxin Contamination in Recombinant Human Heat Shock Protein 70 (Hsp70) Preparation Is Responsible for the Induction of Tumor Necrosis Factor α Release by Murine Macrophages. *J. Biol. Chem.* **2003**, *278*, 174–179. [CrossRef]

101. Suri, R.M.; Austyn, J.M. Bacterial lipopolysaccharide contamination of commercial collagen preparations may mediate dendritic cell maturation in culture. *J. Immunol. Methods* **1998**, *214*, 149–163. [CrossRef]

102. Commission, E.P. *European Pharmacopoeia 7.0*; Council Of Europe: European Directorate for the Quality of Medicines and Healthcare: Strasbourg, France, 2010; pp. 171–175, 520–523.

103. *Guidance for Industry Pyrogen and Endotoxins Testing: Questions and Answers*; U.S. Department of Health and Human Services Food and Drug Administration: Washington, DC, USA, 2012.

104. Gorbet, M.B.; Sefton, M.V. Endotoxin: The uninvited guest. *Biomaterials* **2005**, *26*, 6811–6817. [CrossRef] [PubMed]

105. Zhang, Z.-P.; Shen, C.-C.; Gao, F.-L.; Wei, H.; Ren, D.-F.; Lu, J. ; Isolation, Purification and Structural Characterization of Two Novel Water-Soluble Polysaccharides from *Anredera cordifolia*. *Molecules* **2017**, *22*, 1276. [CrossRef] [PubMed]

106. Li, J.; Chase, H.A. Applications of membrane techniques for purification of natural products. *Biotechnol. Lett.* **2010**, *32*, 601–608. [CrossRef] [PubMed]

107. Azzam, T.; Eliyahu, H.; Makovitzki, A.; Linial, M.; Domb, A.J. Hydrophobized dextran-spermine conjugate as potential vector for in vitro gene transfection. *J. Control. Release* **2004**, *96*, 309–323. [CrossRef] [PubMed]

108. Kim, S.-H.; Chu, C.-C. Synthesis and characterization of dextran–methacrylate hydrogels and structural study by SEM. *J. Biomed. Mater. Res.* **2000**, *49*, 517–527. [CrossRef]

109. Kumar, A. ; Deepak; Sharma, S.; Srivastava, A.; Kumar, R. Synthesis of xanthan gum graft copolymer and its application for controlled release of highly water soluble Levofloxacin drug in aqueous medium. *Carbohydr. Polym.* **2017**, *171*, 211–219. [CrossRef]

110. Oudshoorn, M.H.M.; Rissmann, R.; Bouwstra, J.A.; Hennink, W.E. Synthesis of methacrylated hyaluronic acid with tailored degree of substitution. *Polymer* **2007**, *48*, 1915–1920. [CrossRef]

111. Palma, S.I.C.J.; Rodrigues, C.A.V.; Carvalho, A.; Morales, M.D.; Freitas, F.; Fernandes, A.R.; Cabral, J.M.S.; Roque, A.C.A. A value-added exopolysaccharide as a coating agent for MRI nanoprobes. *Nanoscale* **2015**, *7*, 14272–14283. [CrossRef]

112. Tang, Y.J.; Sun, J.; Fan, H.S.; Zhang, X.D. An improved complex gel of modified gellan gum and carboxymethyl chitosan for chondrocytes encapsulation. *Carbohydr. Polym.* **2012**, *88*, 46–53. [CrossRef]

113. Theilacker, C.; Coleman, F.T.; Mueschenborn, S.; Llosa, N.; Grout, M.; Pier, G.B. Construction and characterization of a *Pseudomonas aeruginosa* mucoid exopolysaccharide-alginate conjugate vaccine. *Infect. Immun.* **2003**, *71*, 3875–3884. [CrossRef]

114. van Dijk-Wolthuis, W.N.E.; Franssen, O.; Talsma, H.; van Steenbergen, M.J.; Kettenes-van den Bosch, J.J.; Hennink, W.E. ; Synthesis, Characterization, and Polymerization of Glycidyl Methacrylate Derivatized Dextran. *Macromolecules* **1995**, *28*, 6317–6322. [CrossRef]

115. Ratner, B.D. The Biocompatibility Manifesto: Biocompatibility for the Twenty-first Century. *J. Cardiovasc. Transl. Res.* **2011**, *4*, 523–527. [CrossRef] [PubMed]

116. Sousa, A.; Neves, S.C.; Gonçalves, I.C.; Barrias, C.C. cells. In *Characterization of Polymeric Biomaterials*; Tanzi, M.C., Farè, S., Eds.; Woodhead Publishing: Sawston, Cambridgeshire, England, 2017; pp. 285–315.

117. Technical Committee ISO/TC 194. Part 5: Tests for in vitro cytotoxicity. In *Biological Evaluation of Medical Devices*, 3rd ed.; International Organization for Standardization: Geneva, Switzerland, 2009; ISO 10993-5; p. 34.

118. Doak, S.H.; Manshian, B.; Jenkins, G.J.S.; Singh, N. *In vitro* genotoxicity testing strategy for nanomaterials and the adaptation of current OECD guidelines. *Mutat. Res. Genet. Toxicol. Environ. Mutagen.* **2012**, *745*, 104–111. [CrossRef] [PubMed]

119. Cowie, H.; Magdolenova, Z.; Saunders, M.; Drlickova, M.; Carreira, S.C.; Kenzaoi, B.H.; Gombau, L.; Guadagnini, R.; Lorenzo, Y.; Walker, L.; et al. Suitability of human and mammalian cells of different origin for the assessment of genotoxicity of metal and polymeric engineered nanoparticles. *Nanotoxicology* **2015**, *9*, 57–65. [CrossRef] [PubMed]

120. Tchobanian, A.; Van Oosterwyck, H.; Fardim, P. Polysaccharides for tissue engineering: Current landscape and future prospects. *Carbohydr. Polym.* **2019**, *205*, 601–625. [CrossRef]

 International Journal of
Molecular Sciences

Article

Self-Association of Antimicrobial Peptides: A Molecular Dynamics Simulation Study on Bombinin

Peicho Petkov [1], Elena Lilkova [2], Nevena Ilieva [2,3,*] and Leandar Litov [1]

[1] Faculty of Physics, Atomic Physics Department, Sofia University "St. Kliment Ohridski", 5 J. Bouchier Blvd, 1164 Sofia, Bulgaria

[2] Institute of Information and Communication Technologies at the Bulgarian Academy of Sciences, Acad. G. Bonchev Str., Block 25A, 1113 Sofia, Bulgaria

[3] Institute of Informatics and Mathematics at the Bulgarian Academy of Sciences, Acad. G. Bonchev Str., Block 8, 1113 Sofia, Bulgaria

* Correspondence: nevena.ilieva@parallel.bas.bg

Received: 2 October 2019; Accepted: 28 October 2019; Published: 1 November 2019

Abstract: Antimicrobial peptides (AMPs) are a diverse group of membrane-active peptides which play a crucial role as mediators of the primary host defense against microbial invasion. Many AMPs are found to be fully or partially disordered in solution and to acquire secondary structure upon interaction with a lipid membrane. Here, we report molecular dynamics simulations studies on the solution behaviour of a specific AMP, bombinin H2. We show that in monomeric form in water solution the peptide is somewhat disordered and preferably adopts a helix-loop-helix conformation. However, when more than a single monomer is placed in the solution, the peptides self-associate in aggregates. Within the aggregate, the peptides provide each other with an amphipathic environment that mimics the water–membrane interface, which allows them to adopt a single-helix structure. We hypothesise that this is the mechanism by which bombinin H2 and, possibly, other small linear AMPs reach the target membrane in a functional folded state and are able to effectively exert their antimicrobial action on it.

Keywords: antimicrobial peptides; self-association; aggregation; promotion of folding

---

1. Introduction

Antimicrobial peptides (AMPs) are a crucial part of the nonspecific innate immunity of all eukaryotes to microbial invasion [1–3]. They are a diverse group of compounds, displaying various types of structures, including α-helices, β-sheets or cyclic structures. Nonetheless, AMPs share some general characteristics: they consist of 6–100 amino acids, usually are cationic and have an amphiphilic nature. Most importantly, AMPs are generally membrane active peptides that interact with target membranes and can cause cell death through various mechanisms [1,4,5]: they can disturb the membrane by inducing thinning, altering its curvature or fluidity, modifying the transmembrane electrochemical gradient, and inducing pore formation [5,6].

Experimental data and theoretical studies have shown that many AMPs are partially or fully disordered in solution and acquire their functional secondary structure upon interaction with the amphiphilic membrane–solvent interface [6–10]. The issue of whether secondary structure content is necessary for successful attack on the bacterial membrane is still controversial [11–14]. In addition, the activity of many AMPs depends on the peptide concentration and it has been shown that AMPs self-organise and cooperatively form pores upon interaction with a lipid bilayer [11,15–23]. AMP aggregation in solution and in the absence of a membrane is not well studied, while this certainly is the first stage of their activity pathway whose significance is that way largely underestimated. This is all the more important in view of recently reported observations [24] about the decrease in AMPs activity upon aggregation due to increased membrane-embedding free-energy costs. In the same time, in a very recent paper [25], the authors demonstrated that not only was α-helical structure necessary for the antimicrobial action of a specific AMP—halictine-1—but also the *"mechanism of the peptide mode of action probably involves formation of peptide assemblies (possibly membrane pores), which disrupt bacterial membrane and, consequently, allow membrane penetration"*.

We consider the issue of whether aggregation is beneficial or unfavourable for AMPs efficient action is still controversial and probably depends on the type of AMP. The design of AMPs with predefined properties requires a detailed and precise understanding of their mechanism of action that allows identification of crucial for each step of this mechanism residues to be targeted for optimisation. Therefore, the process of peptide aggregation has to be especially taken into account in designing potent AMPs for therapeutic purposes, with the attention focused on aggregation-prone amino acids and amino-acid motifs.

In recent papers [26,27], we discussed the process of peptide aggregation and its effect on the monomer's secondary structure in the case of one intrinsically disordered AMP (indolicidin) and another linear α-helical AMP, magainin 2. Here, we report molecular dynamics (MD) simulations studies on the solution behaviour of a specific AMP, bombinin H2 (amino acid sequence IIGPVLGLVGSALGGLLKKI), secreted by the skin of the European *Bombina variegata* frog species. It is active against both Gram-positive and Gram-negative bacteria, and also fungi. In addition, bombinin H2 peptides display hemolytic activity at relatively low concentrations [28]. These peptides are rich in glycine (25%), which allows them to adopt different conformations [29]: α-helical, partially disordered and even β-sheet structures [28]. However, at physiological salt concentrations and pH levels bombinin H2 peptides usually form classical amphiphilic α-helices at the lipid bilayer [29].

In this work, we demonstrate that the monomeric bombinin H2 peptide in water solution is somewhat disordered and preferably adopts a helix-loop-helix conformation. When multiple peptide chains are present in the solution, they rapidly self-associate in aggregates. Aggregation promotes further folding of bombinin H2 by mimicking the water–membrane amphipathic interface. Individual monomers adopt a single-helix structure and are stabilised in this conformational state.

2. Results

2.1. Dynamics of the Bombinin H2 Monomer in Solution

We first focus on the dynamics of an isolated bombinin H2 monomer in water solution. We started the production MD simulation from a helix-loop-helix conformation of the peptide. The initial experimental all α-helical structure rapidly adopts this state within the 10 ns NPT equilibration simulation. The bend is at $\text{Gly}^{10}\text{-Ser}^{11}$.

The peptide explores various conformations multiple times during the first 1.2 μs of the simulation. During the last 800 ns of the simulation, bombinin H2 stabilises in the collapsed helix-loop-helix conformation.

The cluster analysis reveals six clusters, accommodating about 90% of the conformations, and a number of smaller clusters without statistical significance, each of them containing less than 2% of the structure (Figure S1). In Figure S2, the cluster-size distribution and the centroids of the four largest clusters are shown. The first one, with 1132 conformation states (almost 57% of all states), corresponds to the initial structure—a rather regular V-shaped helix-loop-helix conformation. The fully folded (single-helix) conformations form the second-largest cluster, encompassing some 15% of the structures. Clusters 3 and 4 again correspond to V-shaped helix-loop-helix conformations with a tendency towards unfolding of the N-terminal helical part. Clusters 5 and 6 (not shown) follow the same tendency and the remaining clusters, summing up to about 9% of all conformations, have no statistical significance as each of them does not exceed 2% of the states. There are numerous transitions between the two main conformational states—a single classical linear α-helix and helix-loop-helix. This can be clearly seen from the evolution of the gyration radius of the peptide (Figure 1a).

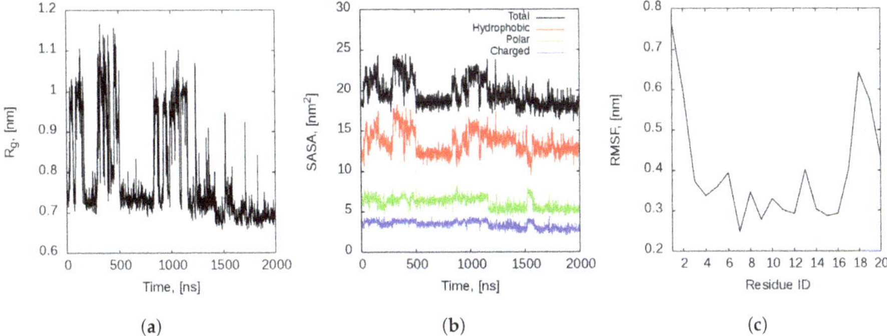

Figure 1. Time evolution of the peptide: (**a**) gyration radius; (**b**) SASA; and (**c**) root mean square fluctuations per residue of monomeric bombinin H2 in water.

The peptide behaviour seems to be driven by the hydrophobic effect. Examination of the solvent accessible surface area (SASA) of the peptide (Figure 1b) reveals that, while the charged and polar amino acid sidechains are solvent-exposed in both states, the compact helix-loop-helix conformation reduces the solvent exposure of the hydrophobic residues in the middle of the peptide (Val[5], Leu[6], Val[9], Leu[13] and Leu[17]).

The most flexible part of the peptide molecule are the two termini, and especially the N-terminus (residues Ile[1]–Val[5]). This is reflected in the plot of the root mean square fluctuations (RMSF) per amino acid residue (Figure 1c). The higher flexibility of these amino acid residues is consistent with the experimental data by Zangger et al. [29]. They observed that, even in a lipid bylayer, bombinin H2 has a well defined α-helical structure only between residues Val[5] and Lys[17]. Henceforth, this is the amino acid range that we use to determine if the peptide is in the single-helix or helix-loop-helix state.

The secondary structure plot (Figure 2a) also demonstrates the multiple transitions between the two main conformational states. Not only does the single helix break at Gly[10]–Ser[11], but the peptide does explore some very disordered conformations, where the whole N-terminal or C-terminal part of the molecule is not folded (e.g., the intervals 120–500 ns, 960–975 ns and 1520–1532 ns). Population of such disordered states is in agreement with experimental data [29].

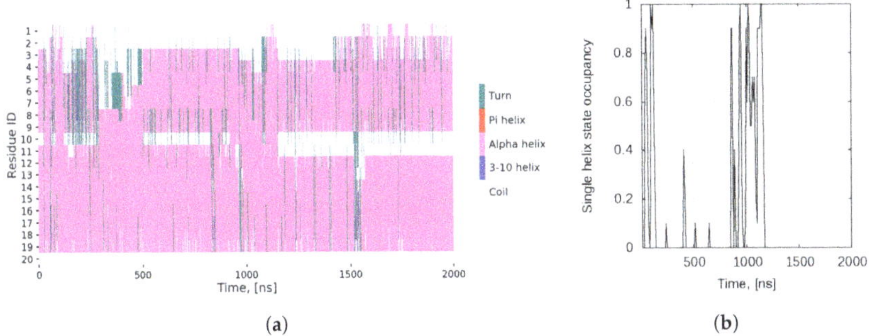

Figure 2. (**a**) Secondary structure; and (**b**) occupancy of the single α-helix conformational state, averaged over 10 ns windows, of monomeric bombinin H2 in water.

The single-helix state is fairly regularly visited, but apparently it is not very stable, since the peptide does not remain in it for long intervals. Figure 2b shows the occupancy of this state, averaged over 10 ns windows. After the 1157th ns, this conformation is no longer adopted at all and the structure transitions permanently to the helix-loop-helix conformational basin. On average, the peptide resides in a classical linear helix conformation in 11.7% of the trajectory frames.

Cartesian principal component analysis (PCA) on the backbone of the peptide confirms that the main mode of motion is indeed the bending of the linear helix (Figure 3a). The projection of the eigenvector with the largest eigenvalue (Principal Component 1 (PC1)) correlates perfectly with the gyration radius of the molecule (Figure 3b). In the first half of the simulation, the peptide visits conformations in both basins—the linear single-helix ($R_g \in [0.9, 1.1]$ nm, $PC_1 \in [-4, -2]$ nm) and the compact helix-loop-helix ($R_g \in [0.55, 0.80]$ nm, $PC_1 \in [-0.5, 2.5]$ nm). In the second half of the simulation, the peptide permanently transitions to the second basin. This demonstrates that PC1 corresponds entirely to the compactification of the peptide, in order to reduce the solvent exposure of hydrophobic residues.

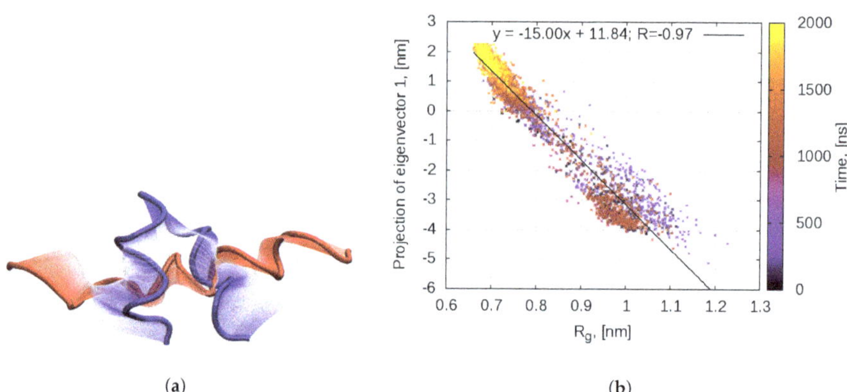

(a)   (b)

Figure 3. (a) Bombinin H2 peptide backbone motion along the first principal component; and (b) correlation between the projection of first principal component and the gyration radius of the peptide.

2.2. Peptide Self-Assembly in Water Solution

Our working hypothesis is that small linear AMPs self-associate into larger compounds (aggregates) prior to attacking the bacterial membrane. Therefore, we studied the evolution of a set of 27 bombinin H2 monomers in water, the peptides being placed in a cubic simulation box sized $15 \times 15 \times 15$ nm^3 (see Section 4). The box size meets the requirement for a minimal distance of 5 nm between two neighbouring monomers, leading to a simulated system of over 325,000 atoms, with a bombinin H2 concentration of 13 mM. The simulated evolution was 2.2 μs. In a somewhat similar study [30], multiple compositions of tetrapeptides were simulated in a cubic box for 100 ns per composition but at about two times higher concentration.

We observed two processes that take place simultaneously when more than one single bombinin H2 peptide chain is placed in the simulation box: a gradual self-association process of the peptides into larger aggregates with an increasing number of monomers and a decreasing number of aggregates, respectively, until all 27 monomers formed a single aggregate that remained stable until the end of the simulation, and a self-organisation within this aggregate that promotes folding of the individual peptide chains.

2.2.1. Aggregation

Although each of the peptide chains has a +2e net positive charge, the peptides do not repulse each other, but rather start to very rapidly aggregate. Figure 4a shows the number of aggregates that are formed in the simulation box and the number of peptide chains participating in the largest aggregate. The number of aggregates starts at 27, as we have 27 peptide chains separated in space.

Then, it decreases very quickly as smaller aggregates lump together and the maximal aggregate size grows.

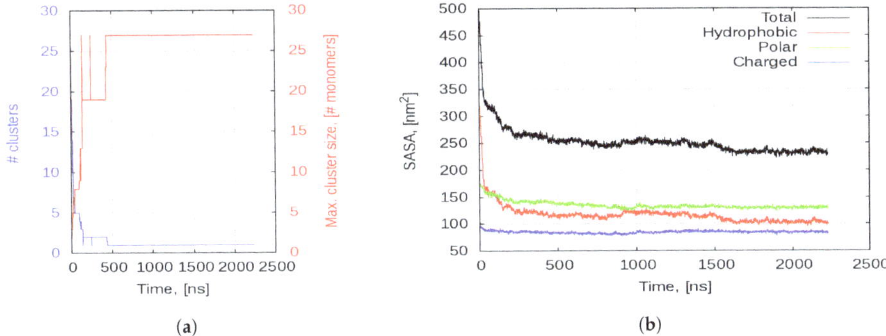

Figure 4. (**a**) Number of separate clusters/aggregates in the solution (blue curve) and maximal size of the clusters (red curve); and (**b**) SASA of all peptides in the solution.

Within the first 4–5 ns of the trajectory, the first few dimers and trimers are formed (Figure 5A). At 15–16 ns, almost all of the peptides are part of a dimer or a trimer and the first tetramer appears (Figure 5B). This is followed by the formation of pentamers at 18 ns. By the 28 ns, there are no monomers and two hexamers assemble. Within the next 50–60 ns, the smaller oligomers aggregate further into medium sized aggregates and stabilise. At about 120 ns, there are only three aggregates—a hexamer, an octamer and a 13-mer—and 20 ns later the hexamer joins the 13-mer to form a 19-mer (Figure 5C,D, respectively). By the first 0.5 μs, all of the peptides are forming one very large aggregate, which remains stable to the end of the simulation but changes in shape and becomes more compact and globular (Figure 5E,F).

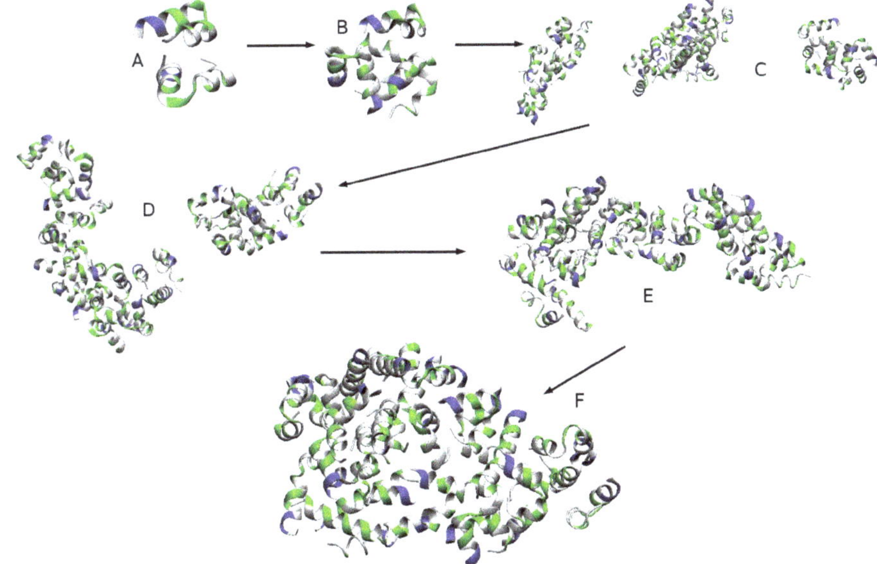

Figure 5. Time evolution of the self-assembly process of bombinin H2 peptides. (Charged residues are coloured in blue, polar in green, and hydrophobic in silver.)

The self-association process is entirely driven by the hydrophobic effect. As evident in Figure 4b, the total SASA of the peptides drops sharply with the formation of the first aggregates and continues to decrease as they consolidate into larger and larger structures. This behaviour is almost fully accounted for by the decrease in hydrophobic SASA. As shown in Figure 5A,B, the individual chains associate in such a way as to orient their hydrophobic surfaces to face each other. This reduces their exposure to the solvent. As smaller aggregates join together into larger ones, the total SASA gradually drops to about 250–260 nm^2 and remains at that level for the next microsecond. Then, after 1.5 μs, it descents further to about 230 nm^2 as the aggregate compactifies into a more globular shape. This is also associated with the same decrease of about 20 nm^2 in the hydrophobic SASA. The SASA of the polar amino acid residues does not change drastically, except for the first few ns. After that, there is a slight decrease in this property as individual peptide chains undergo conformational changes within the structure of the aggregates. After the first microsecond, the polar SASA remains unchanged. Moreover, the SASA of the charged lysines stays practically constant through the simulation at a level 27 times the charged SASA of a monomer in water. This means that virtually all lysines are at the aggregate surface and are completely solvent exposed.

Different amino acid residues exhibit different propensity towards aggregation. As demonstrated in the present study and also observed by Kuroda et al. [30], the aggregation is driven by the hydrophobic effect—the interplay of the Van der Waals interactions between the amino acids plus the entropic contribution to the solvent on the one hand and the Coulomb repulsion on the other hand. The balance between these different forces determines the maximal size of the aggregate. However, exploring this issue is beyond the scope of the present work. In [30], it was reported that Ile, Leu, Val and Met, along with the aromatic Phe, Tyr and Trp, tend to aggregate very quickly into large amorphous clusters. Bombinin H2 does not contain any aromatic residues, but has five leucines, three isoleucines and two valines, i.e., half of all amino acid residues are highly aggregation-prone. Apparently, until reaching certain (saturation) number of monomers in the aggregate, the attractive Van der Waals interactions among these ten hydrophobic residues tend to take over the electrostatic repulsion between the net positive charges provided by the two lysines in each monomer.

2.2.2. Aggregation-Driven Folding

Almost from its very beginning the aggregation is accompanied by another remarkable process—some peptide chains transition from the compact helix-loop-helix to the linear single-helix state. Indeed, when cartesian backbone PCA is performed on the trajectory of all 27 peptides, the principal mode of motion corresponds to straightening and bending of the linear helix at Gly^{10}-Ser^{11} (Figure 6b). This mode matches PC1 of the monomeric bombinin H2 dynamics (Figure 6a). The RMS fluctuations per C$_\alpha$ atom, generated by the first eigenvector for the two trajectories, are shown on Figure 6a. As evident, the two first PCs involve motion of the same atoms.

The first transitions to a linear helix state happen shortly after the aggregation onset—chains *V* and *a* that form a dimer spontaneously adopt the single-helix state within the first 25 ns of the simulation (Table 1). Shortly thereafter, a third helix straightens (chain *W*). For the next more than 330 ns, no further transitions are observed, until chain *F* snaps from the helix-loop-helix state to the single-helix state. Another 330 ns later, a fifth peptide (chain *E*) adopts this conformation, followed by the folding of three more chains (chains *A*, *B* and *D*) at about 900 ns. By 1.5 μs, the last chain (chain *J*) transitions to the single-helix state.

It should be noted that, in the aggregates, once a peptide transitions from a helix-loop-helix to the linear single-helix conformation, it remains stable in this state. This can be observed in Figure S4, which shows the occupancy of the single-helix state for each of the 27 peptide chains. Therein, one sees that, during the 2 μs simulation, all monomers but two (chain *U*, and—with small exceptions—chain *Z*) explore linear conformations towards the single-helix state. Comparing chains *A*, *B*, *D*, *E*, *F*, *J*, *V*, *W*, and *a* with the behaviour of the single monomer in solution in Figure 2b, it becomes obvious that the straightening of the helices in the aggregates has irreversible character. Even when the single-helix state occupancy drops sharply (i.e., in chains *E* or *J* right after 1.5 µs), the peptide chains remain in a linear state. The helix occupancy decreases, because the helix bends somewhat at Leu^6-Gly^7, Val^9-Gly^{10} or Leu^{13}-Gly^{14} or the helix turns widen a little bit. However, as seen in Figure S5, the gyration radius remains stably in the linear conformation domain, $R_g \in [0.9 : 1.05]$ nm.

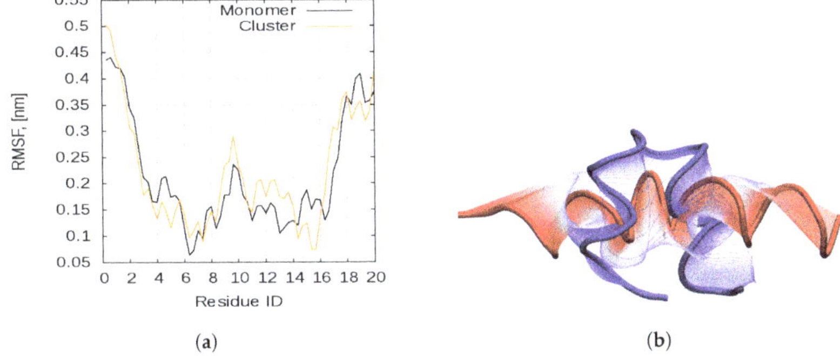

(a)	(b)

Figure 6. (**a**) Backbone RMSF along the first eigenvector of the PCA analysis of the bombinin H2 peptides in monomeric form (blue curve) and in a concentration/aggregate (red curve); and (**b**) peptide backbone motion along the first principal component for the simulation in concentration.

Table 1. Transition of individual bombinin H2 chains from a helix-loop-helix to the single-helix state. The first column gives the chain ID; the second one shows the first moment, when this conformation was adopted; the third column enlists the IDs of the chains, which were within 5 Å of the linear helix at that moment; and the last column gives information on what type of an oligomer the transition took place in.

Chain ID	Time [ns]	Neighbours	Aggregate
A	903	E, F, M, W, Y	27-mer
B	917	G, H, K, N, P, S, Z	27-mer
D	931	H, O, P	27-mer
E	735	A, K, M, R, S, T	27-mer
F	404	A, C, E, I, M, R, U, W, Y	19-mer
J	1516	N, U, V, a	27-mer
V	25	a	Dimer
W	70	A, F, Y	Tetramer
a	22	V	Dimer

The aggregation is associated with the formation of intermolecular contacts between the amino acid residues in different peptide monomers. In Figure S3, the number of intermolecular contacts per frame for each of the amino acid residues in the bombinin H2 molecule is shown for the single-helix and the helix-loop-helix states. In general, the single-helix structured monomers tend to build more contacts (about 19% more) with the neighbouring peptides—this is the case for 13 out of 20 residues in the single-helix peptides. For two residues—Leu^{17} and Lys^{19}—the number of contacts is the same, and only three residues build more contacts with the neighbouring structures while in a helix-loop-helix state. Note that the tendency towards an increase of the contacts number in a single-helix state is particularly pronounced in Leu/Ile and Gly residues: in five out of eight Leu/Ile residues, the single-helix state contacts are 30–300% more than those in the helix-loop-helix state, and in four out of five Gly residues, this increase is even stronger. These are exactly the residues associated with AMPs' antimicrobial activity. Cationic Gly-Leu-rich peptides are hemolytic and very potent against microorganisms [31]. In the design of peptide analogues with higher antimicrobial activity, the increase of the net positive charge and of the hydrophobicity are often targeted through Lys and Leu substitutions (see, e.g., [32]). This, together with the observed correlations between the aggregation propensity and the antimicrobial activity gives one more reason for a detailed research on AMPs' aggregation as an important and possibly decisive part of their antimicrobial action.

3. Discussion

The vast majority of AMP research focuses on the interaction of the peptides with target membranes and here molecular simulations play a particularly important role. However, studying the behaviour of AMPs in water solution, prior to their interaction with the membrane, is in our assessment an undervalued problem when trying to understand the AMP mechanism of action. It was shown recently that MD simulations of the interaction of AMPs and lipid bilayers are very sensitive to the initial simulation setup, including the initial conformation of the AMP and its placement relative to the membrane [11]. In their work, Wang et al. studied the interaction of a synthetic AMP, CM15, with a neutral POPC and a negatively charged POPG:POPC membranes. They performed multiple MD simulations starting from different initial conformations of the CM15 peptide—either a random-coil or a pre-folded α-helical conformation. Somewhat unexpectedly, they found that, when the AMP was pre-folded, its binding and insertion in both membranes was reduced, compared to when the simulation starts from a random-coil conformation of the peptide. Their results demonstrate the significance of the initial conformation of the AMP when simulating its interaction with a target membrane.

The experimental evidence suggests that the functional state of bombinin H2 and, in general, of the predominant part small linear AMPs is that of a single α-helix. Our investigations show that this conformation is supported in a solution only within a self-assembled aggregate, with a gradual increase of the monomers that adopt it, as depicted in Figure 7a. We observe that, at the beginning, the process is very fast—first monomers adopt a single-helix conformation within the first 100 ns, while in small aggregates (dimers or a tetramer). Next, but much later, about 400 ns, a peptide within a 19-mer straightens. The next four transitions happen around 700 and 900 ns, all within the already formed 27-mer. The process saturates at about 1.5 µs, on the level of 1/3 of the monomers (9 out of 27), and is reasonably well approximated by a sigmoid-type curve

$$N_h(t) = \frac{h_{max}}{1 + e^{\lambda(t - t_{1/2})}},\tag{1}$$

where $N_h(t)$ is the number of straightened monomers as a function of time; h_{max} is the asymptotic value corresponding to the maximal number of such monomers; $t_{1/2}$ is the half-saturation time, i.e. the time when half of the asymptotic value is reached; and λ is a shape parameter. The values of these parameters for the investigated dynamics are given in the data-box in Figure 7a.

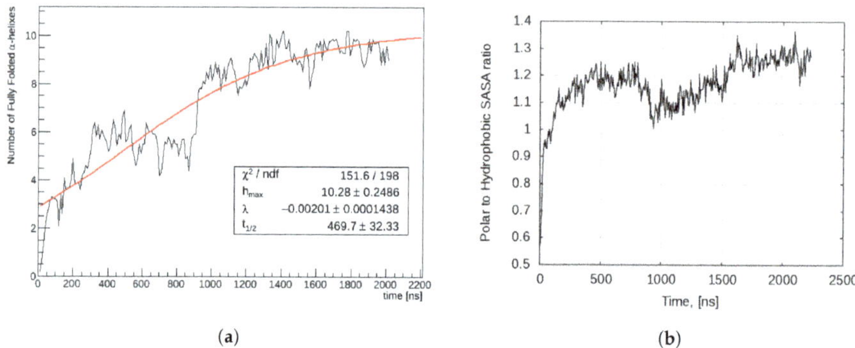

(a)   (b)

Figure 7. (a) Number of monomers in the single-helix state; and (b) polar to hydrophobic SASA ratio.

Thus, the bigger is the aggregate, the slower does the conformational transition occur; however, several transitions might then follow within a short interval. Apparently, the conformational transition from a more compact to a less compact state of a monomer within the aggregate has not only local consequences but is also associated with large-scale rearrangements of individual peptide chains within the aggregate. One might speculate that such a behaviour is rooted in the associated free energy changes. A direct verification of this hypothesis would be very involved if at all possible. However, some insight might be gained by examining the different SASAs behaviour and, in particular, the ratio between polar and hydrophobic ones, $\sigma_{p/hphb}(t) = A_{polar}(t)/A_{hydrophobic}(t)$ (Figure 7b). Note that, while these two SASA curves appear rather smooth and (almost) monotonic, their ratio proves very sensitive to even small but coincidental fluctuations in the respective values.

There is a clear correlation between the evolution of the number of single-helical monomers and $\sigma_{p/hphb}(t)$ (Figure 7a,b). Each act of monomer straightening is actually preceded by a noticeable decrease (drop) in the hydrophobic SASA and a soft decrease in the polar one and is then followed by a continuing polar SASA decrease, together with a local increase in the hydrophobic one. This is manifested in the nonmonotonic character of $\sigma_{p,hphb}(t)$ (Figure 7a). When this transition happens within a small aggregate (events in the first 100 nanoseconds and around 400 ns), the original positive slope is rapidly re-gained, while by the transition around the 700 ns within the final 27-mer this happens only partially, to be succeeded by a very pronounced drop, associated with the almost simultaneous conformational transition of three more monomers, within the already rather compact aggregate. After that, $\sigma_{p,hphb}(t)$ becomes an increasing function again and we see no indications for further transitions or major rearrangements.

This all can be understood as an interplay between the self-assembly and folding-promotion processes, which contribute differently to the formation of the various SASA figures. As a result, due to large-scale rearrangements of the monomers the larger aggregates, though mimicking the membrane amphiphilic environment, effectively resist against monomer straightening.

N_h is actually the number of peptides that are in a fully functional fold. In that sense, the dependence in Figure 7a shows the development of the effective peptide concentration with time and can provide some theoretical background for the experimentally observed sigmoidal dependence of AMP's activity on the concentration (see, e.g., [33], where the authors not only confirmed the importance of AMPs aggregation prior to their interaction with the membrane but also showed the necessity of some additional mechanism for explaining the aforementioned sigmoidal activity dependence).

It has been shown in multiple studies (e.g., [16]) that AMPs action depends on a threshold concentration, below which the peptides are unable to affect the target membrane. However, a number of authors suggested that the action of AMPs depends on their local surface, and not bulk concentration (see, e.g., [34,35]). Several different models for the mechanism of action of AMPs have been proposed, including the carpet model, the toroidal pores model and barrel stave pores model. They all depend on a threshold local surface concentration of the AMPs. Sengupta et al. [34] also demonstrated that aggregation at or near the membrane provides this critical local concentration and is necessary for the formation of transmembrane toroidal pores. However, their starting conformation included peptides, placed *"in the water phase close to one of the leaflets of an equilibrated DPPC bilayer"*. In our work, we demonstrate that this aggregation takes place very quickly, right after the AMPs are secreted and before reaching the membrane surface. The existence of localised isolated clusters and not isolated monomers in the bodily liquids prior to AMPs embedding in the bacterial membrane by no means contradicts the observed low concentration (in particular, of bombinin H2) in solution—it is only that the clusters need to be sparser than the isolated monomers.

Within aggregates, the peptides provide each other with an amphipathic environment mimicking the water–membrane interface that promotes further folding towards the biologically active shape—a single-helix structure, contrary to the case of isolated monomers. The latter might be viewed as representing a low peptide concentration situation. Identifying the critical concentration at which peptide assembly and functional folding promotion occur requires substantial computational resources and will be attempted in a separate study.

The above results support our hypothesis that it is the self-assembly process accompanied by aggregation-driven conformational changes into the biologically active fold that allow the AMPs to reach the target membrane in a fully functional state and to effectively exert their antimicrobial action.

4. Materials and Methods

4.1. Input Structures

We used as a starting model the first frame in the NMR structure of bombinin in DPC micelles with PDB ID 2AP8 [29] (Figure 8a). There, the peptide is not in water solution, but in a DPC micelle environment where it adopts an all α-helical conformation. Once in the absence of a membrane, the peptide rapidly transitions to a helix-loop-helix state within the 10 ns of NPT equilibration simulation (Figure 8c).

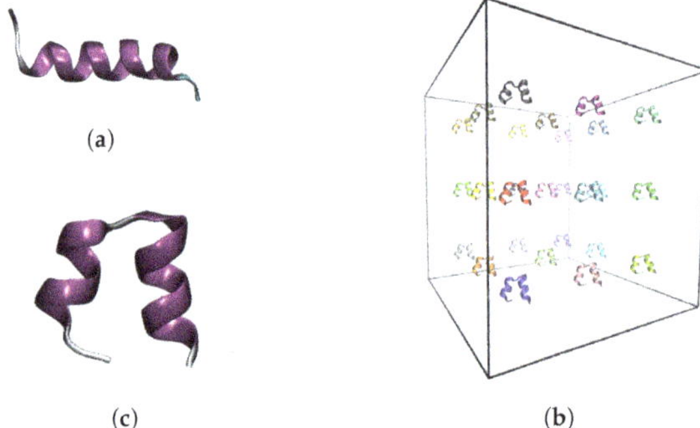

(a)

(c)   (b)

Figure 8. (**a**) Starting experimental structure of Bombinin H2. Input structures for the production simulation of: (**c**) monomeric bombinin H2; and (**b**) 27 bombinin H2 peptides in solution.

This equilibrated structure was used in the isolated-monomer studies and also to build a solution of 27 bombinin H2 peptides in a cubic simulation box with an edge length of 15 nm. The distance between two peptides was 5 nm (Figure 8b).

4.2. MD Simulation Protocol

All simulations were performed with the MD simulation package GROMACS 2016.8 [36]. The CHARMM 27 force field was used for parameterisation of the peptides [37] in combination with the modified TIP3PS water model for the solvent [38]. The peptides were solvated in cubic boxes with a minimal distance to the box walls of 1.2 nm under periodic boundary conditions. Sodium and chlorine ions with a 0.15 mol/L concentration were added to all systems to neutralise their net charge and to ensure physiological salinity of the solution. The systems were energy minimised using the steepest descent with a maximum force tolerance of 10 kJ/(mol nm), followed by short 50 ps position-restraint simulations to equilibrate the solvent. Then, 10 ns isothermal-isobaric simulations were performed, in which the temperature was gradually increased to 310 K using v-rescale thermostat [39] with a coupling constant of 0.1 ps and pressure was equilibrated at 1 atm using a Parrinello–Rahman barostat [40,41] with a coupling constant of 2 ps.

The production MD simulations were also performed in the NPT-ensemble using the same thermo- and barostat parameters. The leapfrog integrator [42] was used with a time-step of 2 fs, whereas constraints were imposed on bonds between heavy atoms and hydrogens using the PLINCS algorithm [43]. Van der Waals interactions were smoothly switched off from a distance of 1.0 nm and truncated at 1.2 nm. Electrostatic interactions were treated using the smooth PME method [44] with a direct PME cut-off of 1.2 nm. Neighbour lists were constructed every 10 ps. Each production simulation had a duration of 2 μs. Trajectory frames were recorded every 100 ps.

4.3. Data Analysis

The trajectories were postprocessed and analysed using the standard GROMACS postprocessing and analysis tools, in particular those for RMSD, RMSF, SASA and R_g calculations and PCA and cluster analysis. The cluster and PC analyses were performed after all global translational and rotational movements were removed by least-square fitting to the starting conformation. Cluster analysis was performed using the gromos algorithm, with a cut-off of 0.3 nm. PCA was carried out in Cartesian coordinates on all 60 backbone atoms (N, Ca, and C). For secondary structure assignment the STRIDE algorithm [45] as implemented by the visualisation and manipulation package VMD [46] was used. All structural figures were also generated by VMD.

5. Conclusions

Understanding in detail the mechanism of action of AMPs is a crucial prerequisite for their optimisation and successful application in the clinical fight against multidrug-resistant bacteria. We consider that studying the behaviour of AMPs right after their secretion in the bodily fluids (water solution), prior to their interaction with the membranes of pathogenic cells, is a largely overlooked first step in this mechanism.

Here, we used molecular dynamics simulations to study the behaviour of a single and multiple bombinin H2 peptides in solution without the presence of a target membrane. We found that in monomeric form bombinin H2 preferably adopts a compact helix-loop-helix conformation and only occasionally visits the classical linear single-helix state. This simulation corresponds to a very low AMP concentration.

At higher concentrations, the bombinin H2 peptides self-associate into aggregates. In addition, the aggregation process drives a significant portion of the peptide chains to permanently transition from the compact helix-loop-helix to the linear single-helix conformational state, by providing the necessary amphypathic environment mimicking the membrane–solvent interface.

The simulations results confirm our initial hypothesis that bombinin H2 and probably other AMP in general do not exist in solution as isolated monomers that assemble into clusters upon interaction with a target membrane to form pores. They rather self-assemble in the solvent into aggregates that deliver a large portion of the peptides into a folded state to the cell membrane and so provide the critical local concentration of peptides in a fully functional form to exert their action of the lipid bilayer.

Supplementary Materials: Supplementary materials can be found at http://www.mdpi.com/1422-0067/20/21/5450/s1.

Author Contributions: P.P. and L.L. developed the concept and methodology; P.P. performed MD simulations and key parts of formal analysis and visualisation; E.L. performed some MD simulations, part of the formal analysis and visualisation, contributed concepts to evaluation and wrote the initial draft of the manuscript; N.I. contributed concepts to evaluation, wrote parts of the manuscript, performed manuscript review and editing, as well as funding acquisition, computational resources and project coordination; and L.L. supervised the study. All authors contributed equally to data curation, investigations and evaluation of the results.

Funding: This research was supported in part by the Bulgarian Science Fund (Grant KP-06-OPR 03-10/2018) and by the Bulgarian Ministry of Education and Science (Grant D01-217/30.11.2018) under the National Research Programme "Innovative Low-Toxic Bioactive Systems for Precision Medicine (BioActiveMed)" approved by DCM # 658/14.09.2018).

Acknowledgments: The computational resources for this study were provided by the HPC Cluster at the Atomic Physics Department of Physics Faculty at Sofia University "St. Kl. Ohridski", Sofia (Bulgaria) and by CI TASK (Centre of Informatics—Tricity Academic Supercomputer & networK), Gdansk (Poland). We thank M. Rangelov (IOCCP–BAS) for bringing our attention to this particular representative of the AMP family—the bombinin H2 peptide—and for the stimulating discussions, and A. Liwo (University of Gdansk) for the interest in our research. We also acknowledge the instructive discussions and ideas exchange on IDPs within COST Action 17139, EUTOPIA. E.L. acknowledges the support by the Bulgarian Ministry of Education and Science under the National Programme "Young researchers and post-doctoral students" approved by DCM # 577/17.08.2018.

Conflicts of Interest: The authors declare no conflict of interest. The funders had no role in the design of the study; in the collection, analyses, or interpretation of data; in the writing of the manuscript, or in the decision to publish the results.

References

1. Ageitos, J.; Sánchez-Pérez, A.; Calo-Mata, P.; Villa, T. Antimicrobial peptides (AMPs): Ancient compounds that represent novel weapons in the fight against bacteria. *Biochem. Pharmacol.* **2017**, *133*, 117–138. [CrossRef]

2. Yount, N.; Yeaman, M. Emerging themes and therapeutic prospects for anti-infective peptides. *Annu. Rev. Pharmacol. Toxicol.* **2012**, *52*, 337–360. [CrossRef]

3. Hancock, R.E.W.; Sahl, H.G. Antimicrobial and host-defense peptides as new anti-infective therapeutic strategies. *Nat. Biotechnol.* **2006**, *24*, 1551–1557. [CrossRef]

4. Boman, H.G. Peptide Antibiotics and Their Role in Innate Immunity. *Annu. Rev. Immunol.* **1995**, *13*, 61–92. [CrossRef]

5. Fjell, C.D.; Hiss, J.A.; Hancock, R.E.W.; Schneider, G. Designing antimicrobial peptides: form follows function. *Nat. Rev. Drug Discovery* **2012**, *11*, 37–51. [CrossRef]

6. Kang, H.K.; Kim, C.; Seo, C.H.; Park, Y. The therapeutic applications of antimicrobial peptides (AMPs): A patent review. *J. Microbiol.* **2017**, *55*, 1–12. [CrossRef]

7. Stöckl, M.; Fischer, P.; Wanker, E.; Herrmann, A. Alpha-synuclein selectively binds to anionic phospholipids embedded in liquid-disordered domains. *J. Mol. Biol.* **2008**, *375*, 1394–1404. [CrossRef]

8. Sato, H.; Feix, J.B. Peptide–membrane interactions and mechanisms of membrane destruction by amphipathic α-helical antimicrobial peptides. *Biochim. Biophys. Acta (BBA) Biomembr.* **2006**, *1758*, 1245–1256. [CrossRef] [PubMed]

9. Bhargava, K.; Feix, J.B. Membrane binding, structure, and localization of cecropin-mellitin hybrid peptides: A site-directed spin-labeling study. *Biophys. J.* **2004**, *86*, 329–336. [CrossRef]

10. Manzini, M.C.; Perez, K.R.; Riske, K.A.; Bozelli, J.C.; Santos, T.L.; da Silva, M.A.; Saraiva, G.K.; Politi, M.J.; Valente, A.P.; Almeida, F.C.; et al. Peptide:lipid ratio and membrane surface charge determine the mechanism of action of the antimicrobial peptide BP100. Conformational and functional studies. *Biochim. Biophys. Acta (BBA) Biomembr.* **2014**, *1838*, 1985–1999. [CrossRef] [PubMed]

11. Wang, Y.; Schlamadinger, D.E.; Kim, J.E.; McCammon, J.A. Comparative molecular dynamics simulations of the antimicrobial peptide CM15 in model lipid bilayers. *Biochim. Biophys. Acta (BBA) Biomembr.* **2012**, *1818*, 1402–1409. [CrossRef] [PubMed]

12. Domingues, T.M.; Perez, K.R.; Miranda, A.; Riske, K.A. Comparative study of the mechanism of action of the antimicrobial peptide gomesin and its linear analogue: The role of the β-hairpin structure. *Biochim. Biophys. Acta (BBA) Biomembr.* **2015**, *1848*, 2414–2421. [CrossRef] [PubMed]

13. Perrin, B.; Scott, J.; Pastor, R.W. Simulations of Membrane-Disrupting Peptides I: Alamethicin Pore Stability and Spontaneous Insertion. *Biophys. J.* **2016**, *111*, 1248–1257. [CrossRef] [PubMed]

14. Chen, C.H.; Wiedman, G.; Khan, A.; Ulmschneider, M.B. Absorption and folding of melittin onto lipid bilayer membranes via unbiased atomic detail microsecond molecular dynamics simulation. *Biochim. Biophys. Acta (BBA) Biomembr.* **2014**, *1838*, 2243–2249. [CrossRef]

15. Leontiadou, H.; Mark, A.E.; Marrink, S.J. Antimicrobial Peptides in Action. *J. Am. Chem. Soc.* **2006**, *128*, 12156–12161. [CrossRef]

16. Huang, H.W. Molecular mechanism of antimicrobial peptides: The origin of cooperativity. *Biochim. Biophys. Acta (BBA) Biomembr.* **2006**, *1758*, 1292–1302. [CrossRef]

17. Almeida, P.F.; Pokorny, A. Mechanisms of Antimicrobial, Cytolytic, and Cell-Penetrating Peptides: From Kinetics to Thermodynamics. *Biochemistry* **2009**, *48*, 8083–8093. [CrossRef]

18. Wadhwani, P.; Strandberg, E.; Heidenreich, N.; Bürck, J.; Fanghänel, S.; Ulrich, A.S. Self-Assembly of Flexible β-Strands into Immobile Amyloid-Like β-Sheets in Membranes As Revealed by Solid-State 19F NMR. *J. Am. Chem. Soc.* **2012**, *134*, 6512–6515. [CrossRef]

19. Lu, N.; Yang, K.; Yuan, B.; Ma, Y. Molecular Response and Cooperative Behavior during the Interactions of Melittin with a Membrane: Dissipative Quartz Crystal Microbalance Experiments and Simulations. *J. Phys. Chem. B* **2012**, *116*, 9432–9438. [CrossRef]

20. Chen, C.; Pan, F.; Zhang, S.; Hu, J.; Cao, M.; Wang, J.; Xu, H.; Zhao, X.; Lu, J.R. Antibacterial Activities of Short Designer Peptides: A Link between Propensity for Nanostructuring and Capacity for Membrane Destabilization. *Biomacromolecules* **2010**, *11*, 402–411. [CrossRef]

21. Buffy, J.J.; Waring, A.J.; Lehrer, R.I.; Hong, M. Immobilization and Aggregation of the Antimicrobial Peptide Protegrin-1 in Lipid Bilayers Investigated by Solid-State NMR. *Biochemistry* **2003**, *42*, 13725–13734. [CrossRef] [PubMed]

22. Li, W.; Sani, M.A.; Jamasbi, E.; Otvos, L.; Hossain, M.A.; Wade, J.D.; Separovic, F. Membrane interactions of proline-rich antimicrobial peptide, Chex1-Arg20, multimers. *Biochim. Biophys. Acta (BBA) Biomembr.* **2016**, *1858*, 1236–1243. [CrossRef] [PubMed]

23. Pino-Angeles, A.; Lazaridis, T. Effects of Peptide Charge, Orientation, and Concentration on Melittin Transmembrane Pores. *Biophys. J.* **2018**, *114*, 2865–2874. [CrossRef] [PubMed]

24. Zou, R.; Zhu, X.; Tu, Y.; Wu, J.; Landry, M.P. Activity of Antimicrobial Peptide Aggregates Decreases with Increased Cell Membrane Embedding Free Energy Cost. *Biochemistry* **2018**, *57*, 2606–2610. [CrossRef]

25. Pazderkova, M.; Malon, P.; Zima, V.; Hofbauerova, K.; Kocisova, E.; Pazderka, T.; Crovsky, V.; Bednarova, L. Interaction of Halictine-Related Antimicrobial Peptides with Membrane Models. *Int. J. Mol. Sci.* **2019**, *20*, 631. [CrossRef] [PubMed]

26. Marinova, R.; Petkov, P.; Ilieva, N.; Lilkova, E.; Litov, L. Molecular Dynamics Study of the Solution Behaviour of Antimicrobial Peptide Indolicidin. *Stud. Comput. Intell.* **2019**, *793*, 257–265.

27. Petkov, P.; Marinova, R.; Kochev, V.; Ilieva, N.; Lilkova, E.; Litov, L. Computational Study of Solution Behavior of Magainin 2 Monomers. *J. Biomol. Struct. Dyn.* **2019**, *375*, 1231–1240. [CrossRef]

28. Mangoni, M.L.; Papo, N.; Saugar, J.M.; Barra, D.; Shai, Y.; Simmaco, M.; Rivas, L. Effect of Natural l-to d-Amino Acid Conversion on the Organization, Membrane Binding, and Biological Function of the Antimicrobial Peptides Bombinins H. *Biochemistry* **2006**, *45*, 4266–4276. [CrossRef]

29. Zangger, K.; Gößler, R.; Khatai, L.; Lohner, K.; Jilek, A. Structures of the glycine-rich diastereomeric peptides bombinin H2 and H4. *Toxicon* **2008**, *52*, 246–254. [CrossRef]

30. Kuroda, Y.; Suenaga, A.; Sato, Y.; Kosuda, S.; Taiji, M. All-atom molecular dynamics analysis of multi-peptide systems reproduces peptide solubility in line with experimental observations. *Sci. Rep.* **2016**, *6*, 19479. [CrossRef]

31. Vanhoye, D.; Bruston, F.; Amri, S.E.; Ladram, A.; Amiche, M.; Nicolas, P. Membrane Association, Electrostatic Sequestration, and Cytotoxicity of Gly-Leu-Rich Peptide Orthologs with Differing Functions. *Biochemistry* **2004**, *43*, 8391–8409. [CrossRef] [PubMed]

32. Park, Y.; Lee, D.G.; Jang, S.H.; Woo, E.R.; Jeong, H.G.; Choi, C.H.; Hahm, K.S. A Leu–Lys-rich antimicrobial peptide: Activity and mechanism. *Biochim. Biophys. Acta* **2003**, *1645*, 172–182. [CrossRef]

33. Chen, F.Y.; Lee, M.T.; Huang, H.W. Sigmoidal Concentration Dependence of Antimicrobial Peptide Activity: A Case Study of Alamethicin. *Biophys. J.* **2002**, *82*, 908–914. [CrossRef]

34. Sengupta, D.; Leontiadou, H.; Mark, A.E.; Marrink, S.J. Toroidal pores formed by antimicrobial peptides show significant disorder. *Biochim. Biophys. Acta* **2008**, *1778*, 2308–2317. [CrossRef] [PubMed]

35. Laadhari, M.; Arnold, A.A.; Gravel, A.E.; Separovic, F.; Marcotte, I. Interaction of the antimicrobial peptides caerin 1.1 and aurein 1.2 with intact bacteria by 2H solid-state NMR. *Biochim. Biophys. Acta* **2016**, *1858*, 2959–2964. [CrossRef]

36. Abraham, M.J.; Murtola, T.; Schulz, R.; Páll, S.; Smith, J.C.; Hess, B.; Lindahl, E. GROMACS: High performance molecular simulations through multi-level parallelism from laptops to supercomputers. *SoftwareX* **2015**, *1–2*, 19–25. [CrossRef]

37. Mackerell, A.D., Jr.; Feig, M.; Brooks, C.L., III. Extending the treatment of backbone energetics in protein force fields: Limitations of gas-phase quantum mechanics in reproducing protein conformational distributions in molecular dynamics simulations. *J. Comput. Chem.* **2004**, *25*, 1400–1415. [CrossRef]

38. MacKerell, A.D.; Bashford, D.; Bellott, M.; Dunbrack, R.L.; Evanseck, J.D.; Field, M.J.; Fischer, S.; Gao, J.; Guo, H.; Ha, S.; et al. All-Atom Empirical Potential for Molecular Modeling and Dynamics Studies of Proteins. *J. Phys. Chem. B* **1998**, *102*, 3586–3616. [CrossRef]

39. Bussi, G.; Donadio, D.; Parrinello, M. Canonical sampling through velocity rescaling. *J. Chem. Phys.* **2007**, *126*, 014101. [CrossRef]

40. Parrinello, M.; Rahman, A. Crystal Structure and Pair Potentials: A Molecular-Dynamics Study. *Phys. Rev. Lett.* **1980**, *45*, 1196. [CrossRef]

41. Parrinello, M.; Rahman, A. Polymorphic transitions in single crystals: A new molecular dynamics method. *J. Appl. Phys.* **1981**, *52*, 7182. [CrossRef]
42. Hockney, R.; Goel, S.; Eastwood, J. Quiet high-resolution computer models of a plasma. *J. Comput. Phys.* **1974**, *14*, 148–158. [CrossRef]
43. Hess, B. P-LINCS: A Parallel Linear Constraint Solver for Molecular Simulation. *J. Chem. Theory Comput.* **2008**, *4*, 116–122. [CrossRef] [PubMed]
44. Essmann, U.; Perera, L.; Berkowitz, M.L.; Darden, T.; Lee, H.; Pedersen, L.G. A smooth particle mesh Ewald method. *J. Chem. Phys.* **1995**, *103*, 8577–8593. [CrossRef]
45. Frishman, D.; Argos, P. Knowledge-based secondary structure assignment. *Proteins Struct. Funct. Genet.* **1995**, *23*, 566–579. [CrossRef]
46. Humphrey, W.; Dalke, A.; Schulten, K. VMD—Visual Molecular Dynamics. *J. Mol. Graphics* **1996**, *14*, 33–38. [CrossRef]

International Journal of
Molecular Sciences

Article

Examination of Adsorption Orientation of Amyloidogenic Peptides Over Nano-Gold Colloidal Particle Surfaces

Kazushige Yokoyama *, Kieran Brown, Peter Shevlin, Jack Jenkins, Elizabeth D'Ambrosio, Nicole Ralbovsky, Jessica Battaglia, Ishan Deshmukh and Akane Ichiki

Department of Chemistry, The State University of New York at Geneseo College, Geneseo, NY 14454, USA; kb6627500@gmail.com (K.B.); PeterJShevlin@gmail.com (P.S.); jackjenk@iu.edu (J.J.); edambros@udel.edu (E.D.); nralbovsky@albany.edu (N.R.); jessicabattaglia120@gmail.com (J.B.); isd1@geneseo.edu (I.D.); ai7@geneseo.edu (A.I.)

* Correspondence: yokoyama@geneseo.edu

Received: 10 September 2019; Accepted: 23 October 2019; Published: 28 October 2019

Abstract: The adsorption of amyloidogenic peptides, amyloid beta 1–40 ($A\beta_{1-40}$), alpha-synuclein (α-syn), and beta 2 microglobulin (β2m), was attempted over the surface of nano-gold colloidal particles, ranging from d = 10 to 100 nm in diameter (d). The spectroscopic inspection between pH 2 and pH 12 successfully extracted the critical pH point (pH_o) at which the color change of the amyloidogenic peptide-coated nano-gold colloids occurred due to aggregation of the nano-gold colloids. The change in surface property caused by the degree of peptide coverage was hypothesized to reflect the ΔpH_o, which is the difference in pH_o between bare gold colloids and peptide coated gold colloids. The coverage ratio (Θ) for all amyloidogenic peptides over gold colloid of different sizes was extracted by assuming $\Theta = 0$ at $\Delta pH_o = 0$. Remarkably, Θ was found to have a nano-gold colloidal size dependence, however, this nano-size dependence was not simply correlated with d. The geometric analysis and simulation of reproducing Θ was conducted by assuming a prolate shape of all amyloidogenic peptides. The simulation concluded that a spiking-out orientation of a prolate was required in order to reproduce the extracted Θ. The involvement of a secondary layer was suggested; this secondary layer was considered to be due to the networking of the peptides. An extracted average distance of networking between adjacent gold colloids supports the binding of peptides as if they are "entangled" and enclosed in an interfacial distance that was found to be approximately 2 nm. The complex nano-size dependence of Θ was explained by available spacing between adjacent prolates. When the secondary layer was formed, $A\beta_{1-40}$ and α-syn possessed a higher affinity to a partially negative nano-gold colloidal surface. However, β2m peptides tend to interact with each other. This difference was explained by the difference in partial charge distribution over a monomer. Both $A\beta_{1-40}$ and α-syn are considered to have a partial charge (especially $\delta+$) distribution centering around the prolate axis. The β2m, however, possesses a distorted charge distribution. For a lower Θ (i.e., $\Theta < 0.5$), a prolate was assumed to conduct a gyration motion, maintaining the spiking-out orientation to fill in the unoccupied space with a tilting angle ranging between 5° and 58° depending on the nano-scale and peptide coated to the gold colloid.

Keywords: amyloidogenic peptides; amyloid beta; alpha synuclein; beta 2 microglobulin; nano-gold colloids; peptide coverage; aggregation; adsorption orientation; spiking-out orientation; gyration

---

1. Introduction

The interface of a solid surface and a protein at the nanoscale level are of interests for many applications including material science, biomedical science (e.g., implantation of artificial bones, heart, organs or blood clotting), industry (e.g., the manufacturing of biosensors, bio separation processes, and drug delivery), and research in the development of new materials. The functionalities of peptide coated nanomaterials have remarkably broad applications in areas where nano-size has a very significant effect [1], including nano-light switching devices [2], disease controlling materials combined with DNA [3], DNA sensors [4], control of human cell activity [5], photo dynamic therapy [6], and optical biosensors that quantify heavy metal pollution in water [7]. However, very little is known regarding how the proteins adhere to nanoscale solid surfaces.

Amyloidogenic peptides, such as $A\beta_{1-40}$ or $A\beta_{1-42}$, α-synuclein (α-syn), and $\beta2$ microglobulin ($\beta2m$) are all regarded as hallmark peptides associated with key onset mechanisms of neurodegenerative diseases, such as Alzheimer's disease or Parkinson's disease. Because of this, the formation process and characterization of amyloid fibrils have been extensively investigated. Fibrils are usually several hundreds of micrometers in size and consist of pre-fibrils, which are built of unit oligomers. Therefore, the formation of unit oligomers from soluble and nontoxic monomers is regarded as a key intermediate process of fibrillogenesis and is considered to be a reversible process. On the other hand, the formation of fibrils or pre-fibrils are considered non-reversible processes. The nuclei-based pre-fibril formation mechanism is considered to be the most reasonable method for interpreting the fibril formation process. A key stage in this fibrillogenesis is the formation of an intermediate oligomer through a reversible process which then leads to one-direction pre-fibril formation. A major conformational change of the monomer to an intermediate oligomer requires significant protein folding, requiring a Gibbs energy of -10 kcal/mol [8–10] based on computational calculations. The most important on-set process of any fibrillogenesis is networking between peptides. Considering that fibrils are formed irreversibly, the networking between peptides must be "effective" and occur due to a strong interaction. However, a direct investigation of this networking process is still lacking.

The PI's group has been investigating the reversible self-assembly process (i.e., reversible networking process) on amyloidogenic peptide-coated gold colloidal nanoparticles. The peptides are relatively small, amphiphilic (i.e., consists of both hydrophilic and hydrophobic segments) peptides and the temperature/pH conditions for folded/unfolded conformations are well studied. The great advantage of this system is that a monomer peptide can be prepared on the nano-surface by orientating each peptide so that it may undergo the most effective networking process. Peptides are adsorbed over the nano-surface and are used to make connections between two nano-particle surfaces by making networks between peptides. Because of the networking between peptides, the assembly to the gold colloidal aggregates results in a drastic change in a spectroscopic feature, meaning that the networking process can thus be spectroscopically probed. Therefore, this work is viewed as the best prototype system to learn how nanoscale surface potentials interact with a peptide and if a specific structure can be selectively constructed [11–13]. Although these peptides eventually form irreversible insoluble amyloid fibrils, initial stages in fibrillogenesis are still reversible processes. We hypothesize that the peptide-peptide networking must be established by an unfolded conformation of each peptide and this unfolded conformation will be strongly enhanced at the nano-particle surface. As observed in negatively charged micelles and Teflon particles, β-sheet formation of $A\beta$ on hydrophobic graphite surfaces [14] or at air–water interfaces [15] indicate an involvement of interfacial surface potential utilized for the conforming intermediate [16–20].

This study aims to clarify (1) an exact attaching sequence or portion of the peptide and (2) orientation of the peptide over the nano-scale surface, and (3) identify the probable conformation of peptides for successful networking. It is well known that the amyloidogenic peptides (e.g., amyloid beta: $A\beta$; beta 2 microglobulin: $\beta2m$; and alpha-synuclein: α-syn) adsorb onto a gold surface through a sulfur atom of a thiol (–SH) group. These amyloidogenic peptides undergo drastic structural changes (protein folding) to form many units of toxic polymers that eventually combine to create a few micron-sized fibers (i.e., amyloid fibrils), which are known to cause neurodegenerative diseases [21–35]. However, $A\beta$ and α-syn do not possess any sequences which contain a thiol group (i.e., Cysteine (Cys, C)).

Contradicting the lack of a Cys sequence, the existence of gold nano-colloids were reported to enhance peptide-peptide networking using Aβ adsorbed on a nano-gold colloid as a "core" of a fiber [36].

There is no clear explanation of why amyloidogenic peptides adsorb onto gold so effectively. Since the detailed structural information of adsorbed peptides at the "core" is not known, contributing factors to the peptide-peptide networking needs to be fully investigated. This study describes a systematic method used to extract both a plausible peptide orientation and which segments interact with the colloidal surface. Based on these data, in Section 4.3.3 we describe a novel systematic procedure to extract the coverage ratio of peptide (Θ) onto the nano-gold colloids. Quite unexpectedly, the surface coverage conditions appeared to depend somewhat on the nano-particle size. While simulation does not explain the nano-size dependence on surface coverage, the observed trend suggests a plausible "packing" formation of the peptide due to a physical surface area condition.

2. Results

2.1. Extraction of Θ

This hypothesis stating a linear relationship between dpH and ΔpH$_o$ was clearly proved to be true when dpH was plotted as a function of ΔpH$_o$ for each tested gold colloidal size and in all three amyloidogenic peptides (Aβ$_{1-40}$, α-syn, and β2m) as shown in Figure 1, while ovalbumin-coated gold colloid did not show any sign of linear relationship [36]. Each data point shown in Figure 1 corresponds to different gold colloidal sizes for a given peptide. It shows that the coverage area is determined by the size of the nano-particle and there must be an equilibrium electrostatic shielding value for a given nano-gold metal surface. The average coverage ratio, $\Theta_{avg.}$, for each amyloidogenic peptide was extracted as: Θ_{avg} (Aβ$_{1-40}$) = 0.6 ± 0.2, Θ_{avg} (α-syn) = 0.6 ± 0.2, and Θ_{avg} (β2m) = 0.7 ± 0.2. By using the reported structural data most suited to the conditions of our work, the axial length a and b ($a < b$) of an approximated prolate for Aβ$_{1-40}$ [17] α-syn [37], and β2m [38] were initially estimated to be: Aβ$_{1-40}$ (a, b) = (2.1 nm, 4.1 nm), α-syn (a, b) = (3.1 nm, 8.5 nm), and β2m (a, b) = (2.5 nm, 4.6 nm).

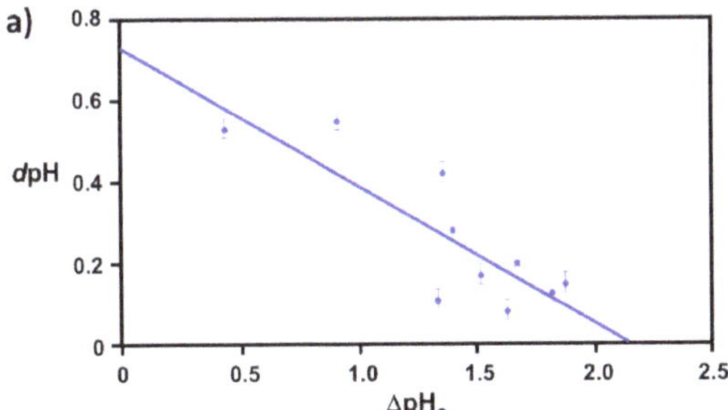

Figure 1. *Cont.*

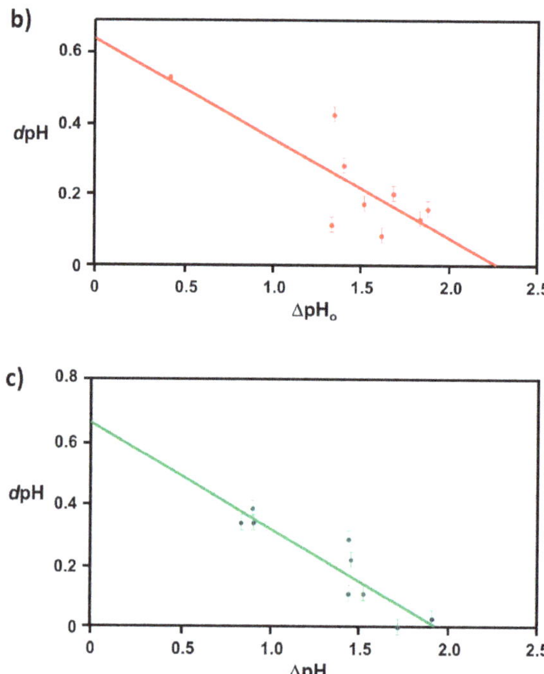

Figure 1. A hypothesized linear relationship between dpH vs. ΔpH$_o$ was plotted for each amyloidogenic peptide-coated nano gold colloid, (**a**). Aβ_{1-40} in blue, (**b**). α-syn in red, and (**c**).β2m in green based on the values obtained from fitting sigmoidal plot with Equation (1). Each linear line was given as a guide for a correlation between dpH vs. ΔpH$_o$. (dpH = m ΔpH$_o$ + b). Supporting information of this figure is available at Supplementary Materials.

2.2. Distance between Colloidal Particles

A representative TEM image of β2m-coated 30 nm gold colloids is shown in Figure 2a. As the magnified image clearly shows, distinct spacing noted as "Δ" between gold colloids were observed. (Figure 2b) The spaces between the gold particles was measured over multiple measurements per image for each size of gold particle and β2m. The morphology of the gold colloidal aggregates coated with Aβ_{1-40}, α-syn, as well as albumin were also studied, and the aggregates formed were more extensive in size and number (size exceeded up to a few microns and the number of gold colloids far exceeded 500 or 1000) than those formed by β2m-coated gold colloid. Therefore, the density of the aggregates by Aβ_{1-40}, as well as albumin coated gold colloid aggregates, extended in the longitudinal direction resulting in preventing the view the section of planar topology. In contrast, β2m formed relatively smaller aggregates, ranging within 500 nm with less than 100 colloids. This allowed us to visualize a two-dimensional view of the aggregate and made it possible to disclose the spacing between two adjacent gold colloids. While the peptide character of the studied peptides is not equivalent, we assume the networking character can be similar. Also, the physical dimension of the networking section compared to the diameter of each colloid can be approximated to be the same. Therefore, the information obtained for β2m will be shown as illustrative for the other two peptide coated gold colloids.

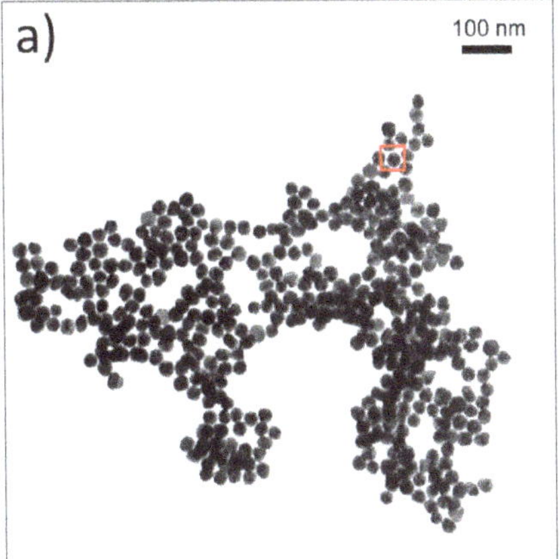
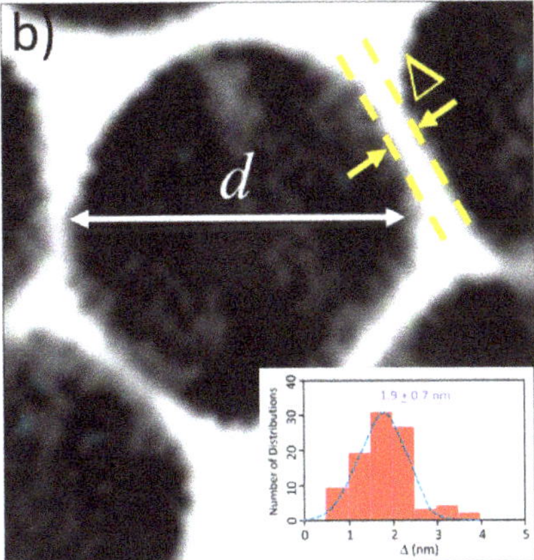

Figure 2. (**a**) A TEM (transmission electron microscopy) image of 30 nm gold colloidal particles coated with β2m at pH 4. The scale bar for 100 nm is shown as a guide. (**b**) A magnified section of the red box showing the diameter of a gold colloid d and the distance between the adjacent colloidal particles (Δ). (Inset): A typical histogram showing the Δ and the observed numbers of the distribution. The distribution was fit by a Gaussian profile and shown by a dotted curve. This histogram is for the β2m-coated 30 nm gold colloid.

The average distances of adjacent nano-gold colloids and the average number of gold nano-particles to form one aggregate for d = 10, 30, 60, and 80 nm particles are summarized in Table 1. We can conclude that the average distance of the adjacent nano-gold colloids was extracted to be 2.0 ± 0.8 nm, indicating insignificant size dependence of the gold colloid and the number of nano-gold colloids forming a cluster.

Table 1. The average distance, (Δ), of adjacent β2m-coated nano-gold colloids and the average number (η) of gold colloids consisting in an aggregate (gold colloid cluster) at pH 4.0 ± 0.3.

Gold Colloidal Size (Diameter d nm)	(Δ)	(η)
10	2.2 ± 0.6	25 ± 39
30	1.9 ± 0.7	123 ± 130
60	2.0 ± 0.7	39 ± 38
80	2.1 ± 0.5	17 ± 13

2.3. Simulation of Θ and Orientation

The procedures explained in Section 4.3.3 was applied to analyze the data points shown in Figure 3. The optimized axial lengths of prolate for three amyloidogenic peptides as a function of colloidal size are listed in Table 2. In order to reproduce extracted Θ, the spiking-out orientation need to be utilized for all cases except for d = 100 nm gold coated with Aβ$_{1-40}$ and α-syn as well as d = 10, 20, 60 nm gold coated with β2m, which exhibited a lie-down orientation. For example, Aβ$_{1-40}$ coated gold 20 nm diameter showed Θ_{obs} = 0.74 for a = 1.4 nm and b = 2.2 nm in order to reach maximum coverage ratio 0.37 under the first layer, $\Theta_{f,cal}$, with total number of attached peptides to be n_{total} = 111. After the second layer was added, it is calculated to have a maximum of $\Theta_{s,cal}$ = 0.72. In order to be consistent with the observed Θ_{obs} = 0.74, a contribution of the second layer, γ ($\Theta_{s,cal}$) should be 0.51, so that Θ_{obs} = 0.74 = $\Theta_{f,cal}$ + γ ($\Theta_{s,cal}$) × $\Theta_{s,cal}$ = 0.37 + 0.51 × 0.72 = 0.37 + 0.37. There was almost no contribution of the second layer when Θ_{obs} < 0.5 for d = 100 nm of both Aβ$_{1-40}$ and α-syn as well as d = 10, 20, and 60 nm for β2m.

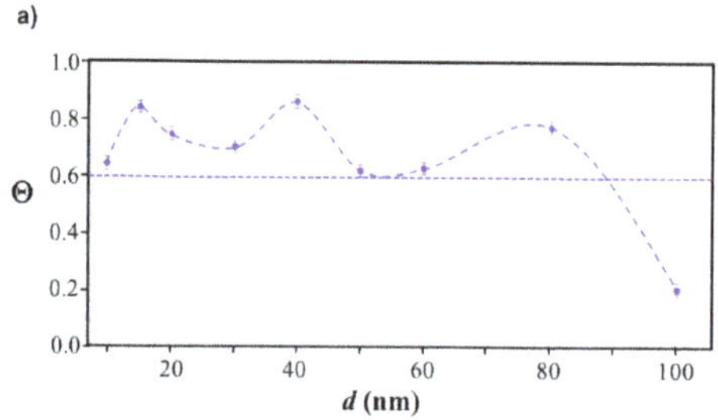

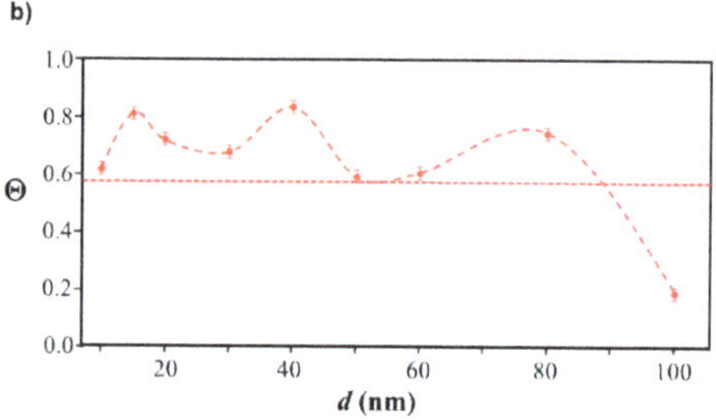

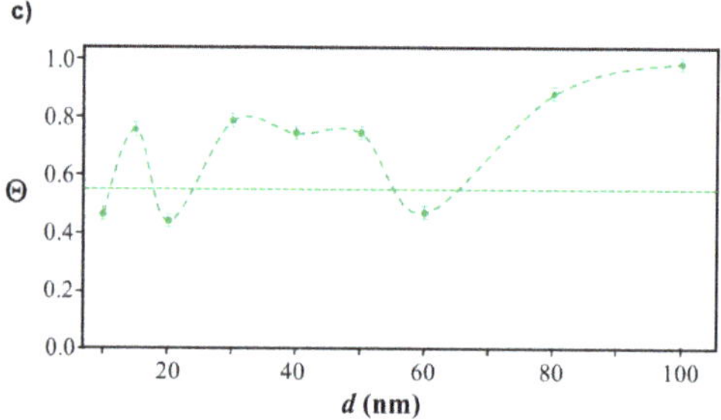

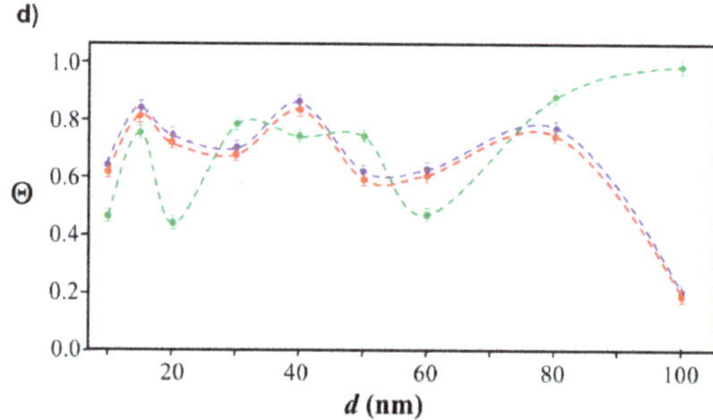

Figure 3. A plot for experimentally obtained Θ for (**a**) Aβ_{1-40} in blue (**b**) α-syn in red, and (**c**) β2m in green as a function of reported gold diameter d (nm). All data points were extracted by using Equation (2). Each dashed curve line is simulated by a method described in Section 4.3.3 with the parameters tabulate in Table 2. The dotted line shows an upper limit of the Θ value obtained by a single layer model. (**d**) The over-laid plots of all (**a–c**) are shown, and indicating (**a,b**) almost overlay each other.

Table 2. In each box, optimized axial length of a prolate (*a* and *b*), the sketches of orientation of adsorption, $n_{f,tot}$ (see Equation (3)), and circular graph indicating % of occupied surface area by adsorption (first layer in red, second layer in blue, and unoccupied in gray) for Aβ$_{1-40}$, α-syn, and β2m as a function of gold size d (and *d*) nm.

d (*d*)	Aβ$_{1-40}$	α-syn	β2m
10 (9.8)	$a = 1.4$ nm, $\Theta = 0.64$; $b = 2.2$ nm, $n_{f,tot} = 39$; 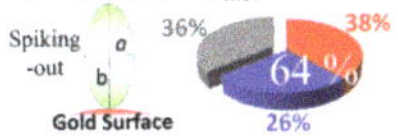 Spiking-out. 36% / 38% / 64% / 26% (Gold Surface)	$a = 4.6$ nm, $\Theta = 0.62$; $b = 7.4$ nm, $n_{f,tot} = 10$; 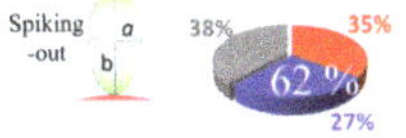 Spiking-out. 38% / 35% / 62% / 27%	$a = 2.7$ nm, $\Theta = 0.47$; $b = 4.0$ nm, $n_{f,tot} = 10$; 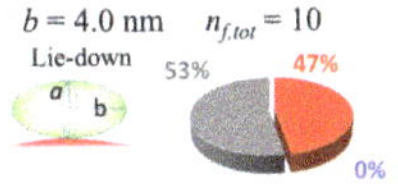 Lie-down. 53% / 47% / 0%
15 (15.2)	$a = 1.4$ nm, $\Theta = 0.83$; $b = 2.2$ nm, $n_{f,tot} = 91$; 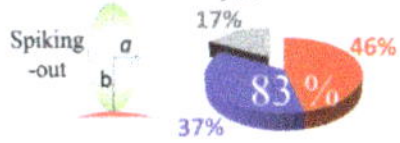 Spiking-out. 17% / 46% / 83% / 37%	$a = 4.6$ nm, $\Theta = 0.81$; $b = 7.4$ nm, $n_{f,tot} = 12$; 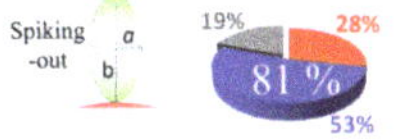 Spiking-out. 19% / 28% / 81% / 53%	$a = 2.5$ nm, $\Theta = 0.75$; $b = 4.6$ nm, $n_{f,tot} = 17$; 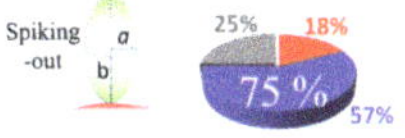 Spiking-out. 25% / 18% / 75% / 57%
20 (19.7)	$a = 1.4$ nm, $\Theta = 0.74$; $b = 2.2$ nm, $n_{f,tot} = 111$; 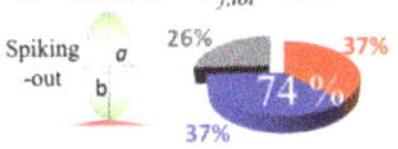 Spiking-out. 26% / 37% / 74% / 37%	$a = 4.6$ nm, $\Theta = 0.72$; $b = 7.4$ nm, $n_{f,tot} = 13$; 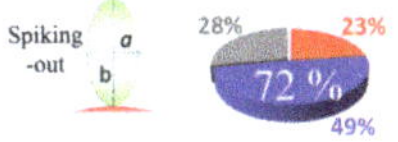 Spiking-out. 28% / 23% / 72% / 49%	$a = 2.7$ nm, $\Theta = 0.44$; $b = 6.4$ nm, $n_{f,tot} = 16$; 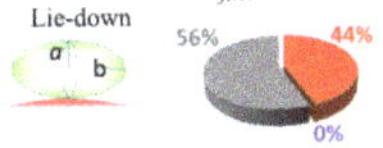 Lie-down. 56% / 44% / 0%
30 (30.7)	$a = 1.4$ nm, $\Theta = 0.70$; $b = 2.2$ nm, $n_{f,tot} = 287$; 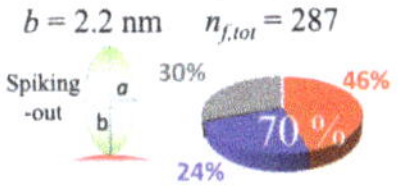 Spiking-out. 30% / 46% / 70% / 24%	$a = 4.6$ nm, $\Theta = 0.68$; $b = 7.4$ nm, $n_{f,tot} = 39$; 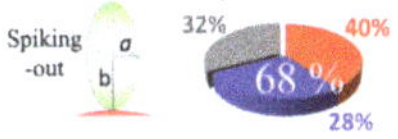 Spiking-out. 32% / 40% / 68% / 28%	$a = 2.5$ nm, $\Theta = 0.79$; $b = 4.6$ nm, $n_{f,tot} = 101$; 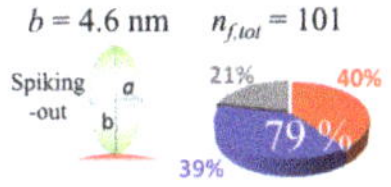 Spiking-out. 21% / 40% / 79% / 39%
40 (40.6)	$a = 1.4$ nm, $\Theta = 0.86$; $b = 2.2$ nm, $n_{f,tot} = 528$; 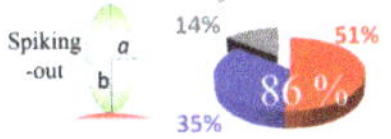 Spiking-out. 14% / 51% / 86% / 35%	$a = 4.6$ nm, $\Theta = 0.83$; $b = 7.4$ nm, $n_{f,tot} = 48$; 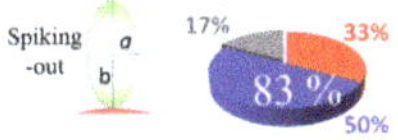 Spiking-out. 17% / 33% / 83% / 50%	$a = 2.5$ nm, $\Theta = 0.75$; $b = 4.6$ nm, $n_{f,tot} = 175$; 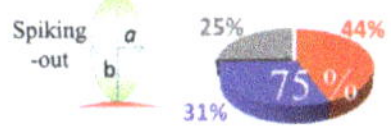 Spiking-out. 25% / 44% / 75% / 31%
50 (51.5)	$a = 1.4$ nm, $\Theta = 0.61$; $b = 2.2$ nm, $n_{f,tot} = 854$; 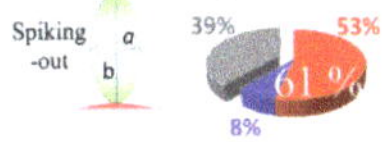 Spiking-out. 39% / 53% / 61% / 8%	$a = 4.6$ nm, $\Theta = 0.69$; $b = 7.4$ nm, $n_{f,tot} = 92$; 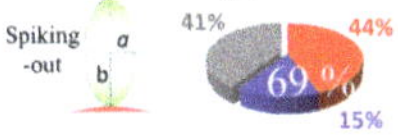 Spiking-out. 41% / 44% / 69% / 15%	$a = 2.5$ nm, $\Theta = 0.75$; $b = 4.6$ nm, $n_{f,tot} = 276$; 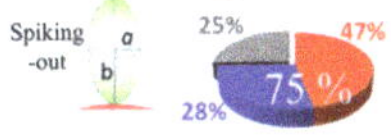 Spiking-out. 25% / 47% / 75% / 28%
60 (60.0)	$a = 1.4$ nm, $\Theta = 0.62$; $b = 2.2$ nm, $n_{f,tot} = 1212$; 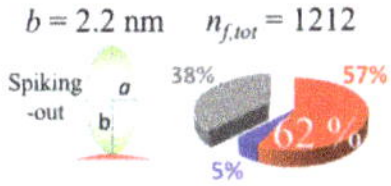 Spiking-out. 38% / 57% / 62% / 5%	$a = 4.6$ nm, $\Theta = 0.60$; $b = 7.4$ nm, $n_{f,tot} = 105$; 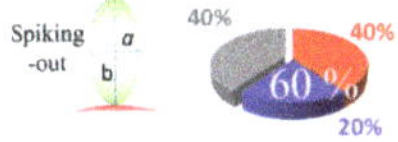 Spiking-out. 40% / 40% / 60% / 20%	$a = 3.0$ nm, $\Theta = 0.47$; $b = 4.3$ nm, $n_{f,tot} = 88$; 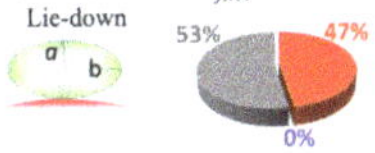 Lie-down. 53% / 47% / 0%
80 (80.0)	$a = 1.4$ nm, $\Theta = 0.61$; $b = 2.2$ nm, $n_{f,tot} = 2038$; 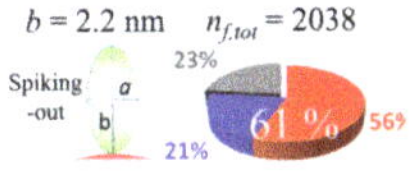 Spiking-out. 23% / 61% / 56% / 21%	$a = 4.6$ nm, $\Theta = 0.75$; $b = 7.4$ nm, $n_{f,tot} = 186$; 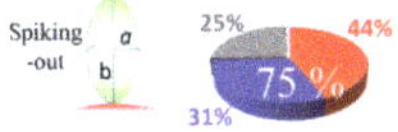 Spiking-out. 25% / 44% / 75% / 31%	$a = 2.5$ nm, $\Theta = 0.88$; $b = 4.6$ nm, $n_{f,tot} = 666$; 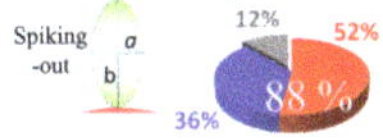 Spiking-out. 12% / 52% / 88% / 36%
100 (99.5)	$a = 0.91$ nm, $\Theta = 0.20$; $b = 3.7$ nm, $n_{f,tot} = 597$; 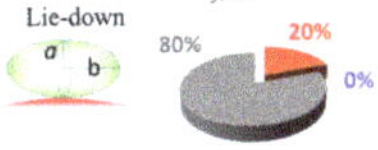 Lie-down. 80% / 20% / 0%	$a = 1.4$ nm, $\Theta = 0.19$; $b = 7.4$ nm, $n_{f,tot} = 190$; 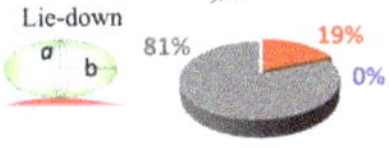 Lie-down. 81% / 19% / 0%	$a = 3.0$ nm, $\Theta = 0.99$; $b = 4.3$ nm, $n_{f,tot} = 1025$; 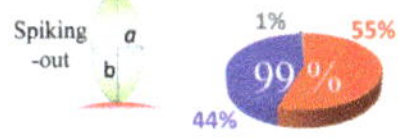 Spiking-out. 1% / 55% / 99% / 44%

In summary, as shown in Figure 4, all studied amyloidogenic peptides interacted with nano-gold surface at the original pH where a sample was prepared (~pH 7) with unfolded condition with spiking out condition.

Figure 4. A picture of an illustrative model and sketch demonstrating a peptide aligning over the surface of a gold colloidal particle with a diameter, *d*. In the image on the **left**, the peptide and gold core are both shown with the same color, and a dotted oval indicates a prolate shaped peptide. In the sketch on the **right**, prolate shaped peptide is shown in blue and the core gold colloid with diameter d is shown in red sphere.

3. Discussions

3.1. Elucidation of Θ and Spectroscopic Measurement

The spectroscopic measurement used in this study exhibits the transition of dispersed peptide-coated gold colloid to aggregated gold colloid through a networking of peptides on the colloidal surface. The surface charge or surface charge potential of each colloid should ideally be neutral, in order to avoid mutual repulsions which would impede aggregation. It is speculated that an aggregation of $A\beta_{1-40}$ coated gold colloid takes place at an isoelectric point, pI, of $A\beta_{1-40}$ (pI = 5.2) [39]. If so, a constant pH_o (~pI) is expected to be observed regardless of the sizes of gold colloid. However the observed pH_o for $A\beta_{1-40}$ coated gold colloid ranging from $d = 10$ nm to $d = 100$ nm scanned between $pH_o = 4.38 \pm 0.06$ for $d = 100$ nm and $pH_o = 6.20 \pm 0.01$ for $d = 40$ nm [40]. The extracted pH_o implies the amount of positive charge, i.e., $[H_3O^+]$, required to neutralize the colloidal surface. This work demonstrated the correlation between ΔpH_o and dpH as a key concept in extracting the change of surface charge potential of the gold colloidal particle. Due to the fact that the bare gold colloid possesses surface plasmon (electrons) over its surface, the excess amount of $[H_3O^+]$ needs to be supplied in order to neutralize the surface. Thus, the ΔpH_o shows the difference of the amount of $[H_3O^+]$ required between bare gold colloid and the peptide-coated gold colloid. Essentially, this means that the ΔpH_o indirectly shows the amount of the negative charges quenched due to the attachment of the peptides over the surface and the changing surface charge potential of the colloidal surface.

The measured quantity dpH is defined in a method described in Section 4.3.1. Since $\lambda_{peak}^{(1)}$ is the first derivative of $\lambda_{peak}(pH)$, it has the dimension of $\Delta\lambda/\Delta pH$ ($\Delta\lambda$ indicates the wavelength change associated to the transition from a dispersed stage to an aggregated stage, and ΔpH indicates the amount of H_3O^+ ion needed for that transition). The marking wavelengths λ_{max} and λ_{min} indicate that the colloid is at the aggregated stage (λ_{max}) or at the dispersed stage (λ_{min}). In practice, the marking wavelengths can be treated as dimensionless numbers or as an index (or with arbitrary units), with the replacement of $\lambda_{peak}^{(1)}$ by $\Delta\Lambda/\Delta pH$, where $\Delta\Lambda$ implies the difference of an index Λ. Overall, dpH indicates an inverse of $\lambda_{peak}^{(1)}$ and it is proportional to ΔpH, if $\Delta\Lambda$ is treated as a constant. Therefore, because $\Delta pH \propto -\log \Omega$ and $dpH \propto \log \Omega$, where Ω is a constant associated with charge. Thus, in principle, there should be a linear correlation between dpH and ΔpH.

A. Wang et. al. reported a pH dependence in protein coverage [41], and it is presumable that the surface charge condition is fully influenced by the residual pH condition. This suggests that the peptides start occupying and aggregating at the gold surface as the pH point gets closer to pH_o. If this is the case, peptides may not adsorb on the gold surface with our reported Θ values at $pH > pH_o$. However, our work is designed to determine pH_o, where the coverage of amyloidogenic peptide was already completed, our approach allows us to extract Θ only at pH_o, and it limits a quantitative conclusion regarding the pH dependence of Θ.

3.2. Orientation of the Peptide over the Surface of Gold Colloidal Surface

Relatively high Θ (i.e., $\Theta > 0.5$) can be accommodated by filling the surface area with a greater number of smaller unit surface area. Thus, the most supported orientation of the prolates is a spiking-out orientation, as sketched in Figure 5. While a "lie-down" orientation was highly expected to establish more interaction between peptide and the gold surface, the higher coverage was only established by creating a larger amount of smaller contacting areas. As supporting evidence, a very similar orientation was reported by Stellaci's group for bipolar polymer spiking out of gold nanoparticle sphere [42–48]. Considering that the gold colloid has a partially negative surface charge, any positively charged sequence can interact with it electrostatically. Since $A\beta_{1-40}$ coated gold colloid is dissolved in an aqueous solution, hydrophobic segments of $A\beta_{1-40}$ (sequences 23–40, C-terminal side) must be used for contacting the gold colloidal surface, causing hydrophilic segments of $A\beta_{1-40}$ (sequences 1–22, N-terminal side) to face outside, making it soluble in water. Among the hydrophobic sequences (23–40), only ^{28}Lysine (^{28}Lys, ^{28}K) can be positively charged at neutral conditions. Therefore, it is hypothesized that $-N^+-$ part of the ^{28}Lysine is responsible for contacting on the gold colloidal surface as shown Figure 5.

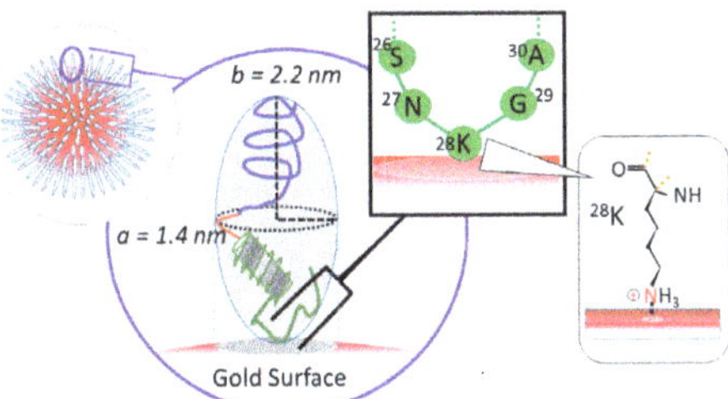

Figure 5. The proposed attachment structure of $A\beta_{1-40}$ over the surface of a gold colloidal particle. At the left top sketch shows the proposed peptide orientation adsorbed over gold nano particle. In the middle the blow up of each peptide with a prolate shape is shown. Inside the prolate, a sketch of $A\beta_{1-40}$ is shown within a prolate of $a = 1.4$ nm and $b = 2.2$ nm. On the top, a blow up and hypothesis of sequences responsible for the adsorption on the gold surface are shown and ^{28}K was speculated to be in a direct contact with gold surface. At the further right, a structure of Lysine (K) is shown and $-NH_3^+$ group is estimated to be a central point for an adsorption.

Due to more complexities in structure, identification of binding sites for α-syn and β2m were less clear. Even so, a similar approach and rough estimation of the site is possible. For α-syn, residues 61–95, or the so-called NAC (Non-Aβ Component) region [49,50], is highly likely to be the region where the peptide is bound to the nano-gold surface. More specifically, ^{80}K, ^{96}K, and/or ^{97}K are candidate residues that could be responsible for the colloidal attachment.

In a similar way, however, with a more complex situation, β2m is considered to possess hydrophobic (and aromatic-rich) region in residues 62–70 implying ^{63}R (^{63}Arginine), ^{66}K, or ^{69}H to be a plausible binding sites to the nano-surface [51]. If we assume multiple concurrent contacting spots are available, the mobility of β2m must be reduced and this can separate the binding property of β2m different from the other two peptides. In order to explain a negative correlation observed in

Θ vs. S_d plot (Section 3.4), β2m was speculated to be a prolate with negative side facing outward. It is quite likely that the sequence from 50 to 58 are the section responsible for the above-mentioned section since the negatively charged section of ^{52}D (^{52}Aspartic Acid), ^{54}E (^{54}Glutamic Acid), and ^{56}D are located therein.

The current model used for extracting Θ was tested for chicken ovalbumin as an example of a globular protein. There was no correlation found between dpH and ΔpH, and we assume that the model can be applied only for amyloidogenic peptides that clearly exhibits folded and unfolded conformations which are drastically different. Also the section of adsorption site has to be clearly determined no matter which size of the gold colloid was applied, otherwise clear mapping of charge distribution due to peptide (as explained in Section 3.4) cannot be obtained. The prolate shape localizes a partial charge in a relatively smaller region, which can become an adsorption point. In contrast, a globular protein takes a spherical shape creating a broader and more homogeneous partial charge distribution. This results in a less sensitive response in aggregation as a function of pH change, which results in a poorly defined dpH value in Equation (1), and reducing a correlation between ΔpH$_o$ and dpH.

3.3. Networking of the Peptide at an Interfacial Area

While the aggregation of Aβ_{1-40} coated gold colloid develops highly condensed networking which results in the mutual overlapping of gold colloids in both horizontal and latitudinal directions, the aggregation of β2m-coated gold colloid was less condensed, and TEM images enabled us to observe the spacing between adjacent gold colloids ($\Delta = 1.9 \pm 0.7$ nm), particularly around the edge area of the aggregates. Since Aβ_{1-40} and β2m both approximate as prolate tops with similar dimensions, the information on this spacing between adjacent gold colloid can be extrapolated to these two peptide systems. This result implies that the peptide layer covered the gold colloidal surface at a thickness of 0.95 nm [52]. If we consider a monolayer of peptide with spiking-out orientation, the adjacent distance between two gold colloid should correspond to $2b$. This assumption strongly contradicts the extracted Δ (~2 nm) since calculated $2b$ value are 4.4 nm for Aβ_{1-40}, 14.8 nm for α-syn, and 9.2 nm for β2m based on the values shown in Table 2. In order to allow peptides to be constrained within 2 nm spacing, the most probable conformation allows the peptides to be bent or spiraled around each other at the interface. Since amyloidogenic peptides studied in this work are all regarded as disordered proteins, it is reasonable to assume that disordered regions are flexible enough to take best suited configuration including a bent form in order to fit in 2 nm inter-colloidal surfaces. The process of gold colloidal aggregation is summarized as the process of a mixture (interaction) between monomer and gold colloidal surface, followed by the adsorption of each monomer over the nano-colloidal surface, and under acidic pH, the networking of peptides forming the gold colloid aggregation (Figure 6).

Based on the fact that all experimental observation needs to involve a second layer, we deduce that the first layer is responsible for the coverage of nano-gold surface and the second layer is the result of networking to the first layer of each peptide coated gold colloid. Due to the spiking-out orientation of the first layer, this would leave accessible another site for further networking as the peptide conformation becomes unfolded. The networking between dual peptides at an interface matches with a speculation of a dimer formation concluded in our previous work [53].

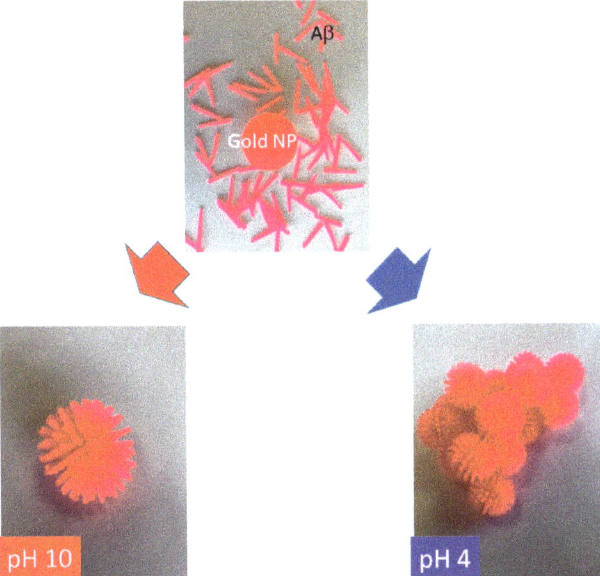

Figure 6. A demonstrative picture of peptides being adsorbed over a surface of gold nano-particle (NP) and adsorption of peptide with spiking out orientation under pH 10 (bottom left) and forming a network with each other in order to form gold colloid aggregates at pH 4 (bottom right). The color of peptide and gold colloid are shown in the same color.

3.4. Verification of the Relationship between Physical Displacement and Coverage Ratio

Our first instinct was that Θ was dominated by the molecular interaction between gold surface and peptide's terminus responsible for an electrostatic interaction. Therefore, the surface field reflecting from the surface curvature would be proportional to molecular interaction and may influence the surface interaction and coverage. However, by assuming that simple term of curvature is proportional to an inverse of radius, we did not see any correlation between a curvature and Θ, as also implied by the complex relationship between Θ and d shown in Figure 4. This implies that at least obvious chemical interaction does not explain the intrinsic reasoning of Θ and its nano-size dependence. While no correlation between the gold colloidal size and its coverage ratio of peptide (Θ) was found in our study, we attempted to find justification of Θ for a given nano-size of gold colloid by using a mathematical approach without involving intermolecular forces. The most simplified explanation of higher or lower coverage is gained from calculating how much space is wasted by a given unit adsorbent. However, the coverage ratio cannot be simply predicted as a function of surface area. For example, if the area to be covered increases, the unit area of an adsorbent may not utilize given space without leaving an unoccupied area, and so the coverage ratio may not increase. The equatorial belt area was used as an index of the effectiveness of space usage, and the spacing between each prolate (S_d) should be correlated with the coverage ratio (Θ). For example, a prolate of $A\beta_{1-40}$ ($a = 1.4$ nm and $b = 2.2$ nm) covering a 40 nm ($d = 40.6$ nm) has $\Theta = 0.86$), and 100 nm ($d = 99.5$ nm)has Θ of 0.20. When maximum prolate with dimension of ($a = 1.4$ nm and $b = 2.2$ nm) was distributed equatorial belt of each gold colloid, $S_d = 0.051$ nm for 40 nm with $n_{eq} = 50$ and $S_d = 0.012$ nm for 100 nm with $n_{eq} = 175$ demonstrating that the larger the S_d, the higher the Θ. A clear correlation between Θ and S_d was confirmed for $A\beta_{1-40}$ and α-syn as shown in Figure 7a,b as a positive slope for β2m as shown in Figure 7c as a negative slope. The finalized fitting parameters and fitting procedures will be further detailed in a report by Yokoyama and Ichiki [54].

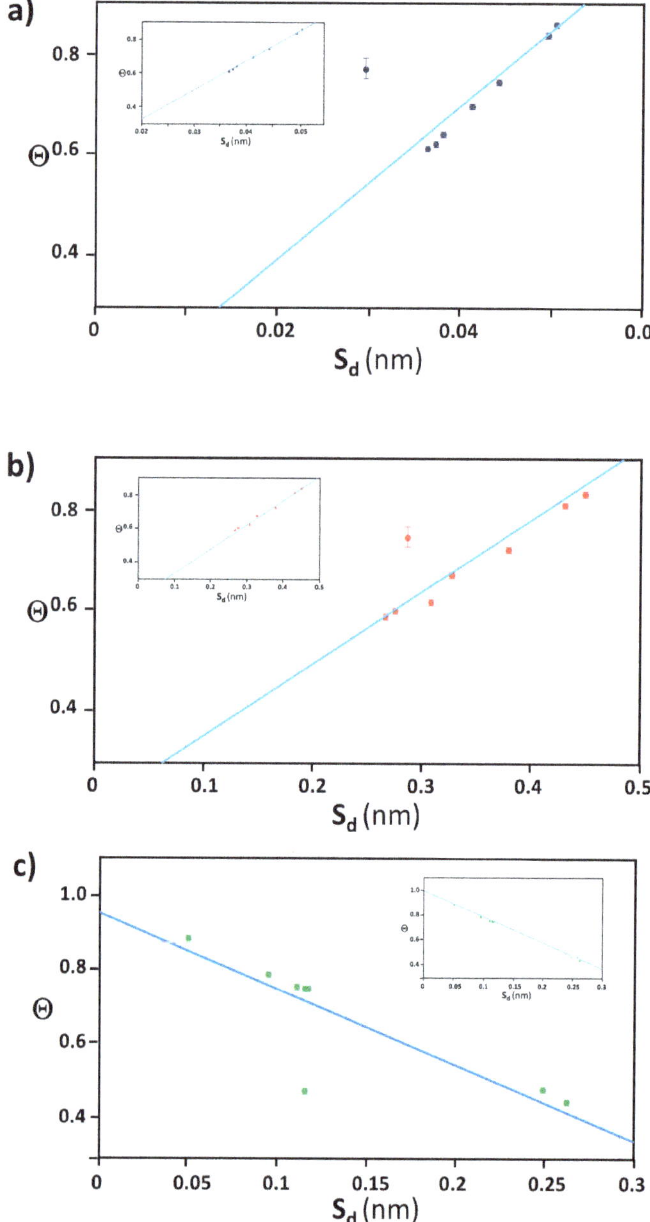

Figure 7. The best optimized plot of Θ vs. S_d for (**a**) $A\beta_{1-40}$ in blue, (**b**) α-syn in red, and (**c**) β2m in green, where fitting values for the linear relationship $\Theta = \Phi\, S_d + \varepsilon$. In each plot, there was always one deviating data point from the linear trend ($d = 80$ nm for $A\beta_{1-40}$ and α-syn, $d = 60$ nm for β2m), and the insets show the plot when each outlier point was removed. Supporting information of this figure is available at Supplementary Materials.

A positive linear relationship between Θ and S_d is explained by considering that both $A\beta_{1-40}$ (Figure 8(a-1)) and α-syn (Figure 8(a-2)) are simplified as a prolate with $\delta+$ region at the adsorption side and opposite side (i.e., exposing side to the outward) as sketched in Figure 8(b-1), respectively. As the prolate attaches onto the gold surface through the $\delta+$ region of a prolate, it also creates the $\delta+$ region on the gold surface as indicated in Figure 8(b-1,b-2). So that an extra peptides are more invited for adsorption as the gold surface possesses more $\delta-$ region when S_d is longer. On the other hand, if S_d is relatively small, not enough $\delta-$ region is available for further adsorption of peptides causing the Θ decreased resulting in the positive linear relationship between Θ and S_d (Figure 8c). As it was speculated before, $A\beta_{1-40}$ may be adsorbing on to the surface through ^{28}K and α-syn adsorbs on to the surface through ^{80}K or $^{96}K^{97}K$. Since those sites are located at relatively close to the N-terminal, it is speculated that $\delta+$ portion of C-terminal side must be exposing outward and away from the colloid surface. As for $A\beta_{1-40}$, we speculate that $^5R^6H$, $^{13}H^{14}H$, or ^{16}K are responsible for distributing $\delta+$ region. The speculated region with $\delta+$ and $\delta-$ are indicated by color coded areas in a prolate and bars in sequences as $\delta+$ in blue and $\delta-$ in red, respectively (Figure 8(a-1)) As for α-syn, all lysines in the C-terminal region (i.e., ^6K, ^{10}K, ^{12}K, ^{21}K, ^{23}K, ^{32}K and ^{34}K) are speculated to be exposing toward the outside and away from the colloidal surface side. While much more information is needed, crude estimation of charge distribution was shown in Figure 8(a-2). In a similar manner as shown in Figure 8(a-1), region with $\delta+$ and $\delta-$ are indicated by color coded areas in a prolate and bars in sequences as $\delta+$ in blue and $\delta-$ in red, respectively.

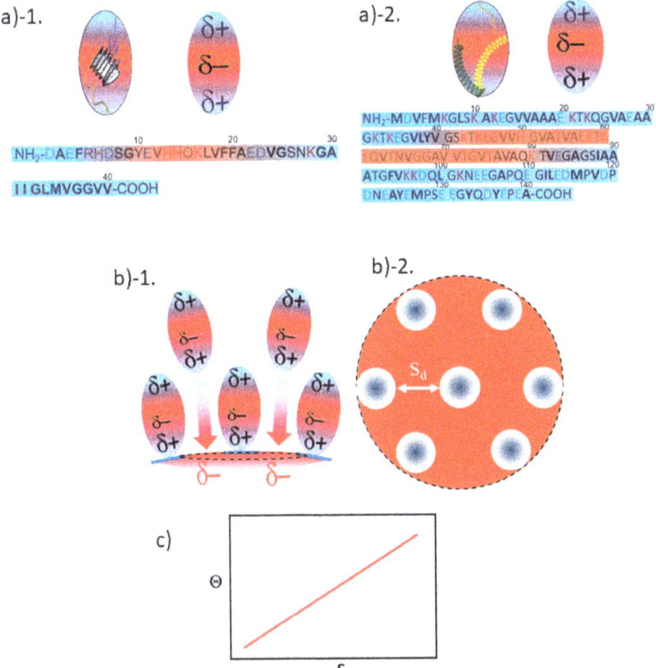

Figure 8. A sketch explaining a positive linear proportionality between Θ vs. S_d, (**a**) simulation of prolate and charge distribution of $A\beta_{1-40}$ (**(a)-1**) and of $\beta2m$ (**(a)-2**). The sequences of each peptide are shown with the colored bar indicating $\delta-$ (in red) or $\delta+$ (in blue). (**(b)-1**). A side view of a prolate top peptide with a partially positive side ($\delta+$) of dipole attaching to the partially negative surface ($\delta-$) of gold colloid. An extra prolate dipole attracted for the space of $\delta-$, if S_d has enough length to let an extra prolate in. (**(b)-2**) The birds eye view of the surface showing area appears as $\delta-$ indicated by red is the highly probable are for an extra prolate to be interacted and may lead to an attachment. (**c**). A graph explaining the expected trend between Θ as a function of S_d.

Opposed to what we observed in Aβ_{1-40} and α-syn, β2m exhibited a negative linear slope for Θ vs. S_d plot. (Figure 7c). This is interpreted that as each β2m (sketched in Figure 9(b-1)) adsorbs onto the gold surface with $\delta+$ segment as exposing more $\delta-$ area to the other side of gold surface as sketched in Figure 9(b-1,b-2). Thus, as S_d decreases, it creates a greater effective attraction to the extra β2m resulting in more coverage (i.e., a negative slope for Θ vs. S_d plot) as shown in Figure 9c. Since the adsorption site of β2m can be speculated to be at relatively toward the C-terminal side (i.e., [63]R, [66]K, or [69]H), the exposing side away from the gold surface is speculated to be N-terminal side. Thus, it is estimated that[18]E is responsible for providing $\delta-$ region. In Figure 9a, region of $\delta+$ and $\delta-$ are indicated by color coded areas in a prolate and bars in sequences as $\delta+$ in blue and $\delta-$ in red, respectively.

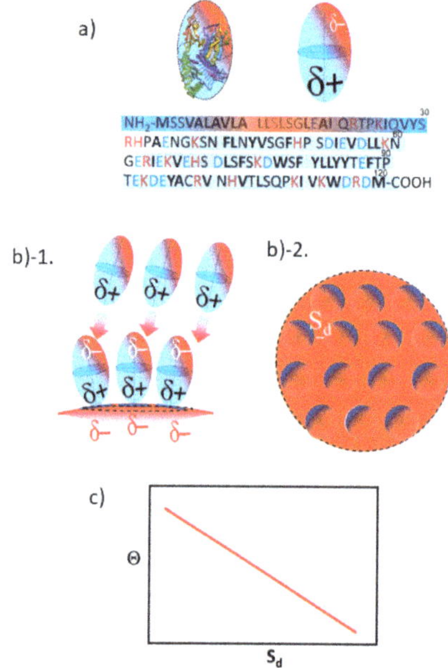

Figure 9. A sketch explaining negative linear relationship between Θ vs. S_d. (**a**) simulation of prolate and charge distribution of β2m. The sequences are shown with the colored bar indicating $\delta-$ (in red) or $\delta+$ (in blue). (**(b)-1**). A side view of a prolate top peptide with $\delta+$ side of dipole attaching to $\delta-$ surface of gold colloid. Because a distribution of $\delta-$ is expected to be spread from the top to the side toward outside, extra prolate is more attracted as more area of $\delta-$ is available. (**(b)-2**). A top view of a focused region in b)-1, where the area appears as $\delta-$ as the prolate locate close by shortening of S_d. (**c**). A graph explaining the expected trend between Θ as a function of $S_{d,}$.

In all three cases explained above, we claim that the spiking-out orientation of the first layer established a corresponding charge distribution seen in each peptide coated gold colloid. If the orientation was lie-down orientation, enhancement of self-adsorption would not take place. For example, a lie-down orientation of prolate dipole in the case of Aβ_{1-40} and α-syn would exhibit a significant amount of $\delta+$ region and not effectively squeeze the prolate dipole with the same orientation. As for the case of β2m, the lie-down orientation exposes a greater amount of $\delta-$ region, resulting in a significant repulsion for the peptide attempts to adsorb with the same lie-down orientation.

3.5. Justification of Lower Coverage Ratio and Associated Prolate Shape

While the overall characteristic of the coverage of amyloidogenic peptides was relatively higher value (i.e., $\Theta \geq 0.6$), there were only five cases when Θ was < 0.5; Aβ_{1-40} coated $d = 100$ nm gold, α-syn coated d = 100 nm gold, β2m-coated d = 10, 20, and 60 nm gold. For all cases, the fit was not optimized with a prolate with spiking-out orientation but with lie-down orientation. Under the estimation that the spiking-out orientation is the best orientation to satisfy coverage stability (i.e., effective packing of the surface) and consistent with most of the coverage orientations observed in this experiment. Thus, it is hypothesized that each prolate takes a spiking-out orientation but tilts over the nano-gold surface as shown in Figure 10a and can rotate around the contact point on the nano-gold surface (Figure 10b) resulting in an occupied area with oblate shape as if it takes a lie-down orientation. In order to explain an oval shape occupying over the surface, a gyration type of motion is considered. So that the contact point of the prolate changes due to a change of tilting angle as it rotates over the surface, is plausible (Figure 10c).

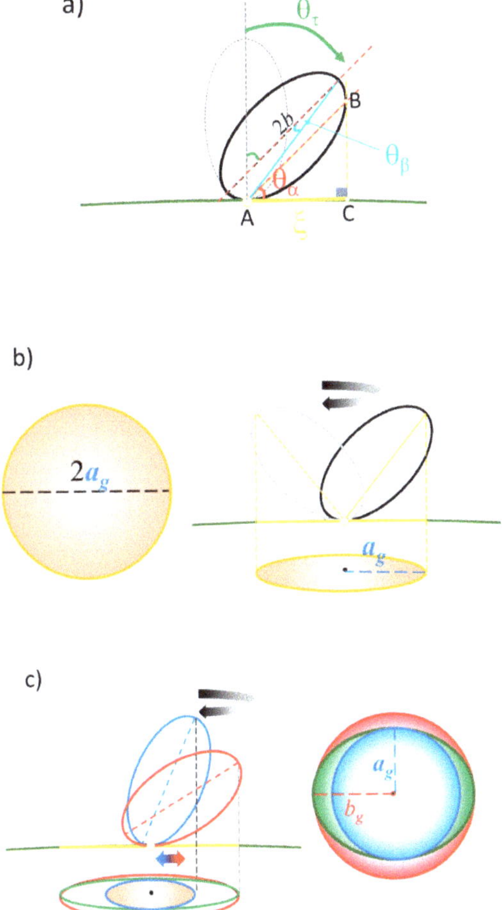

Figure 10. A sketch of the side view of a rotating prolate. (**a**) The tilting of a prolate over the nano-gold surface and approximation for radius ($\overline{AC}$) of the circular plane over the nano surface. Here, θ_α is a tangential angle between prolate axis and the surface line, and the tilting angle of a prolate against surface plane is given by an angle θ_τ. (**b**) A rotational motion of a prolate with a fixed contacting point, resulting in a circular occupied space over the surface. (**c**) A gyration motion of a prolate with a movable contacting point and tilting angle θ_τ, resulting in an oval (in green) occupied space with axial length of a_g (blue circle) and b_g (red circle).

From the geometry shown in Figure 10a, $\overline{AC} = 2bcos\theta_\alpha\theta_\beta$, where θ_α and θ_β are the inner angles as shown in Figure 10a and the length $\overline{AC}$ was approximated as $\overline{AB} \approx 2bcos\theta_\beta$ because $\theta_\beta \ll 1$. A tilting angle of a prolate, $\theta_\tau = 90° - \theta_\alpha$. The extracted θ_α and θ_β are listed in Table 3.

Table 3. The list of extracted tilting angles (θ_τ and θ_β) for the lower coverage for (**a**) Aβ_{1-40}, (**b**) α-syn, and (**c**) β2m. The average tilting angle for each peptide is shown at the bottom for each peptide in (θ_τ).

(a) Aβ_{1-40}	
d (**d**)	99.5 (100) nm
b	2.200 nm
a_g	3.720 nm
θ_τ	57.7°
θ_β	0.155°
b_g	0.905 nm
θ_τ	11.9°
θ_β	0.565°
(θ_τ)	35 ± 2°

(b) α-syn	
d (**d**)	99.5 (100) nm
b	7.400 nm
a_g	7.400 nm
θ_τ	30.0°
θ_β	0.000°
b_g	1.40 nm
θ_τ	5.4°
θ_β	0.127°
(θ_τ)	18 ± 2°

(c) β2m			
d (**d**)	9.80 (10) nm	19.7 (20) nm	60.0 (60) nm
b	4.6 nm	4.6 nm	4.6 nm
a_g	4.03 nm	6.41 nm	5.40 nm
θ_τ	26.0°	44.2°	36.0°
θ_β	0.354°	0.064°	0.326°
b_g	2.70 nm	2.73 nm	4.80 nm
θ_τ	17.1°	17.3°	31.4°
θ_β	0.508°	0.234°	0.060°
(θ_τ)		29 ± 6°	

An example of extracted gyration motion was demonstrated and sketched in Figure 11 for the case of β2m adsorbed over d = 10 nm gold colloid (d = 9.80 nm). Focusing on one unit of prolate as shown in Figure 11a, the tilting angle, θ_τ, changes between 26° and 17° as it rotates, which modulates the surface area. The gyration of the prolate should be taking place simultaneously with the other prolates on the same surface as shown in Figure 11b. We cannot, however, deny that a stationary peptide in an unfolded conformation could occupy the space of the same size calculated by gyration motion. There is a possibility of that the adsorption is more randomized and is an ensemble of multiple orientations. For example, J. A. Yang and et. al., reported that α-syn adsorbs on the poly (allylamide hydrochloride) coated gold nanoparticles with random orientation with an increase in β-sheet and decrease in α-helix structure [55].

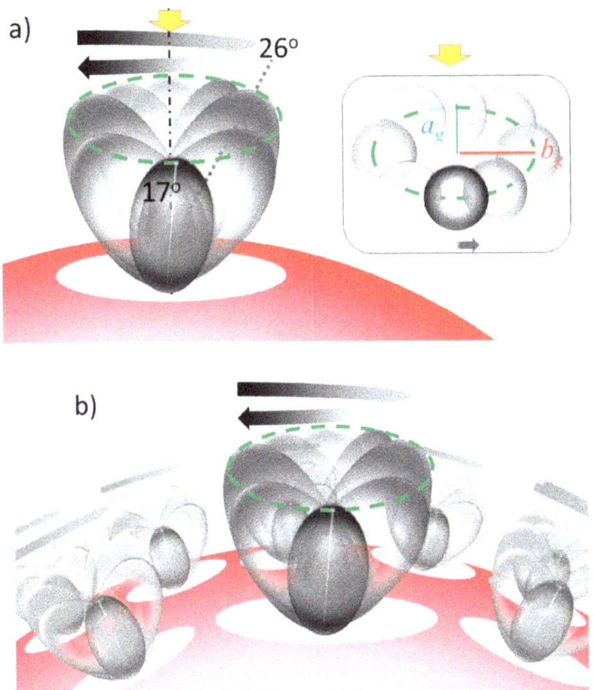

Figure 11. (a) The sketch showing the gyration motion of a prolate (a = 2.7 nm and b = 4.0 nm) representing β2m over a gold nano-particle with a diameter of d = 10 nm, where the prolate major axis tilts between 26° and 17° as it rotates over the surface. It results in an oval occupied space with a_g = 2.7 nm and b_g = 4.0 nm. (See Table 3) (b) The sketch of a gyrating prolate over the nano-gold particle surface.

4. Materials and Methods

4.1. Materials

Lyophilized powder of Aβ$_{1-40}$ peptide (MW; 4.2 kDa, 98% HPLC purity) and α-syn (MW: 14.4 kDa, purity >95% by SDS-PAGE) were purchased from r-Peptide (Bogart, GA, USA). Aqueous 220 μM stock solution of Aβ$_{1-40}$ and 64.2 μM stock solution of α-syn were stored at –80 °C. The β2m (MW: 12 kDa/mol, purity >40% by SDS-PAGE) was purchased from AbD Serotec (Raleigh, NC, USA), and aqueous 77.0 μM stock solution was stored at –20 °C. Gold nanoparticles were purchased from Ted Pella, Inc. (Redding, CA, USA) and have the following estimated diameters (d), reported diameter (d), and particles per mL in O.D. (Optical Density, ϑ) where ϑ = 0.2 at 528 nm: d = 10 nm (d = 9.8 ± 1.0 nm, ϑ = 1.4 × 10^{12} particles mL^{-1}), d = 15 nm (d = 15.2 ± 1.5 nm, ϑ = 2.8 × 10^{11} particles mL^{-1}), d = 20 nm (d = 19.7 ± 1.1 nm, ϑ = 1.4 × 10^{11} particles mL^{-1}), d = 30 nm (d = 30.7 ± 1.3 nm, ϑ = 4.0 × 10^{10} particles mL^{-1}), d = 40 nm (d = 40.6 ± 1.1 nm, ϑ = 1.8 × 10^{10} particles mL^{-1}), d = 50 nm (d = 51.5 ± 4 nm, ϑ = 8.2 × 10^{9} particles mL^{-1}), d = 60 nm (d = 60 ± 1.0 nm, ϑ = 4.3 × 10^{9} particles mL^{-1}), d = 80 nm (d = 80 ± 1.0 nm, ϑ = 2.2 × 10^{9} particles mL^{-1}), and d = 100 nm (d = 99.5 ± 1.3 nm, ϑ = 1.6 × 10^{9} particles mL^{-1}). The residual components in each colloidal particle can be regarded as identical. In order to maintain stability of the nano-gold colloids against salts, deionized and distilled water were used to prepare all aqueous solutions. All sizes of gold colloids were formed by Frens derived citrate reduction method possessing traces of citrate <10^{-5}%, tannic acid <10^{-7}% and potassium carbonate $<10^{-8}$%. Thus, the observed size dependence in this study was not determined by the stabilizer of the gold colloids.[12] The optimized ratio between all peptides and gold nanoparticles was set as 1000:1 so that the concentration of gold nanoparticles was roughly 300 pM [39]. Attachment of peptides to the gold colloidal surface was known to be achieved almost instantaneously and considered to reach equilibrium within a minute. The pH range of the solutions (between pH 2 and pH 12) was achieved by adding either HCl or NaOH to the solution. The UV–Vis absorption spectra were monitored between 200 and 800 nm as the pH value varied by an increment of 0.05 pH to acidic conditions.

4.2. TEM Imaging

The TEM (Transmission Electron Microscopy) experiment was conducted for β2m as well as for ovalbumin, $A\beta_{1-40}$, and α-syn. The β2m samples were prepared with 2.8 µL of β2m stock aqueous solution mixed with 280 µL of gold colloids ranging between 10 nm, 30 nm, 60 nm, and 80 nm in diameter, with pH ranging from 6.5 to 7.5. Before plating on the grid, the sample pH was adjusted to either pH 10 or pH 4 under room temperature. 1 µL of the mixture was then plated onto Formvar Copper Film 400 Grids Mesh. The samples were incubated on the grid for two minutes, after which excess solution was removed from the grid with filter paper. All TEM images were collected on a Morgagni model 268 TEM (FEI Co., Hillsboro, OR, USA) operated at 80 kV and were taken under both 28,000× and 71,000× magnification using a model XR-40 four-megapixel CCD Digital camera. TEM image analysis was performed by converting the image to data consisting of pixel coordinates and corresponding color index using Image J. The threshold in color index was set to recognize the group of pixels corresponding to the gold particles and the average size of the gold particles, the distance between adjacent gold particles, ratio of the area occupied by the gold particles (occupancy, %), and the total numbers of gold particles were calculated. The β2m-coated gold colloid formed relatively small aggregates as opposed to ovalbumin or $A\beta_{1-40}$. The number of gold colloidal particles was extracted by individually counting each particle rather than using the "occupancy" method. Because each gold colloid was easily identified for the β2m-coated gold colloid, in many cases, the space between each gold colloids in each aggregate were observed. In this study, the space between gold particles was focused and its distance was extensively analyzed whenever space between colloids was identified. Using the length of a pixel for calibration, the number of pixels between the colloids was transformed into nanometers. The distribution of the observed length in nm was fit with a Gaussian profile and the average distance was extracted.

4.3. Methods

4.3.1. pH-Dependent UV–Vis Absorption Band

Our group has been investigating the reversible self-assembly process of amyloidogenic peptide-coated colloidal gold nanoparticles extensively. These peptides are relatively small, amphiphilic peptides whose temperature/pH conditions for folded/unfolded conformations are well studied. Therefore, it has been viewed as a useful prototype system to learn how nanoscale surface potentials interact with a peptide, and if a specific oligomeric structure can be selectively constructed [11–13]. Although these peptides eventually form irreversible insoluble amyloids, the initial stage is still a reversible process. In temperature-dependent reversible processes [53], we found that a reversible process between folded and unfolded conformation took place under $A\beta_{1-40}$ coated 20 nm gold colloid as pH externally changed well above or well below a critical pH point (pH_o). The value of pH_o for $A\beta_{1-40}$ coated 20 nm gold was found to be pH_o = 5.45 ± 0.05 at 20 °C, and a reversible process was observed between pH = 4 and pH 10. at 18 ± 0. 2 °C and above. Between 18 ± 0.2 °C and 6 ± 0.2 °C, only $A\beta_{1-40}$ coated 30 nm gold colloid exhibited a reversible process. Under 6 ± 0.2 °C, only $A\beta_{1-40}$ coated 40 nm gold colloid supported a reversible process between folded and unfolded conformations. The results from molecular dynamics (MD) calculations on $A\beta_{10-35}$ suggested the temperature ranges are stable for dimer or trimer formation [56]. For example, the stable dimer formation temperature range matched with the temperature range of reversible process observed for $A\beta_{1-40}$ coated 20 nm gold colloid ($\geq$~18 °C). The trimers were predicted to be stable at the relatively lower temperature range, which reasonably matches temperature ranges for the reversible process over 30 or 40 nm gold colloid's surface (i.e., <18 °C). Also, the stable temperature for trimer formation was in good agreement with the temperature range of the reversible process observed for $A\beta_{1-40}$ coated 30 nm or 40 nm gold colloids. Since unfolded conformation leading to oligomerization is formed at only lower pH value than pH_o, we concluded that this is evidence that oligomeric dimer units over 20 nm gold colloid particles or trimer units over 30 or 40 nm gold colloid particles were produced over a nano-gold colloidal surface at pH 4 [53]. The key intermediate oligomeric form in the reversible process has not been well studied due to its instability. We hypothesize that the activation energy required to form an intermediate oligomer can be gained from the nano-metal surface potential. While metastable folding intermediates (i.e., the oligomer form) for a folding pathway has been suggested and detected in solution [57–71], a direct identification of an exact oligomer has not been shown. Oligomer observed in negatively charged micelles and Teflon particles, β-sheet formation of $A\beta$ on hydrophobic graphite surfaces [14], or at air–water interfaces [15] indicate an involvement of interfacial surface potential utilized for the conforming intermediate [16–20]. Our group has established a way to reproduce and control the reversible self-assembly of $A\beta$ on spherical gold nanoparticles. The average absorption peak shift (λ_{peak}) at room temperature (Figure 12a) is plotted as a function of the continuous operation of an external pH change (Figure 12b). The value of λ_{peak} corresponds to the color of the solution, which in turn corresponds to the morphology of the gold colloid aggregates. When the colloids assemble in an aggregated form at pH 4, the mixture is a blue color with λ_{peak} ~650 nm or above. On the other hand, the gold colloids are widely dispersed at pH 10, and exhibit a reddish color with λ_{peak} ~525 nm. Therefore, a repetitive pH change enables $A\beta_{1-40}$ coated 20 nm gold colloid to exhibit an oscillating feature of λ_{peak} between 525 nm and 625 nm as they reversibly form dispersed and aggregate forms, respectively.

Each absorption spectrum was fit by the "Peak Fit" program in Origin (Version 9.5) and peak positions of i-th band (λ_i) and peak area of each band of i-th component were extracted (A_i). The observed band average peak position is correlated with the surface plasmon resonance (SPR) of gold colloids, and the peak position of the absorption band depends on the conformation of peptide attached on the gold colloidal surface. The folded or unfolded conformation can be prepared by setting the solution to be basic or acidic, respectively. When the solution is acidic, the absorption band commonly has two or more components. Thus, the average peak position, $\lambda_{peak}(pH)$, of the SPR band at given pH is extracted by the weighted average of two components as $\lambda_{peak}(pH) = \sum_i a_i(pH)\lambda_i(pH)$, where $\lambda_i(pH)$ and $a_i(pH)$ are the peak position and fraction of the i-th component band, and the fraction a_i was determined by the fraction of the area (A_i) of the band to the total area of the entire bands as: $a_i = A_i / \sum_j A_j$.

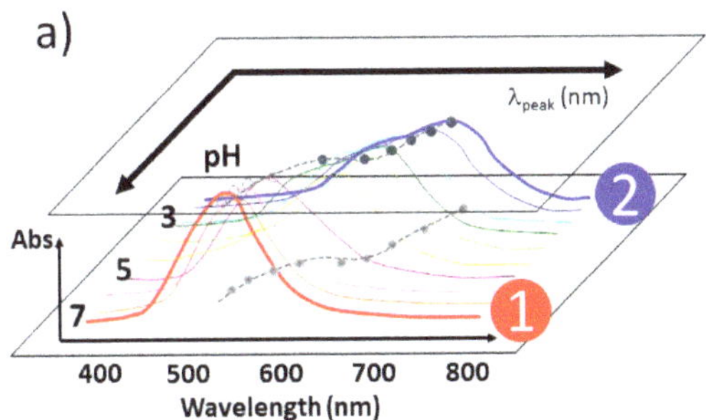

Figure 12. *Cont.*

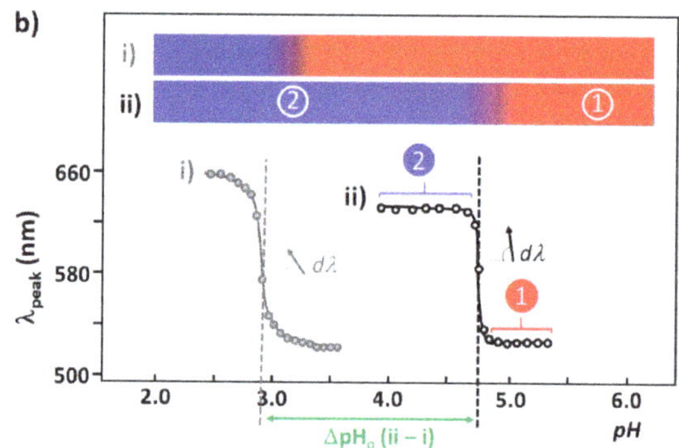

Figure 12. A schematic diagram explaining the spectral analysis and construction of sigmoidal plots. (**a**) The average peak position of SPR, λ_{peak}, was monitored as a function of the pH condition for both bare gold nano-particles and peptide-coated nano gold colloids. The λ_{peak} was extracted by utilizing the method described in Section 4.3.1. Two representative spectrum marked by ① and ② represent that under pH 7 and pH 2, respectively. (**b**) The constructed sigmoidal plot was fit with the Boltzmann formula shown in Equation (1). Both dpH and ΔpH$_o$ were obtained. Here, the sigmoidal plot i indicates that of bare gold colloid, and the sigmoidal plot ii is a typical plot observed for amyloidogenic peptide coated gold nano-particles. The upper colored bar shows the corresponding solution color for regions ① and ②.

Then, the average peak position was surveyed as a function of pH, and the position of the peaks were plotted as a function of pH, as shown in Figure 12b. The constructed sigmoidal plot was then analyzed and fit with a Boltzmann formula (Equation (1)) as shown in Figure 12b.

$$\lambda_{\text{peak}}(\text{pH}) = [\lambda_{\min} - \lambda_{\max}]/\{1 + \exp[(\text{pH} - \text{pH}_0)/d\text{pH}]\} + \lambda_{\max} \tag{1}$$

The $\lambda_{\min}$ and $\lambda_{\max}$ stand for the minimum and maximum of the band peak positions, respectively. Here, pH$_o$ shows the pH where color change takes place, and λ_{peak} (pH$_o$) = ($\lambda_{\min}$ + $\lambda_{\max}$)/2. Also, dpH = ($\lambda_{\max}$ − $\lambda_{\min}$)/4$\lambda_{\text{peak}}^{(1)}$, where $\lambda_{\text{peak}}^{(1)}$ is the first derivative of the λ_{peak}(pH).

Absorption of a collective excitation of the electrons at the interface between a conductor and an insulator is hypothesized to account for the color of suspensions of these particles [72–74]. If the net anionic sites of the metal surface are neutralized by acid, aggregation should be enhanced, resulting in a color change from red to blue. As coverage increases, a shielding effect shifts pH$_o$ to the higher value. This ultimately means that greater coverage of peptide requires a less acidic condition to neutralize the surface. Bare gold colloids change their colors at lower pHs (pH < 4.5) while a peptide-coated colloidal surface shows a color change at pH = 4.5~6 depending on the degree of coverage. This pH$_o$ value change between bare gold and protein coated gold solution is direct evidence of protein adsorption on the metal colloid.

4.3.2. Correlation Relation between ΔpH$_o$ and dpH and Extraction of Coverage Ratio

As the protein covers more of the colloidal surface, a less negative net ionic charge can be achieved. In other words, the negative charge is partially quenched due to the coverage of the peptide over the gold surface. This indicates that a less acidic condition is required to neutralize the colloidal surface.

Assuming a linear relationship (i.e., y = mx = b, where m is a slope and b is an intercept), all data points are fit along the formula given by dpH = m ΔpH$_o$ + b, where dpH corresponds to y and ΔpH$_o$ corresponds to x in the above linear relationship as shown in Figure 13. We hypothesize that ΔpH$_o$ directly relates to the peptide coverage fraction, Θ, since ΔpH$_o$ exhibits the surface character change between peptide coated gold colloid and bare gold colloid. Thus, ΔpH$_o$ = 0 corresponds to Θ = 0 (i.e., no peptide coverage or bare gold colloid), and the x-axis intercept of $dpH = m \, \Delta pH_o + b$ indicates the maximum value of ΔpH_o, which occurs when $\Theta = 1$ at $dpH = 0$ (i.e., λ_{peak} [1] ~∞). The maximum coverage, therefore, can be achieved at the ΔpH_o value given by $\Delta pH_o(max) = -\frac{b}{m}$. By replacing ΔpH_o (max.) with $\Theta = 1$, any Θ values in between $\Delta pH_o = 0$ and ΔpH_o (max.) can then be calculated by

$$\Theta = \frac{\Delta pH_o}{\Delta pH_o(max)} \tag{2}$$

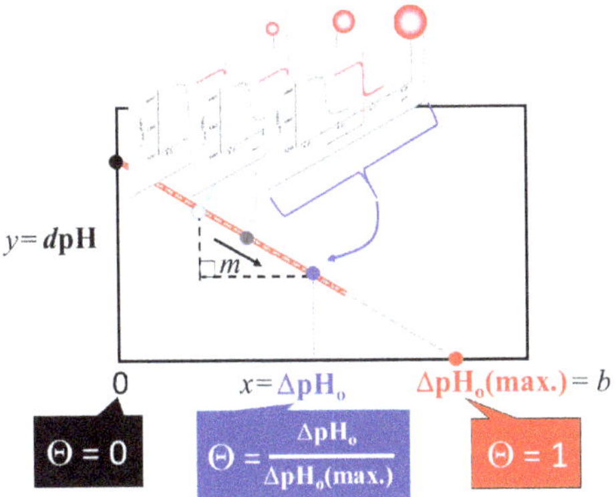

Figure 13. For each amyloidogenic peptide, dpH obtained by Equation (1) by analyzing each sigmoidal plot (See Figure 12b) was plotted as a function of ΔpH_o. The coverage ratio, Θ, for each gold colloidal size was obtained by scaling the data point, based on the x-axis intercept $\Theta = 1$ (See Equation (2)) Each data point corresponds to different nano-gold colloidal size and was obtained by analyzing sigmoidal plot shown in Figure 12b) for bare gold and amyloidogenic peptide coated gold.

4.3.3. Simulation Process for Calculating the Coverage Fraction

A full explanation of a simulation procedure of calculating peptide coverage fraction, Θ, will be described in a report by Yokoyama and Ichiki [54]. In this paper, an essential concepts for calculating Θ are briefly presented.

The adsorption orientation of a prolate (axial length of a and b, $a < b$) was selected from either spiking-out orientation (Figure 14a top) or lie-down orientation (Figure 14a bottom). In spiking-out orientation, b-axis of a prolate contacts tangentially at a nano-gold surface. On the other hand, in lie-down orientation, a-axis of a prolate contacts tangentially at a nano-gold surface. An area projected on the sphere surface as an occupying area (A_{sphere}) is $A_{sphere} = \pi a^2$ for spiking-out orientation and $A_{sphere} = \pi ab$ for lie-down orientation. Once adsorption orientation was selected, the numbers of adsorption points were calculated for the first layer and the second layer.

For the first layer, the total adsorption points ($n_{f,tot}$) is a sum of a top and a bottom spot of a sphere ($n_{f,top}$ and $n_{f,bot}$), equatorial spots ($n_{f,eq}$), and both hemi-sphere's axial position for each equatorial spot j ($n_{f,ax,j}$) as:

$$n_{f,\,tot} = n_{f,top} + n_{f,bot} + n_{f,eq} + 2\sum n_{f,ax,j} \tag{3}$$

As for the second layer, avoiding the adsorption points taken by the first layer, the total adsorption points ($n_{s,tot}$) are given by

$$n_{s,\,tot} = n_{s,eq} + 2\sum n_{s,ax,j} \tag{4}$$

where equatorial spots ($n_{s,eq}$), and both hemi-sphere's axial position for each equatorial spot j ($n_{s,ax,j}$) are counted over a sphere corresponds to a second layer. The axial length a and b ($a > b$) of an approximated prolate for Aβ_{1-40} [17], α-synuclein [37] and $\beta 2m$ [38] are estimated to be: Aβ_{1-40} (a, b)

= (2.1 nm, 4.1 nm), α-syn (a, b) = (2.9 nm, 6.0 nm), and β2m (a, b) = (2.1 nm, 4.6 nm). Each colloidal particle is approximated to be a sphere with diameter d (radius $r = d/2$).

In order to reproduce observed Θ, Θ_{obs} (= Θ_{total}), a combination of the first and second layers were calculated by Equation (5).

$$\Theta_{total} = \Theta_{f,\,total} + \gamma\,\Theta_{s,total} \tag{5}$$

where $\Theta_{f,\,total}$ is a fraction of coverage contributed from the first layer, $\Theta_{s,\,total}$ is a fraction of coverage contributed from the second layer, and γ is an empirical factor indicating the weight of the second layer. Each of $\Theta_{f,\,total}$ and $\Theta_{s,\,total}$ is determined by Equation (6) and Equation (7), respectively.

$$\Theta_{f,total} = \frac{A_{prolate} \times n_{f,tot}}{A_{f,\,sphere}} \tag{6}$$

$$\Theta_{s,total} = \frac{A_{prolate} \times n_{s,tot}}{A_{s,\,sphere}} \tag{7}$$

Here, $A_{prolate}$ indicates an area projected on the sphere surface as an occupying area. The areas to be covered for the first and the second layer are represented by $A_{f,sphere}$ and $A_{s,sphere}$, respectively. The values of $A_{f,sphere}$ and $A_{s,sphere}$ are different depending the adsorption orientation. The total adsorption point for the first layer ($n_{f,tot}$) and the second layer ($n_{s,tot}$) are illustrated in Figure 14b and detailed in Equation (3) and (4), respectively.

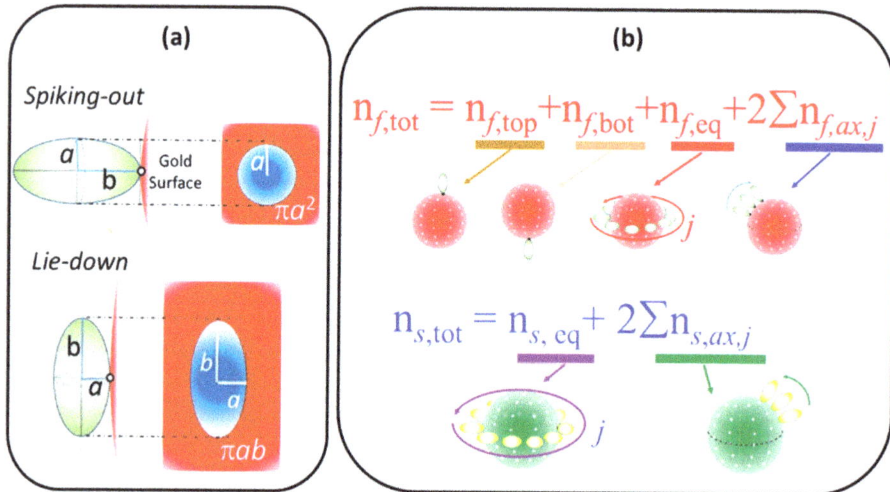

Figure 14. (a) Sketches of two possible orientation of prolate and corresponding occupying area are shown. Top: spiking out orientation and occupying πa^2 on the surface. Bottom: lie-down orientation and occupying πab on the surface. (b) For spiking-out orientation, a procedure of counting total number of adsorption points for the first layer ($n_{f,tot}$) and for the second layer ($n_{s,\,tot}$) are shown. The $n_{f,tot}$ is an addition of top ($n_{f,top}$) and bottom ($n_{f,bot}$) adsorption points of a sphere, maximum adsorption position at equatorial positions ($n_{f,eq}$), and the adsorption point along axial axis $n_{f,\,ax}$ added along each equatorial position j for a semi-sphere. For the second layer, the total number of adsorption points ($n_{s,tot}$) is given by a summation of total number of equatorial adsorption position over a sphere of second layer ($n_{s,,eq}$) and total number of adsorption points associated with axial positions, where total adsorption points of each axial line ($n_{s,ax}$) is counted for each equatorial position of second layer, j.

5. Conclusions

The surface properties of nano-gold colloidal surfaces due to adsorption of amyloidogenic peptides were successfully monitored and characterized by observing the response of spectroscopic features as a function of an external pH change. This surface property change was found to be linearly correlated with the coverage ratio of the peptide onto the colloid, Θ. With the simplification of the space occupied by a peptide into a prolate, the Θ could be extracted through a simplified tessellation logic applied for a sphere. The simulation suggested that a prolate needs to have a spiking-out orientation with representative prolate axial length of $(a, b) = (1.4\ \text{nm}, 2.2\ \text{nm})$ for $A\beta_{1-40}$, $(a, b) = (4.6\ \text{nm}, 7.4\ \text{nm})$ for α-syn, and $(a, b) = (2.5\ \text{nm}, 4.6\ \text{nm})$ for β2m. Of note, these values were similar to the values estimated by the reported protein structural data. However, the above-mentioned prolate dimensions could not reproduce the cases when Θ is less than ~0.50. A lower Θ is required to have less unit coverage area; this increase of unit area was interpreted as a gyration motion of each peptide, which kept a fixed contact spot but changed the tilting angle. The average tilting angle of the prolate was $(\theta_\tau)\ (A\beta_{1-40}) = 35 \pm 2°$, $(\theta_\tau)\ (\alpha\text{-syn}) = 18 \pm 2°$, and $(\theta_\tau)\ (\beta2m) = 29 \pm 6°$, indicating that when the colloid coverage ratio is below 0.5 a prolate can possess high degree of freedom in mobility while still maintaining a high level of interaction with the gold nano-particle surface. At the same time, it indicated many other possibilities of conformation including multiple contacting points or an ensemble of different adsorption orientations. The resulting Θ was fully explained by a relationship between the distances of each unit monomer under a given colloidal area. However, the degree of affinity for a second layer required us to account for the distribution of partially positive charge ($\delta+$) over a peptide. The segment possesses a $\delta+$ that was considered to be highly used when $A\beta_{1-40}$ and α-syn each interacted with a nano-gold colloidal surface. This possesses a distribution of centering around the prolate axis. On the other hand, the $\delta+$ of β2m was used to interact with each monomer, and the charge distribution was spread around with a distortion, resulting in a high exposure for the counter acting monomer. Therefore, it guided us to predict that β2m possesses different charge distributions than $A\beta_{1-40}$ and α-syn, and sequences or sections of peptide corresponding to $\delta+$ or $\delta-$ thus postulated. In closing, we demonstrated that nano-scale geometrical simulation with a simplified protein structure (i.e., prolate) successfully represents peptide adsorption orientation, providing insights into interfacial conformation and indicating the presence of electrostatic intermolecular and interfacial interactions for these pathophysiologic peptide cases.

Supplementary Materials: Supplementary materials can be found at http://www.mdpi.com/1422-0067/20/21/5354/s1.

Author Contributions: Conceptualization, K.Y.; methodology, K.Y. and E.D.; software, K.Y. and J.J.; validation, K.Y., and A.I.; formal analysis, K.Y., K.B., P.S., J.J., E.D., N.R., J.B., I.D., and A.I.; investigation, K.Y., K.B., J.J. and A.I.; resources, K.Y., I.D., and A.I.; data curation, K.Y.; writing—original draft preparation, K.Y.; writing—review and editing, K.Y., J.B., N.R., and A.I.; visualization, K.Y., E.D., and N.R.; supervision, K.Y.; project administration, K.Y.; funding acquisition, K.Y.

Funding: This research received no external funding.

Acknowledgments: We are grateful for the support by Geneseo Foundation at the initial stage of this project. Ishan Deshmukh and Akane Ichiki thank for the gracious support by SUNY Geneseo Chemistry Department Alumni Summer Research Scholarship. We also thank Jonathan Bourne for helpful discussion during preparation of this manuscript.

Conflicts of Interest: The authors declare no conflict of interest.

References

1. Norde, W. BioInterface Perspective-My voyage of discovery to proteins in flatland and beyond. *Colloids Surf. B Biointerfaces* **2008**, *61*, 1–9. [CrossRef] [PubMed]
2. Cook, N.P.; Kilpatrick, K.; Segatori, L.; Martí, A.A. Detection of α-Synuclein Amyloidogenic Aggregates in Vitro and in Cells using Light-Switching Dipyridophenazine Ruthenium(II) Complexes. *J. Am. Chem. Soc.* **2012**, *134*, 20776–20782. [CrossRef] [PubMed]

3. Muangchuen, A.; Chaumpluk, P.; Suriyasomboon, A.; Ekgasit, S. Colorimetric detection of Ehrlichia canis via nucleic acid hybridization in gold nano-colloids. *Sensors* **2014**, *14*, 14472–14487. [CrossRef] [PubMed]

4. Rasheed, P.A.; Sandhyarani, N. Femtomolar level detection of BRCA1 gene using a gold nanoparticle labeled sandwich type DNA sensor. *Colloids Surf. B Biointerfaces* **2014**, *1*, 7–13. [CrossRef] [PubMed]

5. Suarasan, S.; Focsan, M.; Soritau, O.; Maniu, D.; Astilean, S. One-pot, green synthesis of gold nanoparticles by gelatin and investigation of their biological effects on Osteoblast cells. *Colloids Surf. B Biointerfaces* **2015**, *32*, 122–131. [CrossRef] [PubMed]

6. Zhao, T.; Yu, K.; Li, L.; Zhang, T.; Guan, Z.; Gao, N.; Yuan, P.; Li, S.; Yao, S.Q.; Xu, Q.H.; et al. Gold nanorod enhanced two-photon excitation fluorescence of photosensitizers for two-photon imaging and photodynamic therapy. *ACS Appl. Mater. Interfaces* **2014**, *6*, 2700–2708. [CrossRef]

7. Politi, J.; Spadavecchia, J.; Iodice, M.; De Stefano, L. Oligopeptide–heavy metal interaction monitoring by hybrid gold nanoparticle based assay. *Analyst* **2015**, *140*, 149–155. [CrossRef]

8. Heinz, H.; Farmer, B.L.; Pandey, R.B.; Slocik, J.M.; Patnaik, S.S.; Pachter, R.; Naik, R.R. Nature of Molecular Interactions of Peptides with Gold, Palladium, and Pd-Au Bimetal Surfaces in Aqueous Solution. *J. Am. Chem. Soc.* **2009**, *131*, 9704–9714. [CrossRef]

9. Heinz, H.; Vaia, R.A.; Farmer, B.L.; Naik, R.R. Accurate Simulation of Surfaces and Interfaces of Face-Centered Cubic Metals Using 12-6 and 9-6 Lennard-Jones Potentials. *J. Phys. Chem. C* **2008**, *112*, 17281–17290. [CrossRef]

10. Venditti, I.; Hassanein, T.F.; Fratoddi, I.; Fontana, L.; Battocchio, C.; Rinaldi, F.; Carafa, M.; Marianecci, C.; Diociaiuti, M.; Agostinelli, E.; et al. Bioconjugation of gold-polymer core-shell nanoparticles with bovine serum amine oxidase for biomedical applications. *Colloids Surf. B Biointerfaces.* **2015**, *134*, 314–321. [CrossRef]

11. Yokoyama, K. Nanoscale Surface Size Dependence in Protein Conjugation. In *Advances in Nanotechnology*; Chen, E.J., Peng, N., Eds.; Nova Science Publishing: Hauppauge, NY, USA, 2010; pp. 65–104.

12. Yokoyama, K. Modeling of Reversible Protein Conjugation on Nanoscale Surface. In *Computational Nanotechnology: Modeling and Applications with MATLAB*; Musa, S.M., Ed.; CRC Press-Taylor and Francis Group, LLC: Boca Raton, FL, USA, 2011; pp. 381–410.

13. Yokoyama, K. Nano Size Dependent Properties of Colloidal Surfaces. In *Colloids: Classification, Properties and Applications*; Ray, P.C., Ed.; Nova Science Publishing: Hauppauge, NY, USA, 2012; pp. 25–58.

14. Kowalewski, T.; Holtzman, D.M. In situ atomic force microscopy study of Alzheimer's beta-amyloid peptide on different substrates: New insights into mechanism of beta-sheet formation. *Proc. Natl. Acad. Sci. USA* **1999**, *96*, 3688–3693. [CrossRef] [PubMed]

15. Schladitz, C.; Vieira, E.P.; Hermel, H.; Mohwald, H. Amyloid-beta-sheet formation at the air-water interface. *Biophys. J.* **1999**, *77*, 3305–3310. [CrossRef]

16. Kusumoto, Y.; Lomakin, A.; Teplow, D.B.; Benedek, G.B. Temperature dependence of amyloid beta-protein fibrillization. *Proc. Natl. Acad. Sci. USA* **1998**, *95*, 12277–12282. [CrossRef] [PubMed]

17. Coles, M.; Bicknell, W.; Watson, A.A.; Fairlie, D.P.; Craik, D.J. Solution structure of amyloid beta-peptide(1–40) in a water-micelle environment. Is the membranespanning domain where we think it is? *Biochemistry* **1998**, *37*, 11064–11077. [CrossRef] [PubMed]

18. Shao, H.Y.; Jao, S.C.; Ma, K.; Zagorski, M.G. Solution structures of micelle-bound amyloid beta-(1–40) and beta-(1–42) peptides of Alzheimer's disease. *J. Mol. Biol.* **1999**, *285*, 755–773. [CrossRef]

19. Giacomelli, C.E.; Norde, W. Conformational changes of the amyloid beta-peptide(1-40) adsorbed on solid surfaces. *Macromol. Biosci.* **2005**, *5*, 401–407. [CrossRef]

20. Rocha, S.; Krastev, R.; Thunemann, A.F.; Pereira, M.C.; Mohwald, H.; Brezesinski, G. Adsorption of amyloid beta-peptide at polymer surfaces: A neutron reflectivity study. *Chem. Phys. Chem.* **2005**, *6*, 2527–2534. [CrossRef]

21. Moshe, A.; Landau, M.; Eisenberg, D. Preparation of Crystalline Samples of Amyloid Fibrils and Oligomers. *Methods Mol. Biol.* **2016**, *1345*, 201–210.

22. Scarff, C.A.; Ashcroft, A.E.; Radford, S.E. Characterization of Amyloid Oligomers by Electrospray Ionization-Ion Mobility Spectrometry-Mass Spectrometry (ESI-IMS-MS). *Methods Mol. Biol.* **2016**, *1345*, 115–132.

23. Pujol-Pina, R.; Vilapriny-Pascual, S.; Mazzucato, R.; Arcella, A.; Vilaseca, M.; Orozco, M.; Carulla, N. SDS-PAGE analysis of Aβ oligomers is disserving research into Alzheimer's disease: Appealing for ESI-IM-MS. *Sci. Rep.* **2015**, *5*, 14809–14821.

24. El-Shimy, I.A.; Heikal, O.A.; Hamdi, N. Minocycline attenuates Aβ oligomers-induced pro-inflammatory phenotype in primary microglia while enhancing Aβ fibrils phagocytosis. *Neurosci. Lett.* **2015**, *609*, 36–41. [CrossRef] [PubMed]

25. Attanasio, F.; Convertino, M.; Magno, A.; Caflisch, A.; Corazza, A.; Haridas, H.; Esposito, G.; Cataldo, S.; Pignataro, B.; Milardi, D.; et al. Carnosine Inhibits Ab42 Aggregation by Perturbing the HBond Network in and around the Central Hydrophobic Cluster. *ChemBioChem* **2013**, *14*, 583–592. [CrossRef] [PubMed]

26. Liu, G.; Aliaga, L.; Cai, H. α-synuclein, LRRK2 and their interplay in Parkinson's disease. *Future Neurol.* **2012**, *7*, 145–153. [CrossRef] [PubMed]

27. Paslawski, W.; Andreasen, M.; Nielsen, S.B.; Lorenzen, N.; Thomsen, K.; Kaspersen, J.D.; Pedersen, J.S.; Otzen, D.E. High Stability and Cooperative Unfolding of α-Synuclein Oligomers. *Biochemistry* **2014**, *53*, 6252–6263. [CrossRef]

28. Pchelina, N.; Nuzhnyi, E.P.; Emelyanov, A.K.; Boukinac, T.M.; Usenko, T.S.; Nikolaev, M.A.; Salogub, G.N.; Yakimovskii, A.F.; Zakharova, E.Y. Increased plasma oligomeric alpha-synuclein in patients withlysosomal storage diseases. *Neurosci. Lett.* **2014**, *583*, 188–193. [CrossRef]

29. van Rooijen, B.D.; Claessens, M.M.A.E.; Subramaniam, V. Lipid bilayer disruption by oligomeric α-synuclein depends on bilayer charge and accessibility of the hydrophobic core. *Biochim. Biophys. Acta* **2009**, *1788*, 1271–1278. [CrossRef]

30. Vasudevaraiu, P.; Guerrero, E.; Hegda, M.L.; Collen, T.B.; Britton, G.B.; Rao, K.S. New evidence on alpha-synuclein and Tau binding to conformation and sequence specific GC* rich DNA: Relevance to neurological disorders. *J. Pharm. Bioallied Sci.* **2012**, *4*, 112–117.

31. Comellas, G.; Lemkau, L.R.; Zhou, D.H.; George, J.M.; Rienstra, C.M. Structural Intermediates during α-Synuclein Fibrillogenesis on Phospholipid Vesicles. *J. Am. Chem. Soc.* **2012**, *134*, 5090–5099. [CrossRef]

32. Fecchio, C.; De Franceschi, G.; Relini, A.; Greggio, E.; Dalla Serra, M.; Bubacco, L.; de Laureto, P.P. α-Synuclein Oligomers Induced by Docosahexaenoic Acid Affect Membrane Integrity. *PLoS ONE* **2013**, *8*, 1–12. [CrossRef]

33. Esposito, G.; Garvey, M.; Alverdi, V.; Pettirossi, F.; Corazza, A.; Fogolari, F.; Polano, M.; Mangione, P.P.; Giorgetti, S.; Stoppini, M.; et al. Monitoring the Interaction between β2-Microglobulin and the Molecular Chaperone alpha B-crystallin by NMR and Mass Spectrometry: α B-Crystallin dissociates β2-Microglobulin Oligomers. *J. Biol. Chem.* **2013**, *288*, 17844–17858. [CrossRef]

34. Esposito, G.; Corazza, A.; Bellotti, V. Pathological Self-Aggregation of β2-Microglobulin: A Challenge for Protein Biophysics. In *Protein Aggregation and Fibrillogenesis in Cerebral and Systemic Amyloid Disease, Subcellular Biochemistry*; Harris, J.R., Ed.; Springer: Berlin, Germany, 2012; Volume 65, pp. 165–182.

35. Mustata, M.; Capone, R.; Jang, H.; Arce, F.T.; Ramachandran, S.; Lal, R.; Nussinov, R. K3 fragment of amyloidogenic β2-microglobulin forms ion channels: Implication for dialysis related amyloidosis. *J. Am. Chem Soc.* **2009**, *131*, 14938–14945. [CrossRef]

36. Yokoyama, K.; Catalfamo, C.D.; Yuan, M. Reversible Peptide Oligomerization over Nanosclae Gold Surfaces. *Aims Biophys.* **2015**, *2*, 679–695. [CrossRef]

37. Ulmer, T.S.; Bax, A.; Cole, N.B.; Nussbaum, R.L. Strcture and dynamics of micelle-bound human α-synuclein. *J. Biol. Chem.* **2005**, *280*, 9595–9603. [CrossRef] [PubMed]

38. Verdone, G.; Corazza, A.; Viglino, P.; Pettirossi, F.; Giorgetti, S.; Mangione, P.; Andreola, A.; Stoppini, M.; Bellotti, V.; Esposito, G. The solution structure of human beta2-microglobulin reveals the prodromes of its amyloid transition. *Protein Sci.* **2002**, *11*, 487–499. [CrossRef] [PubMed]

39. Yokoyama, K.; Welchons, D.R. The conjugation of amyloid beta protein on the gold colloidal nanoparticles' surfaces. *Nanotechnology* **2007**, *18*, 105101–105107. [CrossRef]

40. Yokoyama, K.; Briglio, N.M.; Sri Hartati, D.; Tsang, S.M.W.; MacCormac, J.E.; Welchons, D.R. Nanoscale size dependence in the conjugation of amyloid beta and ovalbumin proteins on the surface of gold colloidal particles. *Nanotechnology* **2008**, *19*, 375101–375108. [CrossRef]

41. Wang, A.; Perera, Y.R.; Davidson, M.B.; Fitzkee, N.C. Electrostatic Interactions and Protein Competition Reveal a Dynamic Surface in Gold Nanoparticle–Protein Adsorption. *J. Phys. Chem. C* **2016**, *120*, 24231–24239. [CrossRef]

42. Kuna, J.J.; Voïtchovsky, K.; Singh, C.; Jiang, H.; Mwenifumbo, S.; Ghorai, P.K.; Stevens, M.M.; Glotzer, S.C.; Stellacci, F. The effect of nanometre-scale structure on interfacial energy. *Nat. Mater.* **2009**, *8*, 837–842. [CrossRef]

43. DeVries, G.A.; Brunnbauer, M.; Hu, Y.; Jackson, A.M.; Long, B.; Neltner, B.T.; Uzun, O.; Wunsch, B.H.; Stellacci, F. Divalent Metal Nanoparticles. *Science* **2007**, *315*, 358–361. [CrossRef]
44. Jackson, A.M.; Myerson, J.W.; Stellacci, F. Spontaneous assembly of subnanometreordered domains in the ligand shell of monolayer-protected nanoparticles. *Nat. Mater.* **2004**, *3*, 330–336. [CrossRef]
45. Singh, C.; Ghorai, P.K.; Horsch, M.A.; Jackson, A.M.; Larson, R.G.; Stellacci, F.; Glotzer, S.C. Entropy-Mediated Patterning of Surfactant-Coated Nanoparticles and Surfaces. *Phys. Rev. Lett.* **2007**, *99*, 1–3. [CrossRef] [PubMed]
46. Verma, A.; Uzun, O.; Hu, Y.; Hu, Y.; Han, H.-S.; Watson, N.; Chen, S.; Irvine, D.J.; Stellacci, F. Surface-structure-regulated cell-membrane penetration by monolayer-protected nanoparticles. *Nat. Mater.* **2008**, *7*, 588–595. [CrossRef] [PubMed]
47. Yuan, J.; Liu, X.; Akbulut, O.; Hu, J.; Suib, S.L.; Kong, J.; Stellacci, F. Superwetting nanowire membranes for selective absorption. *Nat. Nanotechnol.* **2008**, *3*, 332–336. [CrossRef] [PubMed]
48. Voïtchovsky, K.; Kuna, J.J.; Contera, S.A.; Tosatti, E.; Stellacci, F. Direct mapping of the solid–liquid adhesion energy with subnanometre resolution. *Nat. Nanotechnol.* **2010**, *5*, 401–405. [CrossRef]
49. Davidson, W.S.; Jonas, A.; Clayton, D.F.; George, J.M. Stabilization of alpha-synuclein secondary structure upon binding to synthetic membranes. *J. Biol. Chem.* **1998**, *273*, 9443–9449. [CrossRef]
50. George, J.M.; Jin, H.; Woods, W.S.; Clayton, D.F. Characterization of a novel protein regulated during the critical period for song learning in the zebra finch. *Neuron* **1995**, *15*, 361–372. [CrossRef]
51. Platt, G.W.; Routledge, K.E.; Homans, S.W.; Radford, S.E. Fibril Growth Kinetics Reveal a Region of β2-microglobulin Important for Nucleation and Elongation of Aggregation. *J. Mol. Biol.* **2008**, *378*, 251–263. [CrossRef]
52. D'Ambrosio, E.; Ralbovsky, N.; Yokoyama, K. Direct probing of the reversible selfassembly of amyloid beta peptide oligomers over nanoscale metal colloidal surfaces. In Proceedings of the 251st ACS National Meeting & Exposition, San Diego, CA, USA, March 13–17 2016.
53. Yokoyama, K.; Gaulin, N.B.; Cho, H.; Briglio, N.M. Temperature Dependence of Conjugation of Amyloid Beta Peptide on the Gold Colloidal Nanoparticles. *J. Phys. Chem. A* **2010**, *114*, 1521–1528. [CrossRef]
54. Yokoyama, K.; Ichiki, A. Peptide adsorption orientation. In *Oligomerization*; NOVA Science Publisher; In Preparation.
55. Yang, J.A.; Johnson, B.J.; Wu, S.; Woods, W.S.; George, J.M.; Murphy, C.J. Study of Wild-Type α-Synuclein Binding and Orientation on Gold Nanoparticles. *Langmuir* **2013**, *29*, 4603–4615. [CrossRef]
56. Jang, S.; Shin, S. Computational study on the structural diversity of amyloid beta peptide (Aβ$_{10-35}$) Oligomers. *J. Phys. Chem. B* **2008**, *112*, 3479–3484. [CrossRef]
57. Fraser, P.E.; Nguyen, J.T.; Surewicz, W.K.; Kirschner, D.A. pH-dependent structural transitions of Alzheimer amyloid peptides. *Biophys. J.* **1991**, *60*, 1190–1201. [CrossRef]
58. Shen, C.L.; Scott, G.L.; Merchant, F.; Murphy, R.M. Light scattering analysis of fibril growth from the amino-terminal fragment beta(1-28) of beta-amyloid peptide. *Biophys. J.* **1993**, *65*, 2383–2395. [CrossRef]
59. Inouye, H.; Fraser, P.E.; Kirshner, D.A. Structure of -crystallite assemblies formed by Alzheimer-amyloid protein analogues: Analysis by X-ray diffraction. *Biophys. J.* **1993**, *64*, 502–519. [CrossRef]
60. Lomakin, A.; Chung, D.S.; Benedek, G.B.; Kirshner, D.A.; Teplow, D.B. On the nucleation and growth of amyloid -protein fibrils: Detection of nuclei and quantitation of rate constants. *Proc. Natl. Acad. Sci. USA* **1996**, *93*, 1125–1129. [CrossRef]
61. Vine, H.L. Soluble multimeric Alzheimer (1-40) pre-amyloid complexes in dilute solution. *Neurobiol. Aging* **1995**, *16*, 755–764.
62. Kuo, Y.M.; Emmerling, M.R.; Vigo-Pelfrey, C.; Kasunic, T.C.; Kirkpatrick, J.B.; Murdoch, G.H.; Ball, M.J.; Roher, A.E. Water-soluble A (N-40, N-42) oligomers in normal and Alzheimer disease brains. *J. Biol. Chem.* **1996**, *271*, 4077–4081. [CrossRef]
63. Lambert, M.P.; Barlow, A.K.; Chromy, B.A.; Edwards, C.; Freed, R.; Liosatos, M.; Morgan, T.E.; Rozovsky, I.; Trommer, B.; Viola, K.L.; et al. Diffusible, nonfibrillar ligands derived from A 1-42 are potent central nervous system neurotoxins. *Proc. Natl. Acad. Sci. USA* **1998**, *95*, 6448–6453. [CrossRef]
64. Teller, J.K.; Russo, C.; DeBusk, L.M.; Angelini, G.; Zaccheo, D.; Dagna-Bricarelli, F.; Scartezzini, P.; Bertolini, S.; Mann, D.M.; Tabaton, M.; et al. Presence of soluble amyloid beta-peptide precedes amyloid plaque formation in Down's syndrome. *Nat. Med.* **1996**, *2*, 93–95. [CrossRef]

65. Huang, T.H.J.; Yang, D.S.; Plaskos, N.P.; Go, S.; Yip, C.M.; Fraser, P.E.; Chakrabartty, A. Structural studies of soluble oligomers of the Alzheimer beta-amyloid peptide. *J. Mol. Biol.* **2000**, *297*, 73–87. [CrossRef]

66. Garzon-Rodriguez, W.; Sepulveda-Bacerra, M.; Milton, S.; Glable, C.G. Soluble amyloid A-(1-40) exists as a stable dimer at low concentrations. *J. Biol. Chem.* **1997**, *272*, 21037–21044. [CrossRef]

67. Barrow, C.J.; Zagorski, M.G. Solution structures of peptide and its constituent fragments: Relation to amyloid deposition. *Science* **1991**, *253*, 179–182. [CrossRef] [PubMed]

68. Barrow, C.J.; Yasuda, A.; Kenny, P.T.; Zagorski, M.G. Solution conformations and aggregational properties of synthetic amyloid beta-peptides of Alzheimer's disease. Analysis of circular dichroism spectra. *J. Mol. Biol.* **1992**, *225*, 1075–1093. [CrossRef]

69. Zagorski, M.G.; Barrow, C.J. NMR studies of amyloid beta-peptides: Proton assignments, secondary structure, and mechanism of an alpha-helix—beta-sheet conversion for a homologous, 28-residue, N-terminal fragment. *Biochemistry* **1992**, *31*, 5621–5631. [CrossRef] [PubMed]

70. Walsh, D.M.; Lomakin, A.; Benedek, G.B.; Condron, M.M.; Teplow, D.B. Amyloid -protein fibrillogenesis. *J. Biol. Chem.* **1997**, *272*, 22364–22372. [CrossRef]

71. Crawford, F.; Soto, C.; Suo, Z.; Fang, C.; Parker, T.; Sawar, A.; Frangione, B.; Mullan, M. Alzheimer's -amyloid vasoactivity: Identification of a novel -amyloid conformational intermediate. *FEBS Lett.* **1998**, *436*, 445–448. [CrossRef]

72. Link, S.; El-Sayed, M. Spectral properties and relaxation dynamics of surface plasmon electronic oscillations in gold and silver nanodots and nanorods. *J. Phys. Chem. B* **1999**, *103*, 8410–8426. [CrossRef]

73. Kelly, K.L.; Coronado, E.; Zhao, L.L.; Schatz, G.C. The optical properties of metal nanoparticles: The influence of size, shape, and dielectric environment. *J. Phys. Chem. B* **2003**, *107*, 668–677. [CrossRef]

74. Jensen, T.R.; Schatz, G.C.; Van Duyne, R.P. Nanosphere lithography: Surface plasmon resonance spectrum of a periodic array of silver nanoparticles by ultraviolet–visible extinction spectroscopy and electrodynamic modeling. *J. Phys. Chem. B* **1999**, *103*, 2394–2401. [CrossRef]

International Journal of
Molecular Sciences

Article

Hydroxyapatite Formation on Self-Assembling Peptides with Differing Secondary Structures and Their Selective Adsorption for Proteins

Suzuka Kojima, Hitomi Nakamura, Sungho Lee, Fukue Nagata and Katsuya Kato *

National Institute of Advanced Industrial Science and Technology, 2266-98, Anagahora, Shimo-Shidami, Moriyama-ku, Nagoya, Aichi 463-8560, Japan; suzuka-kojima@aist.go.jp (S.K.); hi.nakamura@aist.go.jp (H.N.); sungho.lee@aist.go.jp (S.L.); f.nagata@aist.go.jp (F.N.)
* Correspondence: katsuya-kato@aist.go.jp; Tel.: +81-52-736-7551; Fax: +81-52-736-7405

Received: 19 July 2019; Accepted: 17 September 2019; Published: 19 September 2019

Abstract: Self-assembling peptides have been employed as biotemplates for biomineralization, as the morphologies and sizes of the inorganic materials can be easily controlled. We synthesized two types of highly ordered self-assembling peptides with different secondary structures and investigated the effects of secondary structures on hydroxyapatite (HAp) biomineralization of peptide templates. All as-synthesized HAp-peptides have a selective protein adsorption capacity for basic protein (e.g., cytochrome c and lysozyme). Moreover, the selectivity was improved as peptide amounts increased. In particular, peptide–HAp templated on β-sheet peptides adsorbed more cytochrome c than peptide–HAp with α-helix structures, due to the greater than 2-times carboxyl group density at their surfaces. It can be expected that self-assembled peptide-templated HAp may be used as carriers for protein immobilization in biosensing and bioseparation applications and as enzyme-stabilizing agents.

Keywords: solid-phase peptide synthesis; hydroxyapatite; peptide; secondary structure; selective protein adsorption; biotemplate

---

1. Introduction

As represented by bone, tooth, pearl, coral, shell, and crustacea, certain organisms have the ability to synthesize inorganic materials with refined structures and superior physical properties that are difficult to imitate. Such biomechanisms are designated as 'biomineralization', which is known as an environmentally-friendly synthesis process of inorganic materials under mild conditions. Synthetic methods of generating bioinspired materials and bioceramics have been reported by many researchers [1–3]. Liu et al. prepared calcium carbonate ($CaCO_3$)-regulated silk fibroin and estimated the drug release of doxorubicin using its vaterite microspheres [4]. DNA-$Cu_3(PO_4)_2$ hybrid nanoflowers were synthesized by Wu et al., and these materials are predicted to employ microRNA detection as captors [5]. He et al. described the synthesis of hematite mesocrystals with hierarchical structures via collagen templates [6]. Biominerals and bioinspired materials continue to be developed and are utilized within a variety of biosensing and biomedical applications.

Even in biomineralization, organic molecules play a crucial role in the formation of the crystalline nucleus and control of crystal polymorphism in vivo, as well as crystal growth and the shaping of the whole inorganic mineral. For example, peptides, which have unique well-ordered structures within their side chains, have been used as organic molecules to synthesize inorganic materials in a variety of methods [7–11]. Wada et al. controlled $CaCO_3$ crystallization within hydrogels by mixing polylysine and polyaspartic acid via double-diffusion methods; furthermore, the influence of peptide in the formation of the composites was elucidated [12]. "End-tethered poly(L-lysine)" monolayer brushes have been employed on silica film mineralization as reported by Wu and colleagues [13]. In summary, small amounts of peptides can affect materials morphologies and surface potentials [14].

Recently, self-assembling peptide-templated inorganic materials have been developed to readily manipulate the functional groups and secondary structures in peptides [15–24]. Lu et al. designed β-folded glutamic acid (leucine–glutamic acid)$_9$ (E(LE)$_9$) peptides, with its peptides working as a key player in the formation of calcium oxalate nanosheets [25]. Xu and colleagues reported the synthesis of inorganic materials based on short peptide self-assembly designed as $I_m K_n$ (e.g., isoleucine–isoleucine–isoleucine–lysine, I_3K), $I_m E_n$ (e.g., isoleucine–isoleucine–isoleucine–glutamic acid, I_3E), and $I_m R_n$ (e.g., isoleucine–isoleucine–isoleucine–isoleucine–arginine–arginine, I_4R_2) [26–29]. Xu et al. revealed that control of morphology and size of nanostructures were observed by using a self-assembling peptide template. In addition, our group previously showed that the morphologies of inorganic materials are controlled on a self-assembling peptide template [30–33] and elucidated that the secondary structures of peptides have a great impact on the resulting particles.

Hydroxyapatite ($Ca_{10}(PO_4)_6(OH)_2$, termed HAp) composes the primary inorganic contents of human tooth and bone and is a representative of biomineralization and bioinspired materials [34–40]. Hadagalli et al. established mineralization of porous HAp scaffolds, in which pores are obtained using organic pore formers, such as wax, wheat flour, or milk powder, and exhibit good cytocompatibility with osteoblasts in vitro [41]. Wei et al. synthesized biomineralized microspheres as follows: an amphipathic poly(L-lactide)-poly(ethylene glycol)-poly(L-lactide) triblock copolymer was coated with gelatin, then the microspheres were immersed in simulated body fluid containing dissolved alendronate. The resulting microspheres exhibited an increased effect on osteogenesis and bone regeneration compared with that of pristine microspheres lacking alendronate [42]. However, studies on HAp mineralization using self-assembled peptide templates could provide additional useful information, and applications based on peptide-template–HAp have rarely been reported.

Our main research is producing adsorbents for biosensing and bioseparation applications, that is, the materials need capable of adsorption selectively. Notably, we investigated not only the impact of calcium phosphate mineralization on peptide templates but also protein and enzyme adsorption performances by using as-synthesized materials [43–45]. It revealed that the morphology of the peptide–HAp hybrid materials included carboxyl groups was influenced from the secondary structures in peptides, and peptide–HAp composites with amino groups carried out for the application as glucose sensors because of its highly selective adsorption ability for proteins. In addition, we previously reported silica biomineralization on self-assembled peptide template using (leucine–lysine–leucine–leucine)$_5$-PEG$_{70}$ and (valine–lysine–valine–valine)$_5$-PEG$_{70}$ [46]. From these, we aimed at the preparation of selective protein adsorption agents using self-assembled peptide templates on HAp mineralization.

Hence, we prepared a peptide–poly(ethylene glycol) (peptide–PEG) block copolymer by solid-phase peptide synthesis using leucine (L), glutamic acid (E), and valine (V) as rich carboxyl groups within peptide side chains (Ac-(LELL)$_5$-PEG$_{70}$ and Ac-(VEVV)$_5$-PEG$_{70}$, Scheme 1). Subsequently, calcium phosphate mineralization using well-arranged peptide templates was attempted. The aim of this study is to provide insight into the influence of peptides on hybrid particle materials and the effect(s) of protein adsorption behavior on particles.

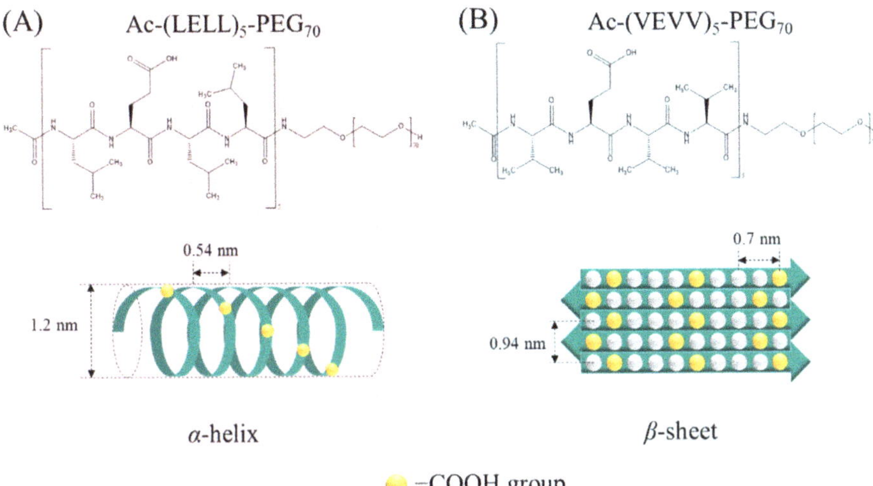

Scheme 1. Structural formula of the peptides. (**A**) Ac-(LELL)$_5$-PEG$_{70}$ and (**B**) Ac-(VEVV)$_5$-PEG$_{70}$ were self-assembled to α-helix and β-sheet conformations, respectively.

2. Results and Discussion

2.1. Peptide–HAp Characterization

The circular dichroism (CD) spectra of Ac-(LELL)$_5$-PEG$_{70}$ (abbreviated as LELL) and Ac-(VEVV)$_5$-PEG$_{70}$ (VEVV) are shown in Figure 1. Two negative peaks at 207 and 220 nm and a positive peak at 191 nm within LELL suggested α-helixes. Conversely, the CD spectra of VEVV showed a positive peak at 195 nm and a negative peak at 215 nm, indicating a β-sheet structure [47,48].

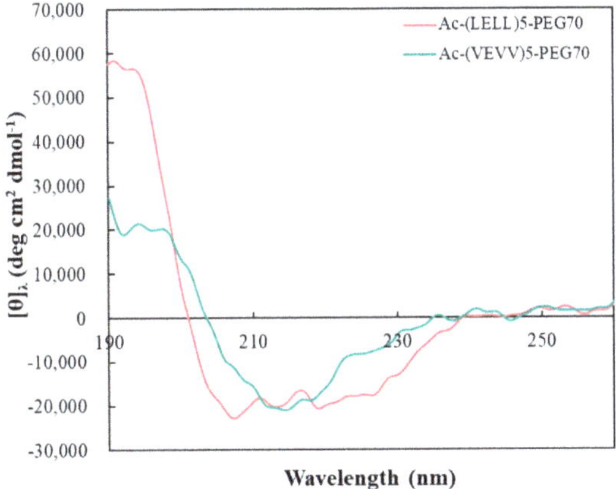

Figure 1. CD spectra of Ac-(LELL)$_5$-PEG$_{70}$ and Ac-(VEVV)$_5$-PEG$_{70}$. Each peptide was dissolved in 10 mM phosphate buffer (pH 7.0) and the concentration of peptide was 1.0×10^{-5} M.

Field-emission scanning electron microscopy (FE-SEM) and transmission electron microscopy (TEM) images of LELL–HAp (1 and 3 mg) and VEVV–HAp (1 and 3 mg) were observed as shown in Figure 2A,B. SEM images of peptide–HAp display nanorods with a length of approximately 60 nm, similar to pristine HAp. Compared with LELL–HAp particles, the morphology of VEVV–HAp exhibited slightly larger plate like particles (Figure 2B).

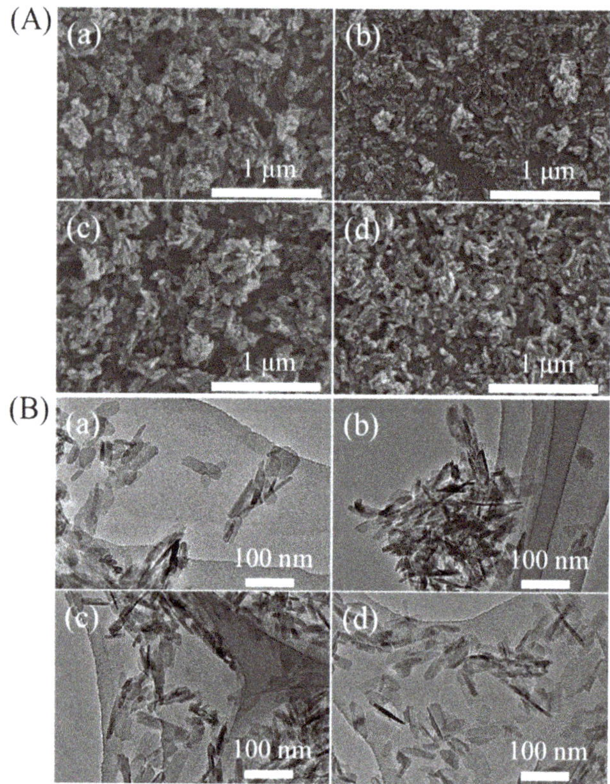

Figure 2. (**A**) FE-SEM and (**B**) TEM images of peptide–HAp composites: (**a**) LELL–HAp (1 mg), (**b**) LELL–HAp (3 mg), (**c**) VEVV–HAp (1 mg), and (**d**) VEVV–HAp (3 mg).

The Brunauer-Emmett-Teller (BET) surface area and pore volume and Barrett-Joyner-Halenda (BJH) pore size distribution of peptide–HAp are shown in Figure 3A,B and Table 1; nitrogen adsorption–desorption isotherms could be classified as a type IV. The specific surface areas of LELL–HAp (1 and 3 mg) and VEVV–HAp (1 and 3 mg) were found to be 106, 101, 101, and 92 m^2 g^{-1}, whereas the pore volumes were 0.81, 0.71, 0.64, and 0.62 cm^3 g^{-1}, respectively. In addition, pore sizes of 30 nm appeared in all samples. We previously observed that pore sizes of 70 nm were not present in any peptide–HAp besides bare HAp, and pore sizes of around 3 nm were confirmed in all samples. However, in the case of peptide–HAp, only α-pLys–HAp (30 and 40 mg) have pore sizes of 30 nm, of which pores may have impacted the enzyme stability of glucose oxidase immobilized on these materials [45]. As-prepared peptide–HAp is predicted to be usable for enzyme immobilization agents in biosensing and bioseparation.

The values of the Ca/P molar ratios are listed in Table 1. The Ca/P molar ratios of LELL–HAp (1 and 3 mg) and VEVV–HAp (1 and 3 mg) were 1.52, 1.50, 1.52, and 1.51, respectively. These ratios exhibited a relatively high degree of similarity to non-peptide–HAp, even though these values were lower than the stoichiometric ratio of HAp of 1.67. As a result, we observed that these calcium phosphates were low-crystallinity HAp or calcium-deficient HAp as composites of HAp and peptides.

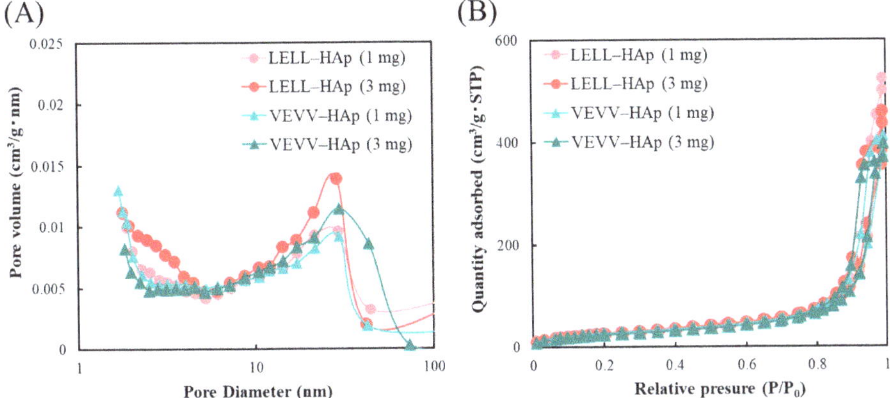

Figure 3. (**A**) Pore size distribution curves, and (**B**) nitrogen adsorption–desorption isotherms of LELL–HAp (1 and 3 mg) and VEVV–HAp (1 and 3 mg).

Table 1. Structural properties, Ca/P molar ratio, amount of peptide, and zeta potential of peptide–HAp composites.

Sample	Surface Area [a] (m^2 g^{-1})	Pore Volume [a] (cm^3 g^{-1})	Ca/P Molar Ratio [b]	Amount of Peptide [c] (mg)	Zeta Potential [d] (mV)
LELL–HAp (1 mg)	106	0.81	1.52	0.43	−19.8
LELL–HAp (3 mg)	101	0.71	1.50	1.4	−12.8
VEVV–HAp (1 mg)	101	0.64	1.52	0.51	−12.5
VEVV–HAp (3 mg)	92	0.62	1.51	1.5	−11.1

[a] The specific surface area, pore volume, and pore size distribution of peptide–HAp were calculated on the basis of nitrogen adsorption–desorption isotherms by the Brunauer-Emmett-Teller (BET) and Barrett-Joyner-Halenda (BJH) methods. [b] To measure the Ca/P molar ratio of peptide–HAp, inductively coupled plasma optical emission spectrometry (ICP-OES) was employed. [c] Peptide amounts within composites were determined by thermogravimetry and differential thermal analysis (TG-DTA). [d] The surface potential of peptide–HAp was measured via electrophoretic light scattering methods, whereas the particles were prepared by dispersion in 10 mM phosphate buffer with pH 7.0 and sonication for 3 min.

Figure 4A shows the powder X-ray diffraction (XRD) patterns of peptide–HAp. The diffraction peaks at $2\theta = 25.9°, 31.8°, 32.2°, 32.8°, 34.0°, 39.7°, 46.7°, 49.5°$, and $53.2°$ correspond to the (002), (211), (112), (300), (202), (310), (222), (213), and (004) planes of hydroxyapatite, respectively, of which broad peaks indicate that all samples synthesized in this study were low-crystallinity HAp (JCPDS card no. 09-0432) [49].

Thermogravimetry and differential thermal analysis (TG-DTA) analysis was performed in order to confirm the peptide content in peptide–HAp composites, and peptide amounts are summarized in Table 1, which were calculated by weight losses from 200 °C to 700 °C that were attributed to peptide loss. The relative peptide amounts of LELL–HAp (1 and 3 mg) and VEVV–HAp (1 and 3 mg) were 0.43, 1.4, 0.51, and 1.5 mg, respectively. In other words, this reveals that the peptide amounts within peptide–HAp were 2.4, 7.5, 2.3, and 6.5 wt % in the particles, respectively.

The zeta potential charge of peptide–HAp was also investigated, as shown in Table 1. The surface potentials were −19.8, −12.8, −12.5, and −11.1 mV for LELL–HAp (1 and 3 mg) and VEVV–HAp (1 and 3 mg), respectively, leading to the independence of the amount of peptides.

Fourier transform infrared (FTIR) data of native peptide and peptide–HAp is shown in Figure 4B. The presence of PO_4^{3-} functional groups in HAp can be observed by the bands at around 560, 600, 960, and 1020 cm^{-1} [49]. The −C=O stretching vibration at 1600–1700 cm^{-1} for amide I could be assessed as the peptide structure. Among these, we focused on two main peaks at around 1650 and 1630 cm^{-1}, attributed to α-helix and β-sheet structure [48,50]. The LELL–HAp (1 and 3 mg) spectra had peaks corresponding to HAp; the PO_4^{3-} bending vibration (O–P–O) at 560 and 600 cm^{-1} and the peaks at around 961 and 1024 cm^{-1} originated from bending modes of the P–O bond in PO_4^{3-}. Moreover, the characteristic of an α-helix in LELL from 1651 to 1653 cm^{-1} was observed. In the case of pure LELL, the band at 1652 cm^{-1} is attributed to an α-helix structure. The characteristic peaks at around 560, 600, 961, and 1022 cm^{-1} for VEVV–HAp (1 and 3 mg) could be designated as PO_4^{3-} groups in HAp. Additionally, the bands registered at 1632 and 1634 cm^{-1} for VEVV–HAp (1 and 3 mg, respectively) were ascribed to β-sheet peaks. For VEVV, the same band (β-sheet structure) was observed at 1626 cm^{-1}. According to these results, the presence of both HAp and peptide in peptide–HAp could be confirmed; furthermore, the peaks derived from each peptide secondary structure were also classified.

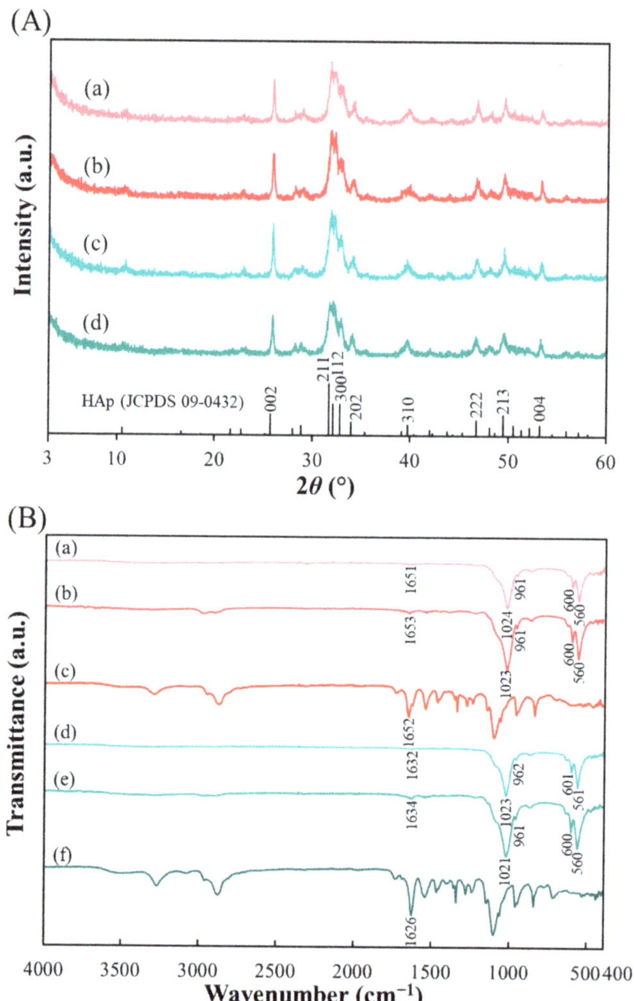

Figure 4. (**A**) X-ray diffraction patterns of (**a**) LELL–HAp (1 mg), (**b**) LELL–HAp (3 mg), (**c**) VEVV–HAp (1 mg), and (**d**) VEVV–HAp (3 mg) (JCPDS card no. 09-0432). (**B**) FTIR spectra of (**a**) LELL–HAp (1 mg), (**b**) LELL–HAp (3 mg), (**c**) pure Ac-(LELL)$_5$-PEG$_{70}$, (**d**) VEVV–HAp (1 mg), (**e**) VEVV–HAp (3 mg), and (**f**) pure Ac-(VEVV)$_5$-PEG$_{70}$.

Secondary structural contents of two peptides and that bound with Ca ions were clarified by FTIR analysis (Table 2). Firstly, native LELL and VEVV had higher contents of α-helixes and β-sheets, respectively, as attributed to the CD data. Moreover, to investigate the secondary structure of peptide in peptide–HAp, we prepared each peptide bond with Ca ions as follows: 3 mg LELL or VEVV was added to 20 mL of prepared $(CH_3COO)_2Ca$ solution (15 mM). After stirring for 2 h at 20 °C, the solid materials were obtained via the freeze-drying process. Most secondary structures were of an α-helix content for LELL bound with Ca ions; meanwhile, VEVV bound with Ca ions not only contained primarily β-sheet structures but also various secondary structural contents (α-helixes, β-turns, and others).

Table 2. Secondary structures (%) of Ac-(LELL)$_5$-PEG$_{70}$ and Ac-(VEVV)$_5$-PEG$_{70}$.

Sample	α-Helix	β-Sheet	β-Turn	Other
LELL	94%	— *	3%	3%
LELL + Ca [a]	>99%	— *	— *	— *
VEVV	— *	96%	1%	3%
VEVV + Ca [a]	2%	47%	24%	27%

* Trace percent. [a] Each peptide (Ac-(LELL)$_5$-PEG$_{70}$ and Ac-(VEVV)$_5$-PEG$_{70}$) (3 mg) was mixed with 20 mL $(CH_3COO)_2Ca$ solution (15 mM) and stirred for 2 h at 20 °C. The resulting product was then freeze-dried.

To investigate the elemental distribution in peptide–HAp particles, especially the peptide, scanning transmission electron microscopy (STEM) images, and energy-dispersive X-ray spectroscopy (EDX) maps were utilized (Figure 5A,B). Nitrogen is attributed to the peptides and calcium, and phosphorous corresponds to HAp. Nitrogen (yellow) is homogeneously distributed throughout nanoparticles; thus, it could be assumed that the peptides are distributed in the particles.

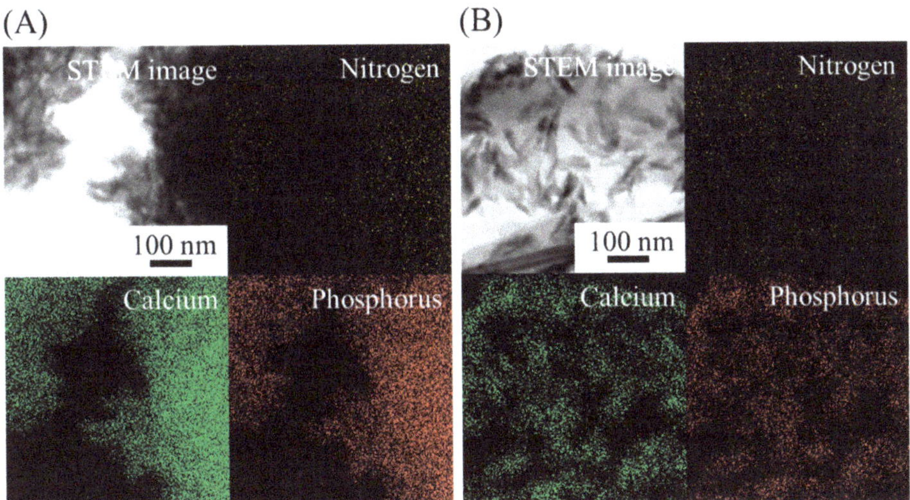

Figure 5. STEM images and EDX maps of elements of (**A**) LELL–HAp (3 mg) and (**B**) VEVV–HAp (3 mg). Nitrogen (yellow), calcium (green), and phosphorous (red) are displayed.

2.2. Protein Adsorption on Peptide–HAp

Three typical proteins with different isoelectric points (i.e., cytochrome c; Cyt c, myoglobin; MGB, and bovine serum albumin; BSA) were adsorbed not only on peptide–HAp but also non-peptide–HAp (Figure 6A). The capacity and tendency for protein adsorption on HAp were similar to the results obtained in a previous report [43]. The amounts of adsorbed Cyt c on LELL–HAp (1 mg) and VEVV–HAp (1 mg) were 94.8 and 78.3 μg mg^{-1}, respectively. In addition, the adsorbed MGB amounts were 41.5 and 0.571 μg mg^{-1} for LELL–HAp (1 mg) and VEVV–HAp (1 mg), and the BSA capacities were 61.8 and 37.9 µg mg^{-1}, respectively. In terms of the amounts of adsorbed Cyt c for LELL–HAp (3 mg) and VEVV–HAp (3 mg), the capacities were 30.4 and 64.8 µg mg^{-1}, respectively, whereas two samples had either no or extremely low adsorption amounts for MGB and BSA. Furthermore, lysozyme (LSZ), conalbumin (ovotransferrin; OVT), and transferrin (TF) was adsorbed on peptide–HAp to confirm its selectivity only for basic protein (Figure 6B). The amounts of adsorbed LSZ on LELL–HAp (1 mg), VEVV–HAp (1 mg), LELL–HAp (3 mg), and VEVV–HAp (3 mg) were 41.9, 69.6, 96.4, and 115 µg mg^{-1}, respectively. For all peptide–HAp, adsorption amounts of OVT and TF were either no or extremely low. From these results, it could be hypothesized that selectivity for protein adsorption on peptide–HAp is due to the presence of glutamic acid (E) within the peptides. Moreover, VEVV–HAp exhibited a high blocking effect itself for the other proteins during the maintenance of Cyt c and LSZ adsorption capacities with increasing peptide amounts.

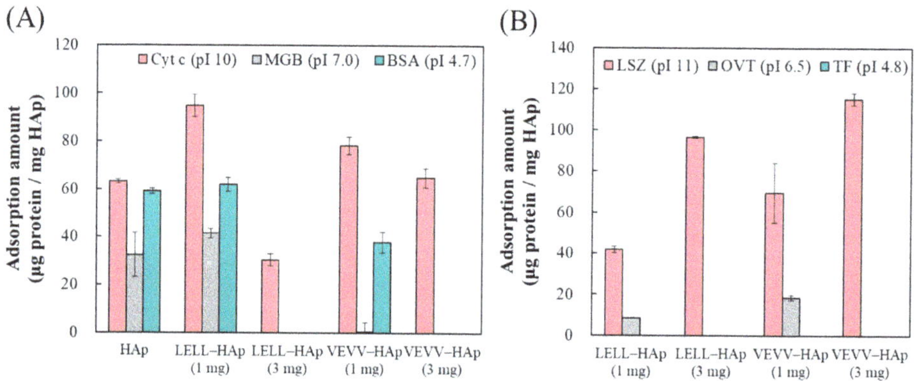

Figure 6. (**A**) Protein (cytochrome c, myoglobin, and bovine serum albumin) adsorption on HAp, LELL–HAp (1 and 3 mg), and VEVV–HAp (1 and 3 mg). (**B**) Protein (lysozyme, conalbumin, and transferrin) adsorption on LELL–HAp (1 and 3 mg) and VEVV–HAp (1 and 3 mg).

2.3. Carboxyl Group Density in Peptide–HAp

The determination of the carboxyl group density in peptide–HAp indicated the cause for selectivity of protein adsorption on peptide–HAp. The densities of carboxyl groups were 2.33, 3.38, 5.97, and 13.0 nmol m^{-2} for LELL–HAp (1 mg), VEVV–HAp (1 mg), LELL–HAp (3 mg), and VEVV–HAp (3 mg), respectively, in which the density of peptide–HAp was improved with increasing amounts of peptide. Moreover, the density of VEVV–HAp (3 mg) was more than twofold higher compared with LELL–HAp (3 mg). From the results, the difference in Cyt c adsorption amounts between VEVV–HAp (3 mg) and LELL–HAp (3 mg) could be explained by the high carboxyl group density in VEVV–HAp (3 mg).

3. Materials and Methods

3.1. Materials

All chemicals were of analytical grade and were used as received without further purification. Calcium acetate monohydrate [(CH$_3$COO)$_2$Ca·H$_2$O] and diammonium hydrogen phosphate [(NH$_4$)$_2$HPO$_4$] were obtained from FUJIFILM Wako Pure Chemical Co. (Osaka, Japan). Cytochrome c from equine heart [Cyt c; isoelectric point (pI) = 10, molecular weight (Mw) = 12,300 Da], myoglobin from equine skeletal muscle (MGB; pI = 7.0, Mw = 17,800 Da), bovine serum albumin (BSA; pI = 4.7, Mw = 67,000 Da), lysozyme from chicken egg white (LSZ; pI = 11, Mw = 14,300 Da), conalbumin from chicken egg white (OVT; pI = 6.5, Mw = 76,000 Da), and transferrin human (TF; pI = 4.8, Mw = 80,000 Da) were purchased from Merck KGaA (Darmstadt, Germany). 1-Ethyl-3-(3-dimethylaminopropyl)carbodiimide hydrochloride (EDC; Mw = 191.7),

N-hydroxysuccinimide (NHS; Mw = 115.1), and 5-aminofluorescein (Mw = 347.3) were obtained from Tokyo Kasei Kogyo Co. (Tokyo, Japan). The Bio-Rad protein assay dye reagent concentrate was purchased from Bio-Rad Laboratories (Hercules, CA, USA).

3.2. Preparation of Peptide–HAp Particles

Two peptides (Ac-(LELL)$_5$-PEG$_{70}$ and Ac-(VEVV)$_5$-PEG$_{70}$) were prepared via solid-phase peptide synthesis according to our previous reports [30–33]. One or 3 mg of Ac-(LELL)$_5$-PEG$_{70}$ or Ac-(VEVV)$_5$-PEG$_{70}$ was added to a 20 mL solution of dissolved (CH$_3$COO)$_2$Ca (52.8 mg), followed by stirring for 30 min at 20 °C. After addition of 20 mL (NH$_4$)$_2$HPO$_4$ solution (23.8 mg), the mixture was heated to 60 °C at a heating rate of 1 °C min^{-1}. The temperature was maintained for 3 h with stirring, and then the precipitant was collected by centrifugation at 6000 rpm for 10 min. The final products were washed twice with deionized water (Milli-Q, Merck KGaA, Darmstadt, Germany) and freeze-dried. To compare protein adsorption behavior, non-peptide–HAp was also synthesized using the same process as peptide–HAp.

3.3. Characterization of Synthesized Peptide–HAp

To analyze the secondary structure of Ac-(LELL)$_5$-PEG$_{70}$ and Ac-(VEVV)$_5$-PEG$_{70}$, the CD spectrum was measured using J-820 (JASCO Co., Tokyo, Japan) in a range scanning of 190–260 nm. The morphologies of all peptide–HAp samples were visualized with FE-SEM (S-4300, Hitachi Ltd., Tokyo, Japan) under 10.0 kV accelerating voltage and TEM (JEM-2010, JEOL Ltd., Tokyo, Japan) at 200 kV accelerator voltage. The specific surface area, pore volume, and pore size distribution were calculated on the basis of nitrogen adsorption–desorption isotherms using a TriStar 3000 (Shimadzu Co., Kyoto, Japan) via the BET and BJH models. For the measurement of the Ca/P molar ratio of synthesized peptide–HAp, inductively coupled plasma optical emission spectrometry (ICP-OES; IRIS Advantage, Thermo Fisher Scientific Inc., Waltham, MA, USA) was employed. The calculated peptide amounts of peptide–HAp were analyzed by TG-DTA (Thermo Plus TG 8120, Rigaku Co., Tokyo, Japan) in the operation range of room temperature to 1000 °C (heating rate of 10 °C min^{-1}). ELSZ-1000 (Otsuka Electronics Co., Tokyo, Japan) was employed to measure the zeta potential of peptide–HAp, with the samples prepared via dispersion in 10 mM phosphate buffer of pH 7.0 with sonication for 3 min. XRD (SmartLab SE/B1, Rigaku Co., Tokyo, Japan) analysis was carried out using CuKα radiation operated at an accelerator voltage of 40 kV and a beam intensity of 30 mA. The XRD patterns were collected at a step size of 2.0° min^{-1} and a 2θ range between 3° and 60°. FTIR spectra in the 400–4000 cm^{-1} range were recorded by FT/IR-4700 (JASCO Co., Tokyo, Japan) with attenuated total reflection. STEM (JEM-2100 Plus, JEOL Ltd., Tokyo, Japan) operated at 200 kV with EDX (Noran System 7, Thermo Fisher Scientific Inc., Waltham, MA, USA) was used to analyze the element distribution of peptide over peptide–HAp.

3.4. Protein Adsorption on Peptide–HAp

Each protein (i.e., Cyt c, MGB, BSA, LSZ, OVT, and TF) was dissolved in 10 mM phosphate buffer (pH 7.0), and the protein solution was prepared at a concentration of 250 μg mL^{-1}. Peptide–HAp (1.5 mg) was mixed with a 1 mL protein solution, followed by stirring overnight at 20 °C. The supernatant was separated from the mixture by centrifugation at 14,000 rpm for 5 min, and excess protein in the supernatant was estimated using the Bradford method by UV–Vis spectroscopy (Infinite F200 PRO, Tecan Group Ltd., Männedorf, Switzerland) at λ = 595 nm. Bio-Rad protein assay dye was employed for the evaluation of protein adsorption performance with the equation

$$Q = Q_0\left(\frac{I_0 - I}{I_0}\right) \tag{1}$$

where Q is the adsorption capacity of protein on peptide–HAp, Q_0 is the initial amount of protein, I_0 is the initial absorbance intensity in the supernatant, and I is the absorbance intensity of the supernatant following adsorption.

3.5. Calculation of Carboxyl Group Density in Peptide–HAp

First, EDC (42.9 mg) and NHS (5.1 mg) were each dissolved in 3 mL of 10 mM phosphate buffer (pH 7.0), and the solutions (500 µL) were mixed together. One mg of peptide–HAp was added to the mixture and stirred for 3 h at 20 °C. The solid materials were separated by centrifugation at 14,000 rpm for 5 min and then washed three times with 10 mM phosphate buffer (pH 7.0). The precipitant was resuspended in the same phosphate buffer (500 µL), and then 500 µL of 5-aminofluorescein solution (8 µg mL^{-1}) was added to the suspension. After stirring in the dark overnight at 20 °C, the solid materials were collected by centrifugation at 14,000 rpm for 5 min and washed with the aforementioned buffer. Finally, the carboxyl group density in the precipitant was redispersed in a 1 mL buffer and measured using a spectrofluorophotometer (RF-5300PC, Shimadzu Co., Kyoto, Japan), of which the excitation and emission wavelengths were 494 and 521 nm, respectively.

4. Conclusions

In summary, we designed two types of self-assembling peptides with different secondary structures: (leucine–glutamic acid–leucine–leucine)$_5$-PEG$_{70}$ (LELL) and (valine–glutamic acid–valine–valine)$_5$-PEG$_{70}$ (VEVV), and these peptides were used as templates for HAp biomineralization. Moreover, we also investigated the effect of secondary structures within peptide-template–HAp on the particles and protein adsorption behavior. It could be shown that as-synthesized peptide LELL or VEVV showed almost entirely α-helix or β-sheet contents within secondary structures, respectively. The morphologies of all peptide–HAp were similar to bare HAp, whereas VEVV–HAp displayed a slightly plate-like structure. Additionally, all peptide–HAp have pore sizes of 30 nm, which may be expected for enzyme stability on enzyme immobilization, as indicated in our previous study. Furthermore, for the adsorption properties of proteins, it was revealed that each peptide–HAp specifically adsorbed basic protein (i.e., Cyt c and LSZ). With increasing amounts of peptide, the blocking effects for proteins, except for basic protein, were also increased. Overall, the reason that VEVV–HAp (3 mg) with β-sheet structures exhibited increased Cyt c adsorption amounts compared with LELL–HAp (3 mg) containing α-helix structures is explained as follows: the carboxyl group density at the surfaces of VEVV–HAp (3 mg) was more than 2-times higher compared with LELL–HAp (3 mg) while the carboxyl group density of peptide–HAp incorporated 1 mg of peptide amount was lower than that of peptide–HAp (3 mg). From these results, it can be stated that synthesized HAp on a self-assembling peptide template could be useful as a carrier for protein immobilization in biosensing and bioseparation applications and as enzyme-stabilizing agents.

Author Contributions: Conceptualization, K.K.; Data curation, S.K. and H.N.; Formal analysis, S.K. and H.N.; Writing—original draft preparation, S.K.; Writing—review and editing, S.L., F.N., and K.K.; Funding acquisition, S.L., F.N., and K.K.

Funding: The financial support of Grants-in-Aid for Scientific Research (C) no.15K06474 from the Japan Society for the Promotion of Science and A-STEP (JPMJTS1624) from the Japan Science and Technology Agency is gratefully acknowledged.

Acknowledgments: The author thanks Hiroyuki Iwata (Aichi Institute of Technology) for his help in the STEM-EDX observation. The authors would like to thank MARUZEN-YUSHODO Co., Ltd. (http://kw.maruzen.co.jp/kousei-honyaku/) for the English language editing.

Conflicts of Interest: The authors declare no conflict of interest.

Abbreviations

HAp	Hydroxyapatite
PEG	Poly(ethylene glycol)
CD	Circular dichroism
FE-SEM	Field-emission scanning electron microscopy
TEM	Transmission electron microscopy
BET	Brunauer-Emmett-Teller
BJH	Barrett-Joyner-Halenda
XRD	Powder X-ray diffraction
TG-DTA	Thermogravimetry and differential thermal analysis
FTIR	Fourier transform infrared
STEM	Scanning transmission electron microscopy
EDX	Energy-dispersive X-ray spectroscopy
Cyt c	Cytochrome c
MGB	Myoglobin
BSA	Bovine serum albumin
LSZ	Lysozyme
OVT	Conalbumin
TF	Transferrin
EDC	1-Ethyl-3-(3-dimethylaminopropyl)carbodiimide hydrochloride
NHS	*N*-Hydroxysuccinimide

References

1. Jiang, W.; Yi, X.; McKee, M.D. Chiral biomineralized structures and their biomimetic synthesis. *Mater. Horiz.* **2019**. [CrossRef]
2. Eichler-Volf, A.; Xue, L.; Dornberg, G.; Chen, H.; Kovalev, A.; Enke, D.; Wang, Y.; Gorb, E.V.; Gorb, S.N.; Steinhart, M. The influence of surface topography and surface chemistry on the anti-adhesive performance of nanoporous monoliths. *ACS. Appl. Mater. Interfaces* **2016**, *8*, 22593–22604. [CrossRef]
3. Cui, Y.; Li, D.; Bai, H. Bioinspired smart materials for directional liquid transport. *Ind. Eng. Chem. Res.* **2017**, *56*, 4887–4897. [CrossRef]
4. Liu, L.; Zhang, X.; Liu, X.; Liu, J.; Lu, G.; Kaplan, D.L.; Zhu, H.; Lu, Q. Biomineralization of stable and monodisperse vaterite microspheres using silk nanoparticles. *ACS. Appl. Mater. Interfaces* **2015**, *7*, 1735–1745. [CrossRef] [PubMed]
5. Wu, T.; Yang, Y.; Cao, Y.; Song, Y.; Xu, L.; Zhang, X.; Wang, S. Bioinspired DNA-inorganic hybrid nanoflowers combined with a personal glucose meter for onsite detection of miRNA. *ACS. Appl. Mater. Interfaces* **2018**, *10*, 42050–42057. [CrossRef] [PubMed]
6. He, M.; Zhang, Y.; Munyemana, J.C.; Wu, T.; Yang, Z.; Chen, H.; Qu, W.; Xiao, J. Tuning the hierarchical nanostructure of hematite mesocrystals via collagen-templated biomineralization. *J. Mater. Chem. B* **2017**, *5*, 1423–1429. [CrossRef]
7. Galloway, J.M.; Staniland, S.S. Protein and peptide biotemplated metal and metal oxide nanoparticles and their pattering onto surfaces. *J. Mater. Chem.* **2012**, *22*, 12423–12434. [CrossRef]
8. Chen, C.; Rosi, N.L. Peptide-based methods for the preparation of nanostructured inorganic materials. *Angew. Chem. Int. Ed.* **2010**, *49*, 1924–1942. [CrossRef]
9. Dickerson, M.B.; Sandhage, K.H.; Naik, R.R. Protein- and peptide-directed synthesis of inorganic materials. *Chem. Rev.* **2008**, *108*, 4935–4978. [CrossRef]
10. Sethi, M.; Pacardo, D.B.; Knecht, M.R. Biological surface effects of metallic nanomaterials for applications in assembly and catalysis. *Langmuir* **2010**, *26*, 15121–15134. [CrossRef]
11. Pan, L.; He, Q.; Liu, J.; Chen, Y.; Ma, M.; Zhang, L.; Shi, J. Nuclear-targeted drug delivery of TAT peptide-conjugated monodisperse mesoporous silica nanoparticles. *J. Am. Chem. Soc.* **2012**, *134*, 5722–5725. [CrossRef]

12. Wada, N.; Horiuchi, N.; Nakamura, M.; Nozaki, K.; Nagai, A.; Yamashita, K. Controlled crystallization of calcium carbonate via cooperation of polyaspartic acid and polylysine under double-diffusion conditions in agar hydrogels. *ACS Omega* **2018**, *3*, 16681–16692. [CrossRef]

13. Wu, J.; Wang, Y.; Chen, C.; Chang, Y. Biomimetic synthesis of silica films directed by polypeptide brushes. *Chem. Mater.* **2008**, *20*, 6148–6156. [CrossRef]

14. Yao, Y.; Dong, W.; Zhu, S.; Yu, X.; Yan, D. Novel morphology of calcium carbonate controlled by poly(L-lysine). *Langmuir* **2009**, *25*, 13238–13243. [CrossRef] [PubMed]

15. Cai, C.; Lin, J.; Lu, Y.; Zhang, Q.; Wang, L. Polypeptide self-assemblies: Nanostructures and bioapplications. *Chem. Soc. Rev.* **2016**, *45*, 5985–6012. [CrossRef] [PubMed]

16. Ghatak, A.S.; Koch, M.; Guth, C.; Weiss, I.M. Peptide induced crystallization of calcium carbonate on wrinkle patterned substrate: Implications for chitin formation in *Molluscs*. *Int. J. Mol. Sci.* **2013**, *14*, 11842–11860. [CrossRef]

17. Bhandari, R.; Pacardo, D.B.; Bedford, N.M.; Naik, R.R.; Knecht, M.R. Structural control and catalytic reactivity of peptide-templated Pd and Pt nanomaterials for olefin hydrogenation. *J. Phys. Chem. C* **2013**, *117*, 18053–18062. [CrossRef]

18. Yuwono, V.M.; Hartgerink, J.D. Peptide amphiphile nanofibers template and catalyze silica nanotube formation. *Langmuir* **2007**, *23*, 5033–5038. [CrossRef]

19. Cui, Y.; Wang, Y.; Liu, R.; Sun, Z.; Wei, Y.; Zhao, Y.; Gao, X. Serial silver clusters biomineralized by one peptide. *ACS Nano* **2011**, *5*, 8684–8689. [CrossRef]

20. Yildirim, A.; Acar, H.; Erkal, T.S.; Bayindir, M.; Guler, M.O. Template-directed synthesis of silica nanotubes for explosive detection. *ACS. Appl. Mater. Interfaces* **2011**, *3*, 4159–4164. [CrossRef]

21. Acar, H.; Garifullin, R.; Guler, M.O. Self-assembled template-directed synthesis of one-dimensional silica and titania nanostructures. *Langmuir* **2011**, *27*, 1079–1084. [CrossRef] [PubMed]

22. Susapto, H.H.; Kudu, O.U.; Garifullin, R.; Yilmaz, E.; Guler, M.O. One-dimensional peptide nanostructure templated growth of iron phosphate nanostructures for lithium-ion battery cathodes. *ACS. Appl. Mater. Interfaces* **2016**, *8*, 17421–17427. [CrossRef] [PubMed]

23. Tomizaki, K.; Kubo, S.; Ahn, S.; Satake, M.; Imai, T. Biomimetic alignment of zinc oxide nanoparticles along a peptide nanofiber. *Langmuir* **2012**, *28*, 13459–13466. [CrossRef] [PubMed]

24. George, J.; Thomas, K.G. Surface plasmon coupled circular dichroism of Au nanoparticles on peptide nanotubes. *J. Am. Chem. Soc.* **2010**, *132*, 2502–2503. [CrossRef] [PubMed]

25. Lu, H.; Schäfer, A.; Lutz, H.; Roeters, S.J.; Lieberwirth, I.; Muñoz-Espí, R.; Hood, M.A.; Bonn, M.; Weidner, T. Peptide-controlled assembly of macroscopic calcium oxalate nanosheets. *J. Phys. Chem. Lett.* **2019**, *10*, 2170–2174. [CrossRef] [PubMed]

26. Tao, K.; Wang, J.; Li, Y.; Xia, D.; Shan, H.; Xu, H.; Lu, J.R. Short peptide-directed synthesis of one-dimensional platinum nanostructures with controllable morphologies. *Sci. Rep.* **2013**, *3*, 2565–2570. [CrossRef] [PubMed]

27. Wang, S.; Xue, J.; Zhao, Y.; Du, M.; Deng, L.; Xu, H.; Lu, J.R. Controlled silica deposition on self-assembled peptide nanostructures via varying molecular structures of short amphiphilic peptides. *Soft Matter* **2014**, *10*, 7623–7629. [CrossRef] [PubMed]

28. Wang, S.; Cai, Q.; Du, M.; Xue, J.; Xu, H. Synthesis of 1D silica nanostructures with controllable sizes based on short anionic peptide self-assembly. *J. Phys. Chem. B* **2015**, *119*, 12059–12065. [CrossRef]

29. Du, M.; Bu, Y.; Zhou, Y.; Zhao, Y.; Wang, S.; Xu, H. Peptide-templated synthesis of branched MnO_2 nanowires with improved electrochemical performances. *RSC Adv.* **2017**, *7*, 12711–12718. [CrossRef]

30. Nonoyama, T.; Kinoshita, T.; Higuchi, M.; Nagata, K.; Tanaka, M.; Sato, K.; Kato, K. Multistep growth mechanism of calcium phosphate in the earliest stage of morphology-controlled biomineralization. *Langmuir* **2011**, *27*, 7077–7083. [CrossRef]

31. Kuno, T.; Nonoyama, T.; Hirao, K.; Kato, K. Influence of the charge relay effect on the silanol condensation reaction as a model for silica biomineralization. *Langmuir* **2011**, *27*, 13154–13158. [CrossRef] [PubMed]

32. Nonoyama, T.; Kinoshita, T.; Higuchi, M.; Nagata, K.; Tanaka, M.; Sato, K.; Kato, K. TiO_2 synthesis inspired by biomineralization: Control of morphology, crystal phase, and light-use efficiency in a single process. *J. Am. Chem. Soc.* **2012**, *134*, 8841–8847. [CrossRef] [PubMed]

33. Murai, K.; Higuchi, M.; Kinoshita, T.; Nagata, K.; Kato, K. Calcium carbonate biomineralization utilizing a multifunctional β-sheet peptide template. *Chem. Commun.* **2013**, *49*, 9947–9949. [CrossRef] [PubMed]

34. Morejón, L.; Delgado, J.A.; Ribeiro, A.A.; Oliveira, M.V.; Mendizábal, E.; García, I.; Alfonso, A.; Poh, P.; Griensven, M.; Balmayor, E.R. Development, characterization and in vitro biological properties of scaffolds fabricated from calcium phosphate nanoparticles. *Int. J. Mol. Sci.* **2019**, *20*, 1790. [CrossRef] [PubMed]

35. Guan, J.; Yang, J.; Dai, J.; Qin, Y.; Wang, Y.; Guo, Y.; Ke, Q.; Zhang, C. Bioinspired nanostructured hydroxyapatite/collagen three-dimensional porous scaffolds for bone tissue engineering. *RSC Adv.* **2015**, *5*, 36175–36184. [CrossRef]

36. Cai, Y.; Li, H.; Karlsson, M.; Leifer, K.; Engqvist, H.; Xia, W. Biomineralization on single crystalline rutile: The modulated growth of hydroxyapatite by fibronectin in a simulated body fluid. *RSC Adv.* **2016**, *6*, 35507–35516. [CrossRef]

37. Zhang, Y.; Li, K.; Zhang, Q.; Liu, W.; Liu, Y.; Banks, C.E. Multi-dimensional hydroxyapatite (HAp) nanocluster architectures fabricated via nafion-assisted biomineralization. *New J. Chem.* **2015**, *39*, 750–754. [CrossRef]

38. Yang, Y.; Cui, Q.; Sahai, N. How does bone sialoprotein promote the nucleation of hydroxyapatite? A molecular dynamics study using model peptides of different conformations. *Langmuir* **2010**, *26*, 9848–9859. [CrossRef] [PubMed]

39. Weiger, M.C.; Park, J.J.; Roy, M.D.; Stafford, C.M.; Karim, A.; Becker, M.L. Quantification of the binding affinity of a specific hydroxyapatite binding peptide. *Biomaterials* **2010**, *31*, 2955–2963. [CrossRef] [PubMed]

40. Mukherjee, K.; Ruan, Q.; Nutt, S.; Tao, J.; Yoreo, J.J.D.; Moradian-Oldak, J. Peptide-based bioinspired approach to regrowing multilayered aprismatic enamel. *ACS Omega* **2018**, *3*, 2546–2557. [CrossRef]

41. Hadagalli, K.; Panda, A.K.; Mandal, S.; Basu, B. Faster biomineralization and tailored mechanical properties of marine-resource-derived hydroxyapatite scaffolds with tunable interconnected porous architecture. *ACS Appl. Bio Mater.* **2019**, *2*, 2171–2184. [CrossRef]

42. Wei, P.; Yuan, Z.; Jing, W.; Huang, Y.; Cai, Q.; Guan, B.; Liu, Z.; Zhang, X.; Mao, J.; Chen, D.; et al. Strengthening the potential of biomineralized microspheres in enhancing osteogenesis via incorporating alendronate. *Chem. Eng. J.* **2019**, *368*, 577–588. [CrossRef]

43. Kojima, S.; Nagata, F.; Kugimiya, S.; Kato, K. Synthesis of peptide-containing calcium phosphate nanoparticles exhibiting highly selective adsorption of various proteins. *Appl. Surf. Sci.* **2018**, *458*, 438–445. [CrossRef]

44. Kojima, S.; Nagata, F.; Inagaki, M.; Kugimiya, S.; Kato, K. Avidin-adsorbed peptide–calcium phosphate composites exhibiting high biotin-binding activity. *New J. Chem.* **2019**, *43*, 427–435. [CrossRef]

45. Kojima, S.; Nagata, F.; Inagaki, M.; Kugimiya, S.; Kato, K. Enzyme immobilisation on poly-L-lysine-containing calcium phosphate particles for highly sensitive glucose detection. *RSC Adv.* **2019**, *9*, 10832–10841. [CrossRef]

46. Kuno, T.; Nonoyama, T.; Hirao, K.; Kato, K. Structural formation ability of peptide secondary structure on silica biomineralization. *Chem. Lett.* **2012**, *41*, 1547–1549. [CrossRef]

47. Qin, S.; Jiang, H.; Peng, M.; Lei, Q.; Zhuo, R.; Zhang, X. Adjustable nanofibers self-assembled from an irregular conformational peptide amphiphile. *Polym. Chem.* **2015**, *6*, 519–524. [CrossRef]

48. Fry, H.C.; Silveira, G.Q.; Cohn, H.M.; Lee, B. Diverse bilayer morphologies achieved via α-helix-to-β-sheet transitions in a short amphiphilic peptide. *Langmuir* **2019**, *35*, 8961–8967. [CrossRef]

49. Stevanović, M.; Đošić, M.; Janković, A.; Kojić, V.; Vukašinović-Sekulić, M.; Stojanović, J.; Odović, J.; Sakač, M.C.; Rhee, K.Y.; Mišković-Stanković, V. Gentamicin-loaded bioactive hydroxyapatite/chitosan composite coating electrodeposited on titanium. *ACS Biomater. Sci. Eng.* **2018**, *4*, 3994–4007. [CrossRef]

50. Suades, A.; Alcaraz, A.; Cruz, E.; Álvarez-Marimon, E.; Whitelegge, J.P.; Manyosa, J.; Cladera, J.; Perálvarez-Marín, A. Structural biology workflow for the expression and characterization of functional human sodium glucose transporter type 1 in Pichia pastoris. *Sci. Rep.* **2019**, *9*, 1203–1213. [CrossRef]

International Journal of
Molecular Sciences

Article

Effect of the Hydrophilic-Hydrophobic Balance of Antigen-Loaded Peptide Nanofibers on Their Cellular Uptake, Cellular Toxicity, and Immune Stimulatory Properties

Tomonori Waku *, Saki Nishigaki, Yuichi Kitagawa, Sayaka Koeda, Kazufumi Kawabata, Shigeru Kunugi, Akio Kobori and Naoki Tanaka

Faculty of Molecular Chemistry and Engineering, Kyoto Institute of Technology, Gosyokaido-cho, Matsugasaki, Sakyo-ku, Kyoto 606-8585, Japan
* Correspondence: waku1214@kit.ac.jp; Tel.: +81-75-724-7811

Received: 22 May 2019; Accepted: 25 July 2019; Published: 2 August 2019

Abstract: Recently, nanofibers (NFs) formed from antigenic peptides conjugated to β-sheet-forming peptides have attracted much attention as a new generation of vaccines. However, studies describing how the hydrophilic-hydrophobic balance of NF components affects cellular interactions of NFs are limited. In this report, three different NFs were prepared by self-assembly of β-sheet-forming peptides conjugated with model antigenic peptides (SIINFEKL) from ovalbumin and hydrophilic oligo-ethylene glycol (EG) of differing chain lengths (6-, 12- and 24-mer) to investigate the effect of EG length of antigen-loaded NFs on their cellular uptake, cytotoxicity, and dendritic cell (DC)-stimulation ability. We used an immortal DC line, termed JAWS II, derived from bone marrow-derived DCs of a C57BL/6 p53-knockout mouse. The uptake of NFs, consisting of the EG 12-mer by DCs, was the most effective and activated DC without exhibiting significant cytotoxicity. Increasing the EG chain length significantly reduced cellular entry and DC activation by NFs. Conversely, shortening the EG chain enhanced DC activation but increased toxicity and impaired water-dispersibility, resulting in low cellular uptake. These results show that the interaction of antigen-loaded NFs with cells can be tuned by the EG length, which provides useful design guidelines for the development of effective NF-based vaccines.

Keywords: peptide; nanofibers; poly(ethylene glycol); antigen delivery; immune stimulation

1. Introduction

Peptide-based synthetic vaccines have attracted a significant amount of attention as a new generation of vaccines, because of their safety benefits and ease of production when compared with that of conventional whole pathogen-based vaccines [1,2]. However, poor immune responses are induced when only minimal antigenic epitopes are used without combining suitable adjuvants (immune stimulants), which are sometimes toxic. Nanocarrier-based delivery systems are a promising approach to overcome those drawbacks of peptide vaccines. In designing the nanocarrier, it is important to consider the interaction between the nanocarrier surface and cells, such as antigen presenting cells (APCs).

Over the past few decades, various nanocarriers have been developed, including liposomes [3–5], polymeric nanoparticles [6–8], and polymeric micelles [9]. In many of these systems, building block molecules for the construction of nanocarriers are first synthesized and then combined with antigenic peptides via several procedures, including nanomaterial formation and loading of antigenic peptides (encapsulation, chemical immobilization or physical adsorption), to give a nano-formulation. Recently, the use of antigenic peptides that are pre-conjugated to self-assembly motifs has attracted attention as an easier and simpler procedure to produce nano-formulations [10,11]. This self-assembly approach ensures highly efficient drug loading without laborious procedures or the use of synthetic components, which sometimes exhibit toxicity. In addition, because the resulting nanostructures consist of a single component, the physicochemical and structural features of these nanostructures can be simply tuned by the design of the building block peptide, and variation in drug loading efficiency among different nanostructures can be eliminated.

Among the various molecular blocks (e.g., lipids [12–14] and hydrophobic polymers [15,16]) used to assemble antigenic peptides into nanostructures, β-sheet-forming-peptides are extremely attractive because: (i) They can assemble in aqueous solution to give nanofibers (NFs) with highly regulated structures, even when functional molecules with a relatively large molecular weight are conjugated; (ii) the resulting well-ordered β-sheet structures allow the integration of antigens at high density; and (iii) they are relatively easy to synthesize and have high biocompatibility. These advantages make NF-vaccines a good alternative to traditional vaccines. Immune induction by NFs formed from antigenic peptides conjugated to β-sheet-forming-peptides have been reported [17–21]. For example, Rudra et al. reported that NFs composed of an antigenic epitope peptide conjugated to self-assembling peptide Q11 were subcutaneously administered, and elicited a strong antibody response [17]. They have also demonstrated that the β-sheet peptide NF system can be applied to various types of antigens, including a malaria epitope [18], a *Staphylococcus aureus* epitope [19] and a tumor-associated antigen MUC1 glycopeptide [20]. However, fundamental studies on how the hydrophilic-hydrophobic balance of NF components affects their cellular interaction—including cellular uptake, cytotoxicity, and immune stimulation response—has not been reported. Recently, studies on other particulate systems reported that surface hydrophobicity is an important factor for determining cellular response [22–32]. In addition to cellular internalization and nontoxicity (i.e., safety), the ability of nanocarriers to stimulate an immune response is an essential property in nanocarrier-based vaccine applications, because uptake of nanocarriers containing an antigen by APCs that do not induce an immune response may lead to unwanted tolerance toward the antigen. Thus, to design nanomaterial-based vaccines that elicit strong immunity without toxicity using a β-sheet assembly system requires a clear understanding of how the hydrophilic-hydrophobic balance of NF components affects their cellular interactions and response.

In previous work, we reported the preparation of antigen-loaded NFs by exploiting the self-assembly of β-sheet peptides [33,34] conjugated to antigenic peptides and hydrophilic chains, such as oligo-ethylene glycol (EG) [35–37]. MHC class I restricted epitope (SIINFEKL) from ovalbumin was selected as a model antigenic peptide. In addition, the structure of the NFs was analyzed in detail by various techniques, including wide-angle X-ray diffraction (WAXD), small-angle X-ray scattering (SAXS), Fourier transform infrared spectroscopy (FT-IR), circular dichroism (CD), transmission electron microscopy (TEM), and atomic force microscopy (AFM). Interestingly, structural analysis revealed that the shape of the NFs is rectangular, rather than a cylinder-like structure observed for filament micelles, possibly because of the lamination structures of β-sheets. Based on this finding, the structural model was proposed as shown in Figure 1b, which shows that the surface of the NFs is not covered with EG chains homogeneously [35,36]. Thus, we hypothesized that the EG chain length is an important parameter for tuning the cellular interactions of NFs, including cellular uptake, cytotoxicity, and immune stimulation response.

In this study, the effect of the EG chain length in building block molecules, which form peptide NFs, on their cellular interaction was investigated. The self-assembling behavior of three kinds of building block peptides with different EG lengths was evaluated by determining their critical aggregation concentration (CAC) and the critical concentration for nanofiber formation (CFC). The structures of the resulting NFs were analyzed by TEM and CD, and their surface hydrophobicity was evaluated using a hydrophobic fluorescence probe. Cellular uptake, cytotoxicity, and immune stimulation ability of the three kinds of NFs were examined in vitro using DCs. In addition, interaction of cells with micelle-like aggregates that were composed of the same building blocks as the NFs were also investigated. Cellular interaction of the NFs was found to be significantly dependent on EG length, whereas that of micelles was independent of EG length. Notably, uptake by DC of NFs composed of EG with a moderate length was effective, and the NFs activated DC without exhibiting significant cytotoxicity. The findings provide useful design guidelines for the development of effective nanofiber-based vaccines.

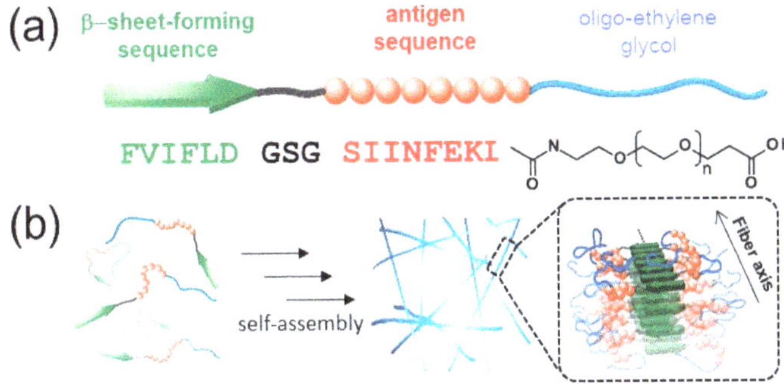

Figure 1. (**a**) Design of the building block peptides (EG$_n$) that are composed of a β-sheet-forming sequence (FVIFLD), a flexible-linker block (GSG), a model antigen sequence (SIINFEKL from OVA), and oligo-ethylene glycol. (**b**) Schematic illustration of the self-assembly process for nanofiber formation and the proposed model of highly antigen-loaded nanofibers based on previous structural study [36]. The schematic illustration was created by modification of Figures 1 and 2 of reference [35].

2. Results

2.1. Self-Assembly Behavior of EG$_n$ Peptides

We have reported previously the preparation of NFs in aqueous solution by heat-treatment of the EG$_{12}$ peptide (EG$_n$, where *n* is the length of the EG chain), which is composed of a β-sheet-forming sequence, an antigen sequence, and a hydrophilic oligo-ethylene glycol (12-mer). In this study, three building block peptide amphiphiles with different EG lengths (6-mer, 12-mer, and 24-mer, which are termed EG$_6$, EG$_{12}$, and EG$_{24}$, respectively) were prepared for assessing the effect of EG length on cellular uptake of NFs, cytotoxicity, and immune stimulatory activities.

Initially, we investigated the effect of EG length on the self-assembly behavior of these peptides by estimating the CFC of each peptide. EG$_n$ peptides were incubated in the presence of the thioflavin T (ThT) dye at 300 µM in phosphate buffered saline (PBS) containing 5% dimethylsulfoxide (DMSO) at 37 °C for 24 h, and a time course of change in ThT fluorescence was measured. The ThT assay is often used to monitor the growth of amyloid-like nanofibers from their component peptides or proteins. A remarkable increase in ThT fluorescence was observed, indicating the formation of NFs by each peptide (Figure 2 and Figure S1). After a certain period of time, the intensity of the ThT fluorescence reached a plateau value. The systems were in a dynamic equilibrium between fibrils and the peptide monomer when the ThT fluorescence intensity reached a plateau value. The concentration of the free peptide monomer at equilibrium corresponds to the CFC [38,39]. The peptide concentration in the supernatant following ultracentrifugation of the EG$_n$ peptide solutions incubated at 300 µM and 37 °C for 24 h provided estimates of the CFC values, which were 96.0 ± 3.6 µM for EG$_6$, 72.4 ± 3.7 µM for EG$_{12}$, and 83.8 ± 1.0 µM for EG$_{24}$.

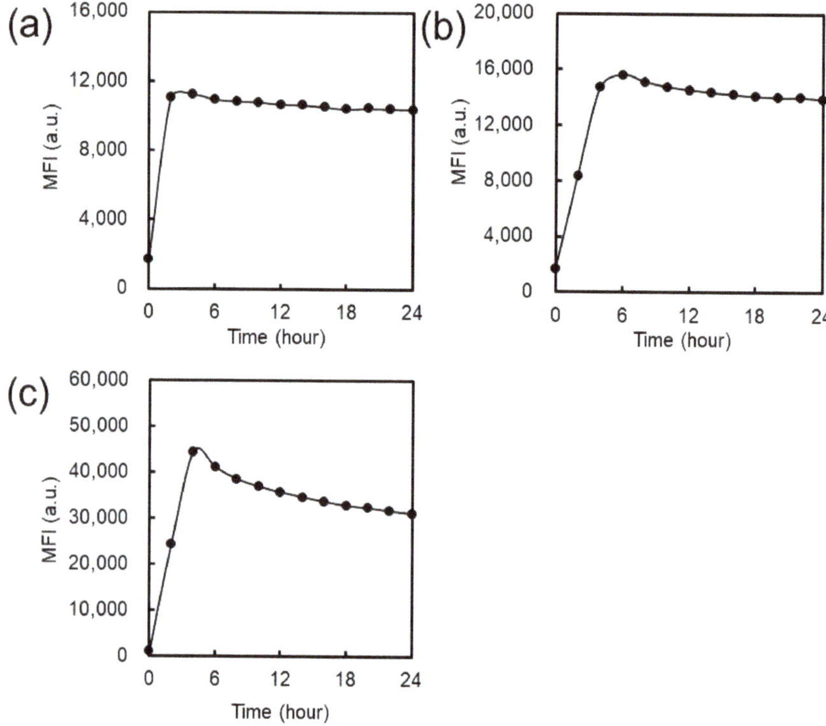

Figure 2. Time-dependent change in the ThT fluorescence intensity of solutions containing the (**a**) EG_6 peptide, (**b**) EG_{12} peptide, and (**c**) EG_{24} peptide when incubated at 37 °C.

Below the CFC, EG_n peptides could either form spherical micelle-like structures or exist as isolated molecules in water because of their amphiphilic structures [40]. The CAC was determined for EG_n peptides in PBS (pH 7.4) using the pyrene 1:3 method to gain information on the association state of EG_n peptides at relatively low concentrations. The CAC was estimated to be 16.6 µM for EG_6, 21.7 µM for EG_{12}, and 29.9 µM for EG_{24} (Figure S2). The CAC value increased as the EG length increased. Because CAC values were smaller than the CFC values, EG_n peptides self-assembled into spherical micelle-like structures over the concentration range between the CAC and the CFC, and existed as monomers at concentrations below the CAC.

2.2. Structural Characterization of EG_n Nanofibers

NFs consisting of EG_n peptides were prepared by incubation of the peptide solution in PBS containing 5% DMSO at 60 °C for 24 h. The incubation was carried out at a higher temperature in this experiment when compared with that of the ThT assay to accelerate NF formation. The resulting NFs were characterized by TEM, CD, ς-potential measurements, and the 8-anilino-1-naphtalene sulfonic acid (ANS) assay.

TEM images revealed that all peptides successfully formed NFs with a homogenous distinct width of ca. 6–8 nm and lengths of several micrometers (Figure 3). The values of the ς-potentials were −33.2 ± 10.7 mV for EG_6 NFs, −33.2 ± 10.7 mV for EG_{12} NFs, and −32.4 ± 9.6 mV for EG_{24} NFs, showing that the surface of the NFs were negatively charged. The ANS assay was performed to obtain information on the surface hydrophobicity of the NFs [41,42]. The peaks shifted toward shorter wavelengths, and the intensities of the signals increased significantly for all NFs (Figure S3). These observations indicate that there are hydrophobic domains on the surface of the peptide NFs. CD was used to examine the secondary structures adopted by peptides in the NFs. The characteristic negative Cotton peak at 217 nm was observed for the three EG_n NFs, showing that the peptides adopt a β-sheet conformation (Figure 4) [43]. These results indicate that the three NFs possess a similar secondary structure regardless of EG length.

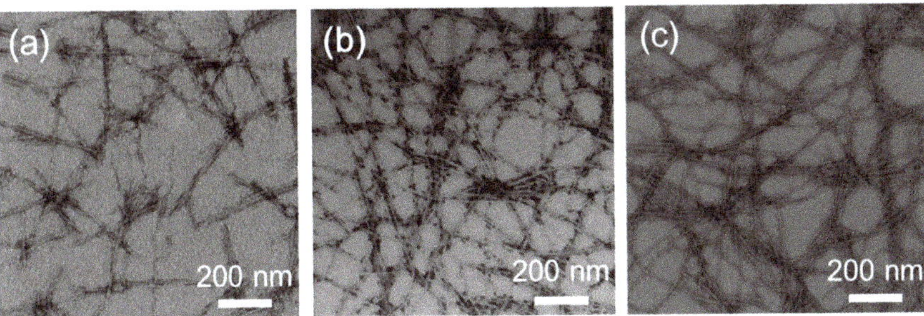

Figure 3. Negatively-stained TEM images of the nanofibers (NFs) obtained by incubation of the (**a**) EG_6 peptide, (**b**) EG_{12} peptide, or (**c**) EG_{24} peptide at a concentration of 300 μM for 24 h in PBS at 60 °C.

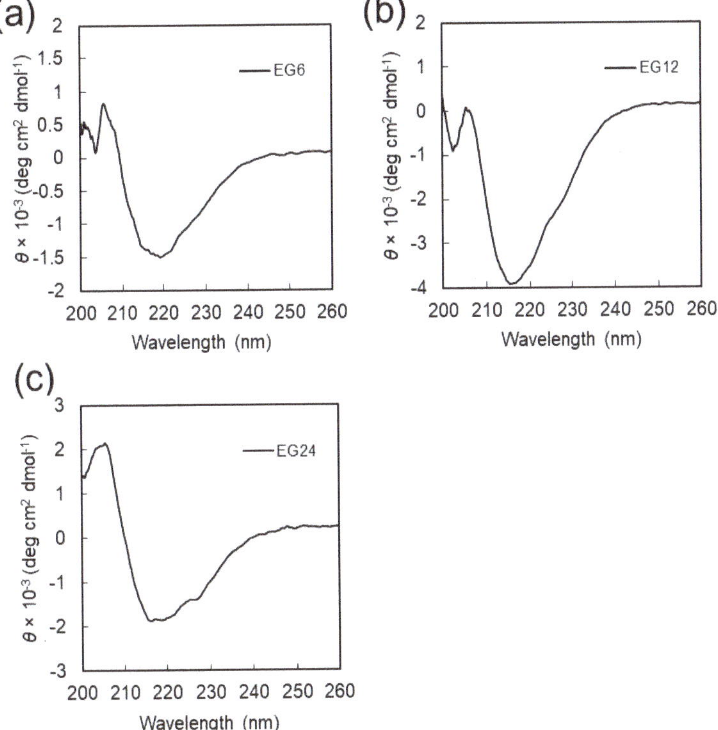

Figure 4. CD spectra of EG_n NFs measured in PBS at room temperature. The EG_n NFs were prepared by incubation of EG_n peptide solutions in PBS at 60 °C for 24 h. (**a**) EG_6 NFs, (**b**) EG_{12} NFs, and (**c**) EG_{24} NFs.

2.3. Cellular Uptake, Cytotoxicity, and Maturation of DCs

We investigated the effect of the EG length of nanofibers on their cell association, cytotoxicity, and DC stimulatory activities. The peptide NFs obtained by incubation at a concentration of 1.5 mM and 60 °C for 24 h were used. The length of the NFs was controlled to be 230–260 nm by an extrusion procedure using a membrane filter with a diameter of 450 nm (Figures S4 and S5). Information about the dispersion state of NFs in water was determined by dynamic light scattering (DLS) measurements of EG_6 NFs, EG_{12} NFs, and EG_{24} NFs. The DLS histograms for EG_{12} and EG_{24} NFs exhibited a unimodal peak, with average diameters of 203.7 ± 119.7 nm for EG_{12} NFs, and 116.6 ± 68.2 nm for EG_{24} NFs. These values are inconsistent with the size estimates from TEM images and this is possibly because of their non-spherical morphology. Nonetheless, these unimodal histograms clearly indicate that EG_{12} and EG_{24} NFs exist as isolated NFs without aggregation in aqueous media (Figure S6). In contrast, the DLS histogram for EG_6 NFs exhibited two peaks with sizes of 135.4 ± 14.5 nm and 3498.3 ± 389.0 nm, indicating that EG_6 NFs formed large aggregates. The secondary aggregation of the NFs may be caused by association of surface-exposed hydrophobic domains on the NFs. Longer EG chains effectively prevent these interactions, yielding highly dispersed, stable EG_{12} and EG_{24} NFs. Conversely, the 6-mer EG is not sufficiently long to prevent secondary aggregation of the peptide NFs, resulting in the observed large aggregates. To compare the behaviors of NFs, the cellular uptake, cytotoxicity, and DC stimulatory activities of non-heat-treated EG_n peptides (non-fiber) were also evaluated. The samples for these experiments were prepared by direct dissolution of the peptides in medium at a given concentration to avoid self-assembly into NFs.

2.3.1. Cellular Association

We evaluated the effect of EG length on the cellular association of EG_n NFs. The fluorescence-labeled EG_n NFs were incubated with JAWS II cells for 1 h at 37 °C. JAWS II cells are an immortalized immature DC line that was established from bone marrow cultures of C57BL/6 p53-knockout mice [44,45]. The amount of NFs associated with the cells was evaluated by flow cytometry (FCM). For comparison, cellular association of non-heat-treated EG_n peptides was also performed.

As shown in Figure 5a, the intensity of fluorescence signals from cells incubated with EG_n peptides (non-fiber) increased as the concentration of the peptides increased. This trend was common to all peptides examined. Comparison of the fluorescence intensity of the three EG_n peptides at the same concentration revealed that they were very similar, indicating that EG length had no effect on cellular association of the peptides. In contrast, cellular association of EG_n NFs was influenced noticeably by EG length (Figure 5b). The amount of associated EG_n NFs was larger as the EG length decreased. This trend was more apparent as the concentration of the peptide increased. This result indicates that longer EG chains may prevent interactions between cells and NFs.

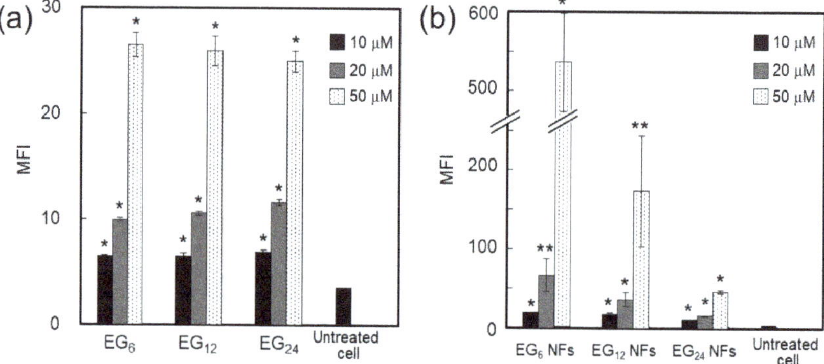

Figure 5. Evaluation of the cellular association of various (**a**) EG_n peptides and (**b**) EG_n NFs labeled with fluorescein using FCM. Mean fluorescence intensity of the treated JAWS II cells is shown. Cellular treatment was performed by incubating cells with peptides or NFs in serum-free medium at 37 °C for 2 h. Each point is the mean ± SD ($n = 3$). * $p < 0.01$, ** $p < 0.01$ compared to untreated cell.

We performed confocal laser scanning microscopic (CLSM) observations of JAWS II cells incubated with various EG$_n$ NFs to evaluate the association of NFs in further detail (Figure 6). The CLSM images of cells incubated with EG$_{12}$ NFs clearly show that the NFs were internalized into cells. The fluorescence signals were observed as dot-like images, indicating that EG$_{12}$ NFs were internalized via endocytosis. In contrast, EG$_{24}$ NFs showed no fluorescence signal, indicating poor cellular uptake of EG$_{24}$ NFs. The confocal images of EG$_6$ NFs-treated cells showed large intensive fluorescence signals on the surface of cells, indicating that some aggregation of NFs were apparently adsorbed onto the surface of cells. This observation indicates that a large proportion of the fluorescence signal from EG$_6$ NFs-treated cells observed in FCM measurements was derived from NFs that had adhered to the surface of cells. The FCM and CLSM results comprehensively showed the efficient uptake of EG$_{12}$ NFs by cells.

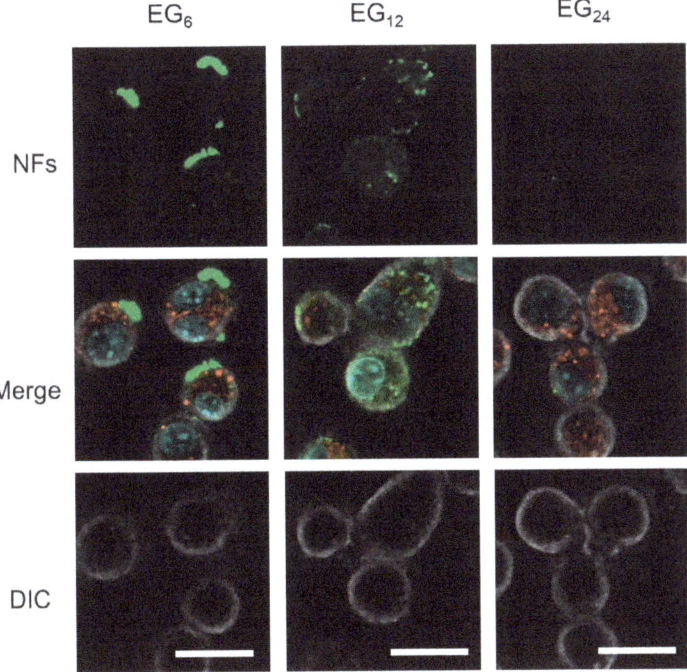

Figure 6. CLSM images of JAWS II cells treated with various EG$_n$ NFs labeled with fluorescein. JAWS II cells were incubated with EG$_n$ NFs at 37 °C for 2 h in serum-free medium. After incubation, the cells were treated with Lyso Tracker Red and Hoechst for staining intracellular acidic compartments and nuclei, respectively. Scale bars represent 20 μM.

2.3.2. Cytotoxicity

The cytotoxicity of EG$_n$ NFs and EG$_n$ peptides (non-fiber) were evaluated. The JAWS II cells were incubated with the EG$_n$ NFs or EG$_n$ peptides at 37 °C for 24 h at different concentrations (10–50 μM), and cell activity was evaluated. Interestingly, cell activity after incubation with all EG$_n$ peptides increased as the peptide concentration increased (Figure 7a). Relative cell activity was essentially 100%, independent of the EG length when peptides were co-incubated with cells at higher concentrations (40–50 μM). Co-incubation of peptides with cells at lower concentrations (10–20 μM) reduced cell activity to 70–90%. In addition, peptides with shorter EG chains showed relatively higher toxicity.

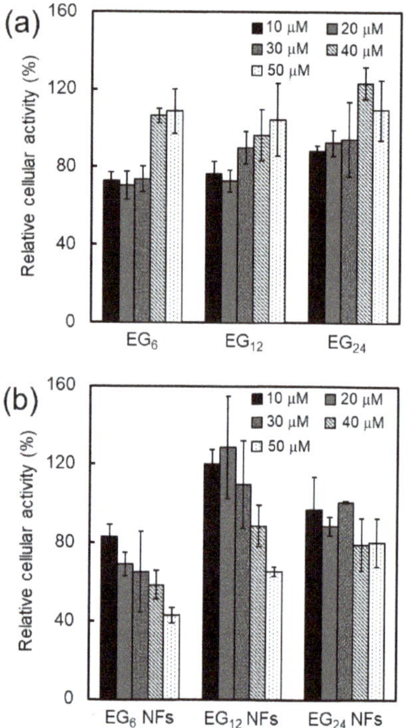

Figure 7. Evaluation of the cytotoxicity of EG_n peptides (**a**) and EG_n NFs (**b**) against JAWS II cells.

In contrast, cell activity was observed to decrease after incubation with each NF, and this reduction in cell activity was concentration-dependent (Figure 7b). In particular, the activity of EG_6 NFs-treated cells was reduced to 40% at a concentration of 50 μM. For EG_{12} and EG_{24} NFs, cell activity was 70% for EG_{12} NFs and 80% for EG_{24} NFs even at a NFs concentration of 50 μM. Thus, NFs with shorter EG chains exhibited higher toxicity.

2.3.3. DC Stimulatory Activity.

It is well known that DC maturation is accompanied by enhanced expression of co-stimulatory molecules (CD40, CD80, and CD86) and by an increase in the secretion of immune-stimulatory cytokines (IL-6, IL-10, IL-12, and TNF-α) [46]. We measured the amount of expressed co-stimulatory molecules (CD86) on JAWS II cells cultured in the presence of EG_n NFs or EG_n peptides for 24 h to determine the effect of EG length on DC maturation. As a positive control, the expression of CD86 on lipopolysaccharides (LPS)-stimulated JAWS II was also measured. The expression of CD86 on JAWS II cells after incubation with EG_n peptides is shown in Figure 8. Even at a peptide concentration of 50 μM, the expression level of CD86 was almost the same as that of untreated JAWS II cells. These results indicate that EG_n peptides do not stimulate DC. In contrast, co-incubation of EG_n NFs with JAWS II cells significantly enhanced the expression of CD86, and this enhancement was dependent on the concentration of the NFs. In particular, when EG_{12} NFs and EG_6 NFs were co-incubated with JAWS II cells, the amount of expressed CD86 was comparable to or larger than that on LPS-stimulated DC.

We also evaluated secretion of immune-stimulatory cytokines. Using enzyme-linked immunosorbent assay (ELISA) methods, we measured the amount of TNF-α and IL-6 contained in the supernatant after 24 h culturing of JAWS II in the presence of EG_n NFs or EG_n peptides (Figure 9). Co-incubation with EG_n peptides did not alter the secretion levels of TNF-α and IL-6. In contrast, interestingly, the secretion of TNF-α and IL-6 was drastically enhanced by co-incubation with EG_6 NFs and EG_{12} NFs, but not by EG_{24} NFs. These results indicate that NFs with relatively short EG chains have an immune-stimulatory effect as adjuvants for DC maturation.

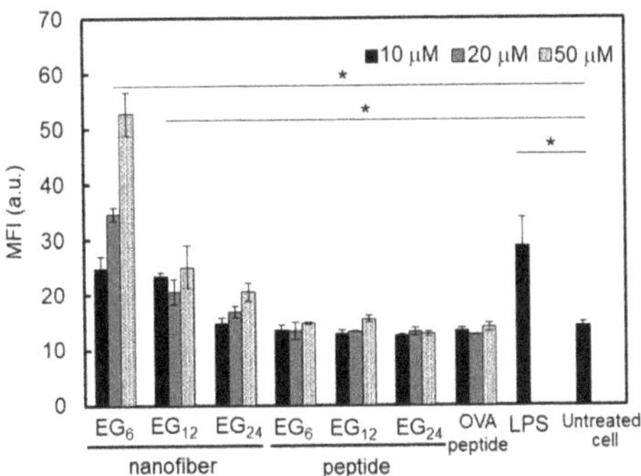

Figure 8. Evaluation of CD86 expressed on the surface of JAWS II cells co-incubated with EG_n NFs (10–50 µM), EG_n peptides (10–50 µM), the OVA peptide (SIINFEKL, 10–50 µM), or 1 µg·mL^{-1} LPS at 37 °C for 24 h. CD86 expression was analyzed by FCM. Each result is the mean ± SD ($n = 3$). * $p < 0.01$. a.u. represents arbitrary unit.

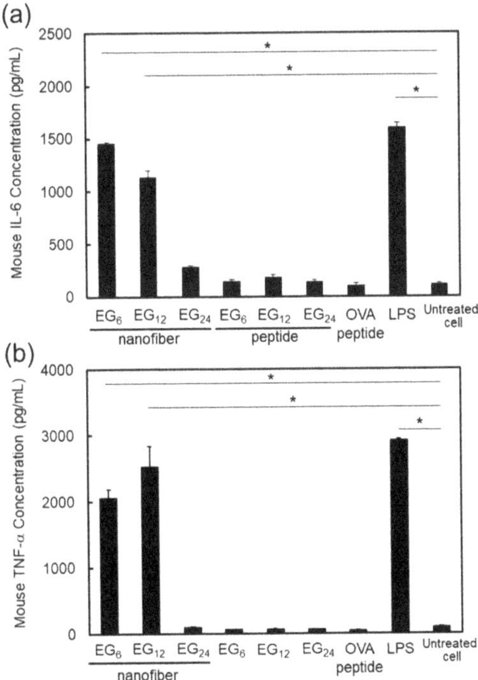

Figure 9. Quantification of immune-stimulatory cytokines secreted from JAWS II cells co-incubated with 10 µM EG_n NFs, 10 µM EG_n peptides, 10 µM OVA peptide (SIINFEKL), or 1 µg·mL^{-1} LPS at 37 °C for 24 h. The cytokine levels (**a,b**) were measured by ELISA. Each result is the mean ± SD ($n = 3$). * $p < 0.01$.

3. Discussion

In this study, we have investigated cellular uptake, cytotoxicity, and DC stimulatory activity of antigen-loaded peptide NFs with different EG lengths and their component peptides. Three building block peptide amphiphiles with different EG lengths (6-mer, 12-mer and 24-mer) were prepared. ThT assay, TEM observation, and CD measurement revealed that all type of peptide amphiphiles are successfully formed β-sheet rich nanofibers with distinct widths (Figures 2–4). The association state of EG_n peptides was dependent on sample concentration. EG_n peptides self-assembled into NFs above the CFC, formed spherical micelles at concentrations between the CFC and CAC, and existed as monomers in solution below the CAC. Based on these findings, we discuss separately the effect of EG length on cellular uptake, cytotoxicity, and immune stimulation for three peptide states: NFs, micelles, and monomers.

3.1. Effect of EG Length of Nanofibers on Their Cellular Uptake, Cytotoxicity, and Immune Stimulation Ability

The EG length of NFs significantly affected their cellular uptake, cytotoxicity, and DC stimulatory activity. Here, we discuss the effect of EG length on these properties of NFs using structural models derived from SAXS, WAXD, CD, and FT-IR data of a previous study [36]. FT-IR, CD, and WAXD results indicate that EG_{12} NFs contain β-sheet structures. In addition, synchrotron X-ray scattering profiles of EG_{12} NFs revealed that the morphology of the NFs is rectangular, and they do not form cylinder structures like filament micelles, presumably because of the laminated structure of β-sheets. In general, amyloid-like nanofibers have a common characteristic cross-β-sheet structure, where tightly packed β-sheets orientate themselves perpendicularly to the fiber elongation axis [47]. By combining these findings, we propose a model of EG_n NFs (Figure S7). β-sheet structures consisting mainly of hydrophobic amino acids form the framework of NFs with EG chains facing outwards to provide water-dispersibility. The surfaces of NFs possess hydrophobic and hydrophilic domains that consist of EG chains based on this model. The ANS assay results support the notion that there are hydrophobic domains on the surface of NFs.

3.1.1. Cellular Association and Internalization of NFs

The amount of NFs associated with cells increased in the order of EG_{24} NFs, EG_{12} NFs, and EG_6 NFs (Figure 5b). Surface hydrophobicity of nanomaterials has been well documented to affect cellular association and uptake by phagocytic cells [24,30–32]. Surface hydrophobicity of nanomaterials facilitates interactions between nanomaterial surfaces and cellular membranes. This may lead to higher cell association of nanomaterials and occasionally increase the chance of recognition by particular receptors involved in cellular uptake. Our results show that cellular association of NFs decreased as the EG chain length increased (Figure 5b), although these NFs commonly possess hydrophobic domains on their surface, as evidenced by the ANS fluorescence assay. These results suggest that longer EG chains inhibit hydrophobic interactions between the NF surface and cell membranes, which can be explained using the model structures presented in Figure 9. The NF skeleton region, consisting mainly of hydrophobic amino acids, may facilitate the interaction with the cell membrane and the EG chain located on the lateral face of the NF may inhibit this interaction.

Results from CLSM observation revealed that EG_{12} NFs were more efficiently internalized by JAWS cells than EG_6 NFs and EG_{24} NFs (Figure 6). Since the surface of EG_{12} NFs is negatively charged, the mechanism for internalization of EG_{12} NFs would be mainly via phagocytosis by scavenger receptor, which recognizes anion species, although further studies using some inhibitor for phagocytosis are required. Thus, the internalization behavior by non-phagocytic cells would be different from that by JAWS II cells. The internalization of EG_6 NFs was low, whereas their association propensity to cells was high. Because the size of nanomaterials can affect cell internalization [48–51], the dispersion state of NFs in aqueous media should be considered in addition to interactions between NFs and the cell surface. The results from DLS indicate that EG_{12} and EG_{24} NFs exist as isolated NFs without aggregation in aqueous media, whereas EG_6 NFs form large aggregates. This observation is consistent with CLSM images showing large aggregates adsorbed onto the cell surfaces. Thus, it is likely that the low efficiency of cellular internalization of EG_6 NFs can be attributed to their apparent size in water. The aggregation of EG_6 NFs is too large for cell uptake. This interpretation is consistent with a previous study that showed that cellular uptake of microparticles with a diameter of a few micrometers or more by phagocytic cells is slow and inefficient [50,51]. Thus, for development of NFs that are efficiently taken up by cells, it is important to design a EG chain length that allows modest interactions with cell membranes while ensuring water-dispersibility.

3.1.2. Cytotoxicity of NFs

Generally, nanomaterials with cationic or hydrophobic surfaces can induce significantly higher toxicities when compared with hydrophilic or anionic nanomaterials [29]. A mechanism of cytotoxicity is cell membrane perturbation, including structural alternation, pore formation, and phase transitions, which cause nonspecific entrance of extracellular components to the cytosol. An increase in hydrophobic interactions between the surface of nanomaterials and cell membranes could perturb the membrane. In the present study, the cytotoxicity of EG_n NFs was found to increase in the order of EG_{24}, EG_{12}, and EG_6 (Figure 7). These results indicate that longer EG chains inhibit the interaction between NFs and cell membranes, which leads to lower cytotoxicity of NFs with long EG chains. It is also possible that the cytotoxicity of NFs may be related to biological stress, e.g., induction of reactive oxygen species (ROS). The detailed mechanism of cytotoxicity by NFs is the subject of ongoing research.

3.1.3. DC Stimulation Ability of NFs

DC activation ability of EG_6 NFs and EG_{12} NFs was much higher than that of EG_{24} NFs. Matzinger and colleagues have proposed that hydrophobic portions in various biomolecules may be involved in the activation of the immune system [22]. Hydrophobic components in molecules are usually masked from the external environment by hydrophilic components. However, when protein denaturation or cell disruption occur, these hydrophobic components become exposed and interact with particular surface receptors of immune cells, which activates the immune system. In agreement with the notion proposed by Matzinger, recently, the relationship between the surface hydrophobicity of nanomaterials and their immune stimulatory activities has been reported [23–28]. For example, Moyano and colleagues reported that the surface hydrophobicity of ligand-modified gold-nanoparticles was correlated with expression of pro-inflammatory cytokine genes in splenocytes from mice in vitro [23]. Shima and colleagues also reported that the activation ability of nanoparticles was significantly affected by the hydrophobicity of polymers constituting the nanoparticles [24]. In the present study, EG_6 NFs and EG_{12} NFs stimulated DC maturation more effectively than EG_{24} NFs, as evidenced by the quantitative evaluation of expressed co-stimulatory molecules (Figure 8) and secreted immune-stimulatory cytokines, IL-6 and TNF-α (Figure 9). Based on these results, it is reasonable to consider that the hydrophobic part of NFs plays an important role in DC activation. Longer EG chains seem to inhibit the recognition of hydrophobic surfaces of NFs by DC surface receptors in a similar manner to that described above. However the mechanisms responsible for DC maturation by EG_n NFs remain unclear and further studies are required. In addition, because IL-6 signaling cannot only promote anti-tumor-adaptive immunity, but also drive malignancy [52], the role of IL-6 in this NFs-based vaccine system should be examined further in vivo.

3.2. Cellular Uptake, Toxicity, and DC Stimulatory Ability of Micelles

In the concentration range where EG_n peptides form micelle-like structures, their cellular association, cytotoxicity, and stimulation ability were not dependent on EG length (Figure 5a, Figure 7a, and Figure 8). These results suggest that the surface components of the micelle-like structures would be almost the same. Poly(ethylene glycol) (PEG) interactions with biological components, including cellular membranes and proteins, are weak because of their nonionic hydrophilicity and high mobility [53]. Thus, a low-level of interaction between the surface of the micelle-like structures and cell membrane components, including receptors involved in cellular uptake and DC maturation, led to lower uptake by DC, lower cytotoxicity, and no DC activation in comparison with NFs.

3.3. Cellular Uptake, Toxicity, and DC Stimulatory Ability of Monomeric Molecules

EG_n peptides exist as monomeric molecules in aqueous media at concentrations below ca. 15–30 μM. The cellular uptake of monomeric peptides was not dependent on EG length (Figure 5a) and this may be attributed to size. Monomeric peptides are too small for efficient uptake by DC regardless to EG length. In addition, the monomeric peptides exhibited some cytotoxicity but no DC activation ability, which is in sharp contrast to the results with NFs (Figure 7a, Figure 8, and Figure 9). These results suggest that monomeric peptides interact with cell membranes, possibly through an N-terminal hydrophobic region, but are not well recognized by receptors involved in DC activation. Recognition by receptors would be dependent on the size of the hydrophobic portion.

3.4. Design of NF-based Vaccines

Recently, various types of NFs for immunotherapy have been reported [54,55]. In particular, NFs formed from antigenic peptides conjugating to β-sheet-forming peptides have been recognized as very promising candidates for next-generation nanoparticle-based vaccines. In the present study, we demonstrated that the hydrophilic-hydrophobic balance of peptide NFs affects their cellular uptake, cytotoxicity, and DC activation ability. NFs consisting of EG with a moderate length (12-mer) showed the most balanced character: Highly efficient cell entry, low cytotoxicity, and high DC activation ability, indicating that the NFs have significant potential as NF-based vaccines, which can be used without additional adjuvants. In general, the relationship between toxicity and DC stimulation ability is a trade-off. It is important to improve the stimulation ability, while simultaneously reducing the cytotoxicity of the NFs. Our results demonstrate that such balance can be simply tuned by the length of the EG. This feature is important for designing safe NF-based vaccines with high immune stimulatory ability. In contrast to NF uptake, the uptake of micelles and monomeric peptides by DC cells inefficient and showed no DC stimulation ability independent of EG length. This result indicates that the assembly style of building block peptide molecules influences the properties of the nanoassembly formed from these building blocks. Finally, to develop NFs with strong immunity-inducing ability, it is necessary to precisely adjust the EG length and introduce intracellular environment-responsive links for efficient release of antigens in cells.

Although we focused on the effect of hydrophilic and hydrophobic balance of nanofibers on their interaction with cells, surface charge of nanofibers is also an important factor in determining the interaction. In general, positively charged nanomaterials are more effectively internalized to cells than negatively charged ones, but they are more toxic. Thus, for design of nanofiber vaccines, it is necessary to address the role of their surface charge. In addition to surface charge, the length of NFs is also an important factor determining their property as a nano-vaccine. In a previous study, we investigated the effect of nanofiber length on their cellular uptake using various NFs with different lengths (40 nm, 120 nm, 280 nm, 800 nm). The study demonstrated that nanofibers with a length of 280 nm were most effectively uptaken by phagocytic cells compared to the others (unpublished data). Based on this finding, we used NFs with a length of 230–260 nm for the cell experiments in the present study. However, other properties—e.g., cytotoxicity and the ability to stimulate immune cells etc.—could exhibit different size-dependencies. Therefore, the optimization of nanofiber length is also required for developing effective nanofiber vaccine.

The important attributes of a vaccine, which are antigen processing, antigen presentation, T-cell stimulation, and successful activation of adaptive immune response against target antigen, should also be evaluated. However, because the NFs used in this study comprised the minimum required block (β-sheet forming peptide, antigenic peptide, oligo(ethylene glycol)) and the antigen could not be released in the cells, the effective antigen presentation via MHC class I pathway and subsequent induction of immunity are not expected. Therefore, we are addressing the development of intracellular environment-responsive NFs for efficient release of antigens in cells and the characterization of their function to induce immunity in vivo.

4. Materials and Methods

4.1. Materials

21-amino-N-(9-fluorenylmethoxycarbonyl)-4,7,10,13,16,19-hexaoxaheneicosanoic acid (Fmoc-N-amido-dPEG$_6$ acid), 39-amino-N-(9-fluorenylmethoxycarbonyl)-4, 7, 10, 13, 16, 19, 22, 25, 28, 31, 34, 37-dodecaoxanonatriacontanoic acid (Fmoc-N-amido-dPEG$_{12}$ acid), and O-[N-(9-fluorenylmethoxycarbonyl)-2-aminoethyl]-O'-(2-carboxyethyl) undecaethyleneglycol (Fmoc-N-amido-dPEG$_{24}$ acid) were purchased from Quanta BioDesign Ltd. (Plain City, OH, USA). 2-chlorotrityl chloride resin, N,N'-diisopropylethylamine (DIPEA), all the L-Fmoc amino acids, 1-[bis(dimethylamino)methylene]-1H-benzotriazolium 3-oxide hexafluorophosphate (HBTU), 1-hydroxybenzotriazole (HOBT), and piperidine were purchased from Watanabe Chemical Industries Ltd. (Hiroshima, Japan). N,N'-dimethylformamide (DMF), isopropanol, methanol, diethyl ether, hexafluoroisopropanol (HFIP), dichloromethane (CH$_2$Cl$_2$), trifluoroacetic acid (TFA), and granulocyte-macrophage colony-stimulating factor (GM-CSF) were purchased from Wako Pure Chemical Industries, Ltd. (Osaka, Japan). Eagle's minimal essential medium (EMEM), penicillin-streptomycin, and lipopolysaccharides from *Escherichia coli* O55:B5 were purchased from Sigma-Aldrich (St. Louis, MO, USA). N^6-[(3',6'-Dihydroxy-3-oxospiro[isobenzofuran-1(3H),9'-[9H]xanthen]-5-yl) carbonyl]-N^2-[(9H-fluoren-9-ylmethoxy)carbonyl]-L-lysine (Fmoc-Lys(5-FAM)-OH) was purchased from AAT Bioquest (Sunnyvale, CA, USA). LysoTracker Red DND-99, and Hoechst 33342, trihydrochloride, trihydrate were purchased from Thermo Fisher Scientific (Waltham, MA, USA). Fetal bovine serum (FBS) was purchased from Biowest (Nuaillé, France). Purified anti-mouse CD16/CD32 (2.4G2) and anti-mouse CD86 (B7-2) PE were purchased from Tonbo Biosciences (San Diego, CA, USA). The ELISA Kit for mouse TNF-alpha and IL-6 were purchased from R&D Systems (Minneapolis, MN, USA).

4.2. Experimental Methods

4.2.1. Synthesis of Building Block Molecules

Loading of resin: Fmoc-N-amido-dPEG$_{12}$ was dehydrated by azeotropy with benzene prior to use. A solution of Fmoc-N-amido-dPEG$_{12}$ acid (0.238 mmol) and DIPEA (0.952 mmol) in CH$_2$Cl$_2$ (2.6 mL) was added to 2-chlorotrityl chloride resin (0.397 mmol, 1.5 mmol/g loading max) for 12 h.

Peptide synthesis: Coupling reactions were performed using a standard Fmoc protocol. The coupling cycle included 3 repeats of Fmoc deprotection (20% piperidine in DMF, 5 min), a wash in DMF, two repeats of amino acid coupling: L-Fmoc amino acids (4 eq.), HBTU (3.6 eq.), HOBt (4 eq.), and DIPEA (8 eq.) for 30 min, and a final DMF wash. After all coupling reactions, the obtained peptides were cleaved from the resin using a solution of H$_2$O/TFA/triisopropylsilane (100:5:2 volume ratio) containing 500 mM phenol for 2 h. The resulting peptides were precipitated in ice cold diethyl ether, filtered, centrifuged, and washed with diethyl ether. The crude peptides were purified by reversed-phase high-performance liquid chromatography (RP-HPLC, SPD-10A and LC-10A, Shimadzu Scientific Instruments, Kyoto, Japan). Other building blocks with different EG lengths were synthesized by a similar procedure. Molecular weight was analyzed by MALDI-TOF mass (autoflex speed system, Bruker, Billerica, MA, USA). MS (MALDI-TOF): EG$_6$; Cald. MASS: 2234.92, Obsd. MASS: 2234.359, EG$_{12}$; Cald. MASS: 2499.12, Obsd. MASS: 2498.11, EG$_{24}$; Cald. MASS: 3027.52, Obsd. MASS: 3027.04.

Fluorescence-labeled building block peptides were synthesized using Fmoc-Lys(5-FAM)-OH as the first amino acid residue by a similar procedure. MS (MALDI-TOF): EG$_6$-FAM; Cald. MASS: 2721.41,

Obsd. MASS: 2720.84, EG_{12}-FAM; Cald. MASS: 2985.61, Obsd. MASS: 2985.26, EG_{24}-FAM; Cald. MASS: 3514.01, Obsd. MASS: 3514.65.

4.2.2. Preparation of Antigen-Loaded Peptide NFs

EG_n peptide was dissolved in HFIP and dried with nitrogen flow to allow film formation. The obtained film was re-dissolved at a concentration of 30 mM in DMSO. The resulting solution (15 μL) was added to PBS (285 μL) to a final concentration of 1.5 mM and incubated at 60 °C for 24 h. Following incubation, the resulting peptide nanofiber dispersion was dialyzed against PBS for 24 h using dialysis membrane (MWCO 8,000, GE Healthcare, Chicago, IL, USA) to remove DMSO and free peptides. For cell-based experiments (cytotoxicity, DC maturation), the length of NFs was controlled by filtration using a syringe filter with a pore size of 0.45 μM (GE Healthcare).

4.2.3. Preparation of Fluorescence-Labeled Antigen-Loaded Peptide NFs

The HFIP-treated mixture of EG_n and EG_n-FAM peptides was dissolved in DMSO; then, 15 μL of the solution was added to 285 μL PBS to give final concentrations of 1.425 mM for EG_n and 0.071 mM for EG_n-FAM. The solution was incubated at 60 °C for 24 h, and then the resulting peptide nanofiber dispersion was dialyzed against PBS for 24 h using dialysis membrane (MWCO 8,000, GE Healthcare) to remove DMSO and free peptides. The length of NFs was controlled by filtration using a syringe filter with a pore size of 0.45 μM (GE Healthcare).

4.2.4. Determination of Critical Aggregation Concentration

CAC was determined using the pyrene 1:3 method [56]. First, a saturated solution of pyrene was prepared by mixing an excess of pyrene with PBS, and using the supernatant to dissolve EG_n peptides at a concentration of 150 mM (the stock solution). A concentration range of EG_n peptides from 2.5 μM to 100 μM was then prepared using serial dilutions of the stock solution with the saturated solution of pyrene. The final concentration of pyrene was equal in each solution. The fluorescence emission of pyrene was monitored using a fluorescence spectrometer (RF5300 PC, Shimadzu Scientific Instruments, Kyoto, Japan) with an excitation wavelength of 335 nm at 37 °C. The ratio of the emission intensities at 376 nm and 392 nm were then plotted as a function of the EG_n peptide concentration (log scale). The CAC was determined from an abrupt change in the slope of the plot using the least-squares fitting technique.

4.2.5. ThT Assay

PBS containing 10 μM Thioflavin T (ThT) was dispensed into a 96-well plate. EG_n peptides solution (6 mM) was prepared and added to the 96-well plate, giving a final concentration of 300 μM. ThT fluorescence intensities at 480 nm (excitation; 440 nm) were monitored using a Genios microplate reader (TECAN, Männedorf, Switzerland) at 37 °C.

4.2.6. Measurement of Surface Hydrophobicity

The surface hydrophobicity of NFs in the solution was determined using an ANS fluorescent probe as previously reported [41,42]. A concentration range of EG_n NFs in PBS from 9.4 μM to 200 μM was prepared. The nanofiber dispersion was mixed with the equivalent volume of PBS containing 20 μM ANS. The intensities of ANS fluorescence ranging from 400 nm to 600 nm (excitation; 370 nm) were monitored using a fluorescence spectrometer (RF5300 PC) at 37 °C.

4.2.7. Cell Culture

JAWS II, a DC line derived from mouse bone marrow, was purchased from the American Type Culture Collection (ATCC, Manassas, VA, USA). The cells were grown in EMEM supplemented with 20% FBS, 5 ng/mL murine GM-CSF, and antibiotics at 37 °C, 5% CO_2.

4.2.8. Evaluation of Cellular Association of Peptide NFs

JAWS II cells were seeded into 12-well plates (2.5×10^5 per well) and cultured for 12 h at 37 °C in a humidified atmosphere (5% CO_2). After 12 h, the cells were washed with PBS and serum-free culture medium. The fluorescence-labeled peptide NF dispersion was added gently to the cells followed by incubation for 2 h at 37 °C. Following incubation, the cells were washed with PBS and 0.2% trypan blue aqueous solution, which was used to quench the flourescence from surface-adsorbed NFs [57]. The cells were then detached using trypsin and subsequently analyzed by FCM (Guava EasyCyte Plus, Millipore, Burlington, MA, USA). As a comparison to peptide NFs, the cellular uptake of the building block peptides without heat treatment was investigated under the same conditions.

4.2.9. CLSM Observation of NF-Treated Cells

JAWS II cells (1.5×10^5) were cultured for 12 h in a 35 mm glass-bottom dish and subsequently washed with PBS and serum-free culture medium. Fluorescein-labeled peptide NFs were gently added to the cells, followed by incubation for 2 h at 37 °C with 5% CO_2. After incubation, the cells were washed with PBS, and then incubated for 5 min with a solution containing LysoTracker Red DND-99 (50 nm) and Hoechst 33342, trihydrochloride, trihydrate (3.24 µM). LysoTracker Red DND-99 and Hoechst 33342 were used to stain the intracellular acidic compartments and nuclei, respectively. After staining, the cells were washed twice with PBS, then observed by CLSM using an FV10i microscope (Olympus, Tokyo, Japan).

4.2.10. Quantitative Expression Analysis of Co-Stimulatory Molecules and Cytokines from NF-Treated Cells

The expression of co-stimulatory molecules and cytokines was evaluated by specific immunostaining, as well as by ELISA. For immunostaining, JAWS II cells (2×10^5) were cultured for 12 h in a 24-well plate followed by washing with PBS containing 3% FBS and 0.05% NaN_3, and then with serum-free culture medium. The DCs were pulsed with peptide NFs for 24 h, and then immunostained with a mouse monoclonal antibody for CD86 (a maturation marker), and subsequently analyzed by FCM to estimate their CD86 expression level. For quantitative analysis of TNF-α and IL-6 expression, the supernatants following the 24 h co-incubation of DCs with peptide NFs were collected and analyzed using an ELISA kit. The maturation of JAWS II cells cultured in medium with and without LPS (1 µg/mL) was evaluated as the positive and negative control, respectively. In addition, to compare the DC-activation ability between the NFs and heat-untreated building block peptides, JAWS II cells cultured with the peptides were also evaluated.

4.2.11. Evaluation of Cytotoxicity of Peptide NFs

The cytotoxicity of peptide NFs was evaluated using a Cell Counting Kit-8 (Dojindo Molecular Technologies, Kumamoto, Japan) according to the manufacturer's instructions. Briefly, JAWS II cells were seeded into 96-well plates (1.0×10^5 per well) and cultured for 12 h at 37 °C in a humidified atmosphere (5% CO_2). After 12 h, the cells were washed with PBS and serum-free culture medium. The nanofiber dispersion was gently added to the cells followed by incubation for 24 h. The cells were washed with PBS three times and the medium was replaced with a solution containing 2-(2-methoxy-4-nitrophenyl)-3-(4-nitrophenyl)-5-(2,4-disulfophenyl)-2H-tetrazolium, monosodium salt (WST-8), and 1-methoxy-5-methylphenazinium methylsulfate at a 10-fold dilution. After a 2 h incubation, the absorbance was measured at 420 nm using a plate reader (Multiskan JX, Thermo Fisher Scientific, Waltham, MA, USA). The relative cellular activity was calculated using the following equation:

$$\% \text{ relative cellular activity} = \frac{A_{420\text{ nm}}\ (\text{NFs} - \text{treated cells}) - A_{420\text{ nm}}\ (\text{blank})}{A_{420\text{ nm}}(\text{untreated cells}) - A_{420\text{ nm}}\ (\text{blank})} \times 100 \qquad (1)$$

where $A_{420\,nm}$ is the absorbance at 420 nm, $A_{420\,nm}$ (untreated cells) is the absorbance at 420 nm after incubation in the absence of peptide NFs, and $A_{420\,nm}$ (blank) is the absorbance of medium containing WST-8 reagent at 420 nm. As a comparison, the cytotoxicity of building block peptides without heat treatment was investigated in a similar manner.

4.3. Other Characterizations

TEM measurements were performed using a JEM-1200EX II (JEOL, Tokyo, Japan) with an acceleration voltage of 85 keV. The samples were negatively stained with 0.1% phosphotungstate. p-potentials of NFs were measured using a Micro-Electrophoresis Zeta Potential Analyzer Model 502 (Nihon Rufuto, Tokyo, Japan). DLS analysis was performed using a particle size analyzer (ELSZ-1000, Otsuka Electronics, Osaka, Japan) at 25 °C. The light source was a He-Ne laser (630 nm) set at an 1ngle of 45°. Experimental data were analyzed using the marquardt provided by the manufacturer. CD spectra were measured using a J-720 spectropolarimeter (Jasco, Tokyo, Japan) at 25 °C. The data were obtained using a 0.1 cm path length cell at a scan speed of 20 nm/min.

5. Conclusions

This study showed that the hydrophilic-hydrophobic balance of antigen-loaded NFs significantly impacted on their cellular uptake, cytotoxicity, and DC stimulation ability, which differs noticeably from the results observed for micelles formed from the same components of NFs. Building blocks consisting of β-sheet-forming peptides conjugated with antigenic peptides and hydrophilic EG with different lengths (6-mer, 12-mer and 24-mer) were found to successfully form NFs with homogenous widths. The uptake of NFs consisting of EG with a moderate length (12-mer) by DC was effective, and these NFs activated DC without exhibiting significant cytotoxicity. Increasing the EG chain length significantly reduced the interactions with cells. Conversely, decreasing the EG chain length enhanced DC activation ability but increased toxicity and impaired water-dispersibility, resulting in low cellular uptake. Thus, since cell entry, cytotoxicity, and the immune stimulation ability of antigen-loaded NFs can be tuned by the length of the EG moiety, the antigen-loaded NFs have potential as NF-based vaccines that can be used without additional adjuvants. In order to achieve efficient immune response in vivo, the development of intracellular environment-responsive NFs is now in progress. We believe the findings obtained in this study contribute to the understanding of the interaction between the surface of one-dimensional assemblies and cells, and provide useful design guidelines for development of effective NF-based vaccines.

Supplementary Materials: Supplementary materials can be found at http://www.mdpi.com/1422-0067/20/15/3781/s1.

Author Contributions: T.W. conceived and designed the experiments; S.N., S.K. (Sayaka Koeda), Y.K., and K.K. performed the experiments; all members discussed the experimental data; T.W. wrote the paper.

Funding: This work was partly supported by the Japan Society for the Promotion of Science (JSPS) KAKENHI Grant number 16K01391.

Acknowledgments: We thank the Edanz Group (www.edanzediting.com/ac) for editing a draft of this manuscript. The authors would like to thank Kaeko Kamei at the Faculty of Molecular Chemistry and Engineering of Kyoto Institute of Technology for technical assistance with the ELISA assay. Also, we thank Kensuke Naka and Hiroaki Imoto at the Faculty of Molecular Chemistry and Engineering of Kyoto Institute of Technology for technical assistance with the DLS measurement.

Conflicts of Interest: The authors declare no conflict of interest.

References

1. Skwarczynski, M.; Toth, I. Peptide-based synthetic vaccines. *Chem. Sci.* **2016**, *7*, 842–854. [CrossRef] [PubMed]

2. Li, W.; Joshi, M.; Singhania, S.; Ramsey, K.; Murthy, A. Peptide vaccine: Progress and challenges. *Vaccines* **2014**, *2*, 515–536. [CrossRef] [PubMed]

3. Yoshizaki, Y.; Yuba, E.; Komatsu, T.; Udaka, K.; Harada, A.; Kono, K. Improvement of peptide-based tumor immunotherapy using pH-sensitive fusogenic polymer-modified liposomes. *Molecules* **2016**, *21*, 1284. [CrossRef] [PubMed]

4. Varypataki, E.M.; Silva, A.L.; Barnier-Quer, C.; Collin, N.; Ossendorp, F.; Jiskoot, W. Synthetic long peptide-based vaccine formulations for induction of cell mediated immunity: A comparative study of cationic liposomes and PLGA nanoparticles. *J. Control. Release* **2016**, *226*, 98–106. [CrossRef] [PubMed]

5. Guan, H.H.; Budzynski, W.; Koganty, R.R.; Krantz, M.J.; Reddish, M.A.; Rogers, J.A.; Longenecker, B.M.; Samuel, J. Liposomal formulations of synthetic MUC1 peptides: Effects of encapsulation versus surface display of peptides on immune responses. *Bioconjugate Chem.* **1998**, *9*, 451–458. [CrossRef] [PubMed]

6. Büyüktimkin, B.; Wang, Q.; Kiptoo, P.; Stewart, J.M.; Berkland, C.; Siahaan, T.J. Vaccine-like controlled-release delivery of an immunomodulating peptide to treat experimental autoimmune encephalomyelitis. *Mol. Pharm.* **2012**, *9*, 979–985. [CrossRef] [PubMed]

7. Chua, B.Y.; Al Kobaisi, M.; Zeng, W.; Mainwaring, D.; Jackson, D.C. Chitosan microparticles and nanoparticles as biocompatible delivery vehicles for peptide and protein-based immunocontraceptive vaccines. *Mol. Pharm.* **2011**, *9*, 81–90. [CrossRef] [PubMed]

8. Ma, W.; Chen, M.; Kaushal, S.; McElroy, M.; Zhang, Y.; Ozkan, C.; Bouvet, M.; Kruse, C.; Grotjahn, D.; Ichum, T.; et al. PLGA nanoparticle-mediated delivery of tumor antigenic peptides elicits effective immune responses. *Int. J. Nanomed.* **2012**, *7*, 1475. [CrossRef]

9. Zeng, Q.; Jiang, H.; Wang, T.; Zhang, Z.; Gong, T.; Sun, X. Cationic micelle delivery of Trp2 peptide for efficient lymphatic draining and enhanced cytotoxic T-lymphocyte responses. *J. Control. Release* **2015**, *200*, 1–12. [CrossRef]

10. Zhao, G.; Chandrudu, S.; Skwarczynski, M.; Toth, I. The application of self-assembled nanostructures in peptide-based subunit vaccine development. *Eur. Polym. J.* **2017**, *93*, 670–681. [CrossRef]

11. Wen, Y.; Collier, J.H. Supramolecular peptide vaccines: Tuning adaptive immunity. *Curr. Opin. Immunol.* **2015**, *35*, 73–79. [CrossRef] [PubMed]

12. Black, M.; Trent, A.; Kostenko, Y.; Lee, J.S.; Olive, C.; Tirrell, M. Self-assembled peptide amphiphile micelles containing a cytotoxic T-Cell epitope promote a protective immune response in vivo. *Adv. Mater.* **2012**, *24*, 3845–3849. [CrossRef] [PubMed]

13. Ghasparian, A.; Riedel, T.; Koomullil, J.; Moehle, K.; Gorba, C.; Svergun, D.I.; Perriman, A.W.; Mann, S.; Tamborrini, M.; Pluschke, G.; et al. Engineered Synthetic Virus-Like Particles and Their Use in Vaccine Delivery. *ChemBioChem* **2012**, *12*, 100–109. [CrossRef] [PubMed]

14. Simerska, P.; Suksamran, T.; Ziora, Z.M.; de Labastida Rivera, F.; Engwerda, C.; Toth, I. Ovalbumin lipid core peptide vaccines and their CD4+ and CD8+ T cell responses. *Vaccine* **2014**, *32*, 4743–4750. [CrossRef] [PubMed]

15. Kakwere, H.; Ingham, E.S.; Allen, R.; Mahakian, L.M.; Tam, S.M.; Zhang, H.; Silvestrini, M.T.; Lewis, J.S.; Ferrara, K.W. Toward personalized peptide-based cancer nanovaccines: A facile and versatile synthetic approach. *Bioconjugate Chem.* **2017**, *28*, 2756–2771. [CrossRef] [PubMed]

16. Skwarczynski, M.; Zaman, M.; Urbani, C.N.; Lin, I.C.; Jia, Z.; Batzloff, M.R.; Good, M.F.; Monteiro, M.J.; Toth, I. Polyacrylate dendrimer nanoparticles: A self-adjuvanting vaccine delivery system. *Angew. Chem. Int. Ed.* **2010**, *49*, 5742–5745. [CrossRef] [PubMed]

17. Rudra, J.S.; Tian, Y.F.; Jung, J.P.; Collier, J.H. A self-assembling peptide acting as an immune adjuvant. *Proc. Natl. Acad. Sci. USA* **2010**, *107*, 622–627. [CrossRef] [PubMed]

18. Rudra, J.S.; Mishra, S.; Chong, A.S.; Mitchell, R.A.; Nardin, E.H.; Nussenzweig, V.; Collier, J.H. Self-assembled peptide nanofibers raising durable antibody responses against a malaria epitope. *Biomaterials* **2012**, *33*, 6476–6484. [CrossRef] [PubMed]

19. Pompano, R.R.; Chen, J.; Verbus, E.A.; Han, H.; Fridman, A.; McNeely, T.; Collier, J.H.; Chong, A.S. Titrating T-Cell Epitopes within Self-Assembled Vaccines Optimizes CD4+ Helper T Cell and Antibody Outputs. *Adv. Health Mater.* **2014**, *3*, 1898–1908. [CrossRef]

20. Huang, Z.H.; Shi, L.; Ma, J.W.; Sun, Z.Y.; Cai, H.; Chen, Y.X.; Zhao, Y.F.; Li, Y.M. A totally synthetic, self-assembling, adjuvant-free MUC1 glycopeptide vaccine for cancer therapy. *J. Am. Chem. Soc.* **2012**, *134*, 8730–8733. [CrossRef] [PubMed]

21. Si, Y.; Wen, Y.; Kelly, S.H.; Chong, A.S.; Collier, J.H. Intranasal delivery of adjuvant-free peptide nanofibers elicits resident CD8+ T cell responses. *J. Control. Release* **2018**, *282*, 120–130. [CrossRef] [PubMed]

22. Seong, S.Y.; Matzinger, P. Hydrophobicity: An ancient damage-associated molecular pattern that initiates innate immune responses. *Nat. Rev. Immunol.* **2004**, *4*, 469. [CrossRef] [PubMed]
23. Moyano, D.F.; Goldsmith, M.; Solfiell, D.J.; Landesman-Milo, D.; Miranda, O.R.; Peer, D.; Rotello, V.M. Nanoparticle hydrophobicity dictates immune response. *J. Am. Chem. Soc.* **2012**, *134*, 3965–3967. [CrossRef] [PubMed]
24. Shima, F.; Akagi, T.; Akashi, M. Effect of hydrophobic side chains in the induction of immune responses by nanoparticle adjuvants consisting of amphiphilic poly (γ-glutamic acid). *Bioconjugate Chem.* **2015**, *26*, 890–898. [CrossRef] [PubMed]
25. Liu, Y.; Yin, Y.; Wang, L.; Zhang, W.; Chen, X.; Yang, X.; Xu, J.; Ma, G. Surface hydrophobicity of microparticles modulates adjuvanticity. *J. Mater. Chem. B* **2013**, *1*, 3888–3896. [CrossRef]
26. Shahbazi, M.A.; Fernández, T.D.; Mäkilä, E.M.; Le Guével, X.; Mayorga, C.; Kaasalainen, M.H.; Salonen, J.J.; Hivonen, J.T.; Santos, H.A. Surface chemistry dependent immunostimulative potential of porous silicon nanoplatforms. *Biomaterials* **2014**, *35*, 9224–9235. [CrossRef] [PubMed]
27. Moyano, D.F.; Liu, Y.; Peer, D.; Rotello, V.M. Modulation of immune response using engineered nanoparticle surfaces. *Small* **2016**, *12*, 76–82. [CrossRef] [PubMed]
28. Gause, K.T.; Wheatley, A.K.; Cui, J.; Yan, Y.; Kent, S.J.; Caruso, F. Immunological principles guiding the rational design of particles for vaccine delivery. *ACS Nano* **2017**, *11*, 54–68. [CrossRef] [PubMed]
29. Saei, A.A.; Yazdani, M.; Lohse, S.E.; Bakhtiary, Z.; Serpooshan, V.; Ghavami, M.; Asadian, M.; Mashaghi, S.; Dreaden, E.C.; Mashaghi, A.; et al. Nanoparticle surface functionality dictates cellular and systemic toxicity. *Chem. Mater.* **2017**, *29*, 6578–6595. [CrossRef]
30. Torres, M.P.; Wilson-Welder, J.H.; Lopac, S.K.; Phanse, Y.; Carrillo-Conde, B.; Ramer-Tait, A.E.; Bellaire, B.H.; Wannemuehler, M.J.; Narasimhan, B. Polyanhydride microparticles enhance dendritic cell antigen presentation and activation. *Acta Biomater.* **2011**, *7*, 2857–2864. [CrossRef]
31. Ulery, B.D.; Phanse, Y.; Sinha, A.; Wannemuehler, M.J.; Narasimhan, B.; Bellaire, B.H. Polymer chemistry influences monocytic uptake of polyanhydride nanospheres. *Pharm. Res.* **2009**, *26*, 683–690. [CrossRef] [PubMed]
32. Chiu, Y.L.; Ho, Y.C.; Chen, Y.M.; Peng, S.F.; Ke, C.J.; Chen, K.J.; Mi, F.L.; Sung, H.W. The characteristics, cellular uptake and intracellular trafficking of nanoparticles made of hydrophobically-modified chitosan. *J. Control. Release* **2010**, *146*, 152–159. [CrossRef] [PubMed]
33. Fukuhara, S.; Nishigaki, T.; Miyata, K.; Tsuchiya, N.; Waku, T.; Tanaka, N. Mechanism of the chaperone-like and antichaperone activities of amyloid fibrils of peptides from αA-crystallin. *Biochemistry* **2012**, *51*, 5394–5401. [CrossRef] [PubMed]
34. Tanaka, N.; Tanaka, R.; Tokuhara, M.; Kunugi, S.; Lee, Y.F.; Hamada, D. Amyloid fibril formation and chaperone-like activity of peptides from αA-crystallin. *Biochemistry* **2008**, *47*, 2961–2967. [CrossRef] [PubMed]
35. Waku, T.; Kitagawa, Y.; Kawabata, K.; Nishigaki, S.; Kunugi, S.; Tanaka, N. Self-assembled β-sheet peptide nanofibers for efficient antigen delivery. *Chem. Lett.* **2013**, *42*, 1441–1443. [CrossRef]
36. Minami, T.; Matsumoto, S.; Sanada, Y.; Waku, T.; Tanaka, N.; Sakurai, K. Rod-like architecture and cross-sectional structure of an amyloid protofilament-like peptide supermolecule in aqueous solution. *Polym. J.* **2016**, *48*, 197–202. [CrossRef]
37. Waku, T.; Tanaka, N. Recent advances in nanofibrous assemblies based on β-sheet-forming peptides for biomedical applications. *Polym. Int.* **2017**, *66*, 277–288. [CrossRef]
38. Inoue, M.; Konno, T.; Tainaka, K.; Nakata, E.; Yoshida, H.O.; Morii, T. Positional effects of phosphorylation on the stability and morphology of tau-related amyloid fibrils. *Biochemistry* **2012**, *51*, 1396–1406. [CrossRef]
39. Lomakin, A.; Chung, D.S.; Benedek, G.B.; Kirschner, D.A.; Teplow, D.B. On the nucleation and growth of amyloid beta-protein fibrils: Detection of nuclei and quantitation of rate constants. *Proc. Natl. Acad. Sci. USA* **1996**, *93*, 1125–1129. [CrossRef]
40. Ghosh, A.; Haverick, M.; Stump, K.; Yang, X.; Tweedle, M.F.; Goldberger, J.E. Fine-tuning the pH trigger of self-assembly. *J. Am. Chem. Soc.* **2012**, *134*, 3647–3650. [CrossRef]
41. Semisotnov, G.V.; Rodionova, N.V.; Razgulyaev, O.I.; Uversky, V.N.; Gripas, A.F.; Gilmanshin, R.I. Study of the molten globule intermediate state by hydrophobic fluorescent probe. *Biopolymers* **1991**, *31*, 119–128. [CrossRef] [PubMed]
42. Yamaguchi, S.; Mannen, T.; Nagamune, T. Evaluation of surface hydrophobicity of immobilized protein with a surface plasmon resonance sensor. *Biotechnol. Lett.* **2004**, *26*, 1081–1086. [CrossRef] [PubMed]

43. Greenfield, N.J.; Fasman, G.D. Computed circular dichroism spectra for the evaluation of protein conformation. *Biochemistry* **1969**, *8*, 4108–4116. [CrossRef] [PubMed]

44. Shaheen, S.M.; Akita, H.; Nakamura, T.; Takayama, S.; Futaki, S.; Yamashita, A.; Katoono, R.; Yui, N.; Harashima, H. KALA-modified multi-layered nanoparticles as gene carriers for MHC class-I mediated antigen presentation for a DNA vaccine. *Biomaterials* **2011**, *32*, 6342–6350. [CrossRef] [PubMed]

45. Jiang, X.; Shen, C.; Rey-Ladino, J.; Yu, H.; Brunham, R.C. Characterization of murine dendritic cell line JAWS II and primary bone marrow-derived dendritic cells in Chlamydia muridarum antigen presentation and induction of protective immunity. *Infect. Immun.* **2008**, *76*, 2392–2401. [CrossRef] [PubMed]

46. Yi, A.K.; Yoon, J.G.; Hong, S.C.; Redford, T.W.; Krieg, A.M. Lipopolysaccharide and CpG DNA synergize for tumor necrosis factor-α production through activation of NF-κB. *Int. Immunol.* **2001**, *13*, 1391–1404. [CrossRef] [PubMed]

47. Kirschner, D.A.; Abraham, C.; Selkoe, D.J. X-ray diffraction from intraneuronal paired helical filaments and extraneuronal amyloid fibers in Alzheimer disease indicates cross-beta conformation. *Proc. Natl. Acad. Sci. USA* **1986**, *83*, 503–507. [CrossRef]

48. Shang, L.; Nienhaus, K.; Nienhaus, G.U. Engineered nanoparticles interacting with cells: Size matters. *J. Nanobiotechnol.* **2014**, *12*, 5. [CrossRef]

49. Foged, C.; Brodin, B.; Frokjaer, S.; Sundblad, A. Particle size and surface charge affect particle uptake by human dendritic cells in an in vitro model. *Int. J. Pharm.* **2005**, *298*, 315–322. [CrossRef]

50. Tabata, Y.; Ikada, Y. Effect of the size and surface charge of polymer microspheres on their phagocytosis by macrophage. *Biomaterials* **1988**, *9*, 356–362. [CrossRef]

51. Ayhan, H.; Tuncel, A.; Bor, N.; Pişkin, E. Phagocytosis of monosize polystyrene-based microspheres having different size and surface properties. *J. Biomater. Sci. Polym. Ed.* **1996**, *7*, 329–342. [CrossRef]

52. Fisher, D.T.; Appenheimer, M.M.; Evans, S.S. The two faces of IL-6 in the tumor microenvironmnt. *Semin. Immunol.* **2014**, *26*, 38–47, Academic Press. [CrossRef] [PubMed]

53. Zheng, M.; Davidson, F.; Huang, X. Ethylene glycol monolayer protected nanoparticles for eliminating nonspecific binding with biological molecules. *J. Am. Chem. Soc.* **2003**, *125*, 7790–7791. [CrossRef]

54. Xing, R.; Li, S.; Zhang, N.; Shen, G.; Möhwald, H.; Yan, X. Self-Assembled Injectable Peptide Hydrogels Capable of Triggering Antitumor Immune Response. *Biomacromolecules* **2017**, *13*, 3514–3523. [CrossRef] [PubMed]

55. Sirc, J.; Hampejsova, Z.; Trnovska, J.; Kozlik, P.; Hrib, J.; Hobzova, R.; Zajicova, A.; Holan, V.; Bosakova, Z. Cyclosporine A Loaded Electrospun Poly(D,L-Lactic Acid)/Poly(Ethylene Glycol) Nanofibers: Drug Carriers Utilizable in Local Immunosuppression. *Pharm. Res.* **2017**, *34*, 1391–1401. [CrossRef] [PubMed]

56. Meng, Q.; Kou, Y.; Ma, X.; Liang, Y.; Guo, L.; Ni, C.; Liu, K. Tunable self-assembled peptide amphiphile nanostructures. *Langmuir* **2012**, *28*, 5017–5022. [CrossRef]

57. Gratton, S.E.; Ropp, P.A.; Pohlhaus, P.D.; Luft, J.C.; Madden, V.J.; Napier, M.E.; DeSimone, J.M. The effect of particle design on cellular internalization pathways. *Proc. Natl. Acad. Sci. USA* **2008**, *105*, 11613–11618. [CrossRef]

International Journal of
Molecular Sciences

Article

Mutational and Combinatorial Control of Self-Assembling and Disassembling of Human Proteasome α Subunits

Taichiro Sekiguchi [1,2,3,4,†], Tadashi Satoh [4,†], Eiji Kurimoto [5], Chihong Song [6], Toshiya Kozai [7], Hiroki Watanabe [3], Kentaro Ishii [3,4], Hirokazu Yagi [4], Saeko Yanaka [1,2,3,4], Susumu Uchiyama [3,8], Takayuki Uchihashi [3,7], Kazuyoshi Murata [6,9] and Koichi Kato [1,2,3,4,*]

1   School of Physical Science, SOKENDAI (The Graduate University for Advanced Studies), Okazaki, Aichi 444-8787, Japan; sekiguchi@ims.ac.jp (T.S.); saeko-yanaka@ims.ac.jp (S.Y.)
2   Institute for Molecular Science, National Institutes of Natural Sciences, 5-1 Higashiyama, Myodaiji, Okazaki, Aichi 444-8787, Japan
3   Exploratory Research Center on Life and Living Systems (ExCELLS), National Institutes of Natural Sciences, 5-1 Higashiyama, Myodaiji, Okazaki, Aichi 444-8787, Japan; hwatanabe@d.phys.nagoya-u.ac.jp (H.W.); ishii@ims.ac.jp (K.I.); suchi@bio.eng.osaka-u.ac.jp (S.U.); uchihast@d.phys.nagoya-u.ac.jp (T.U.)
4   Graduate School of Pharmaceutical Sciences, Nagoya City University, 3-1 Tanabe-dori, Mizuho-ku, Nagoya 467-8603, Japan; tadashisatoh@phar.nagoya-cu.ac.jp (T.S.); hyagi@phar.nagoya-cu.ac.jp (H.Y.)
5   Faculty of Pharmacy, Meijo University, Tempaku-ku, Nagoya 468-8503, Japan; kurimoto@meijo-u.ac.jp
6   National Institute for Physiological Sciences, National Institutes of Natural Sciences, 5-1 Higashiyama, Myodaiji, Okazaki, Aichi 444-8787, Japan; chsong@nips.ac.jp (C.S.); kazum@nips.ac.jp (K.M.)
7   Department of Physics, Nagoya University, Furo-cho, Chikusa-ku, Nagoya 464-8602, Japan; toshiya.kozai@unibas.ch
8   Department of Biotechnology, Graduate School of Engineering, Osaka University, 2-1 Yamadaoka, Suita, Osaka 565-0871, Japan
9   School of Life Science, SOKENDAI (The Graduate University for Advanced Studies), Okazaki, Aichi 444-8787, Japan
*   Correspondence: kkatonmr@ims.ac.jp; Tel.: +81-564-59-5225; Fax: +81-564-59-5224
†   These authors contributed equally to this work.

Received: 27 March 2019; Accepted: 7 May 2019; Published: 9 May 2019

Abstract: Eukaryotic proteasomes harbor heteroheptameric α-rings, each composed of seven different but homologous subunits α1–α7, which are correctly assembled via interactions with assembly chaperones. The human proteasome α7 subunit is reportedly spontaneously assembled into a homotetradecameric double ring, which can be disassembled into single rings via interaction with monomeric α6. We comprehensively characterized the oligomeric state of human proteasome α subunits and demonstrated that only the α7 subunit exhibits this unique, self-assembling property and that not only α6 but also α4 can disrupt the α7 double ring. We also demonstrated that mutationally monomerized α7 subunits can interact with the intrinsically monomeric α4 and α6 subunits, thereby forming heterotetradecameric complexes with a double-ring structure. The results of this study provide additional insights into the mechanisms underlying the assembly and disassembly of proteasomal subunits, thereby offering clues for the design and creation of circularly assembled hetero-oligomers based on homo-oligomeric structural frameworks.

Keywords: proteasome; self-assembly; homo-oligomer; hetero-oligomer; size exclusion chromatography; native mass spectrometry; crystal structure; atomic force microscopy; electron microscopy

---

1. Introduction

Proteins in living systems are often assembled into filamentous and circular oligomers, which exert appropriate biological functions and are deposited as malfunctional aggregates, such as pathological amyloids. Circular assemblages composed of identical protomers give rise to functional barrels or cages, as exemplified by chaperonins and AAA ATPases [1–3]. The design and creation of circular oligomers have been important and challenging issues in protein engineering [4,5]. During evolutionary processes, the building blocks of these homo-oligomers acquire diversity in sequence and structure, maintaining their assembling properties. The proteasome system is one of the best examples demonstrating this concept [6–8].

The proteasome is a huge protein complex harboring a proteolytic chamber termed the 20S core particle. This 20S core particle comprises the α- and β-rings, two types of heptameric rings arranged as a cylindrical, four-layered αββα structure [6–13]. In the archaea, the heptameric α-ring is composed of seven identical α subunits and the heptameric β-ring is composed of one or two kinds of β subunits. A total of 28 subunits are spontaneously assembled into the 20S core particle [14]. In contrast, in eukaryotes, both the α- and the β-rings are heteroheptamers composed of seven different but homologous subunits, i.e., α1–α7 and β1–β7. Assembly of these subunits is not an autonomous process but is assisted by several chaperones operating as molecular matchmakers and checkpoints [8,11,13,15–21].

Among all the human proteasome α subunits, α7 exhibits a unique feature of in vitro self-assembly into a homotetradecameric double-ring structure [22–25], thereby raising the question whether the α7 homotetradecamer is an off-pathway dead-end product of the proteasome formation process. Another possibility is that certain mechanisms exist for disassembling the homo-oligomer of the α7 subunit, thereby resulting in its monomeric form, which is a component of the heteroheptameric α-ring.

We previously determined a crystal structure of the human α7 homotetradecamer and demonstrated that α6, which exists as a monomer, interacts with this assemblage, thereby disrupting its double-ring structure [25,26]. These findings suggest that different α subunits have different assembly properties. We comprehensively characterized the oligomeric states of the α subunits of the human proteasome and examined their possible interplay for obtaining a better understanding of the design principles underlying the self-assembly and disassembly of the proteasomal subunits, and more generally, those behind formation of circularly assembled hetero-oligomers composed of structurally homologous subunits.

2. Results and Discussion

2.1. Oligomeric States of Human Proteasomal α Subunits

To characterize the oligomeric states of the α subunits in solution, we performed size-exclusion chromatography (SEC) and native mass spectrometry (MS). SEC data revealed that the major fractions of α1, α2, α3, α4, α5, and α6 corresponded to monomeric-to-dimeric forms and that α7 formed a significantly larger oligomer, as shown in Figure 1a. These findings were confirmed by native MS, which indicated that α7 existed as a homotetradecamer and that the remaining subunits exhibited one major and one minor ion series, corresponding to the molecular masses of the monomer and dimer, respectively, as shown in Figure 1b. Based on these data and our previously reported data regarding sedimentation velocity analytical ultracentrifugation [25], we concluded that among the seven types of α subunits, only α7 is self-assembled into the homotetradecameric double-ring structure and that the remaining α subunits are under equilibrium between the monomeric and dimeric forms.

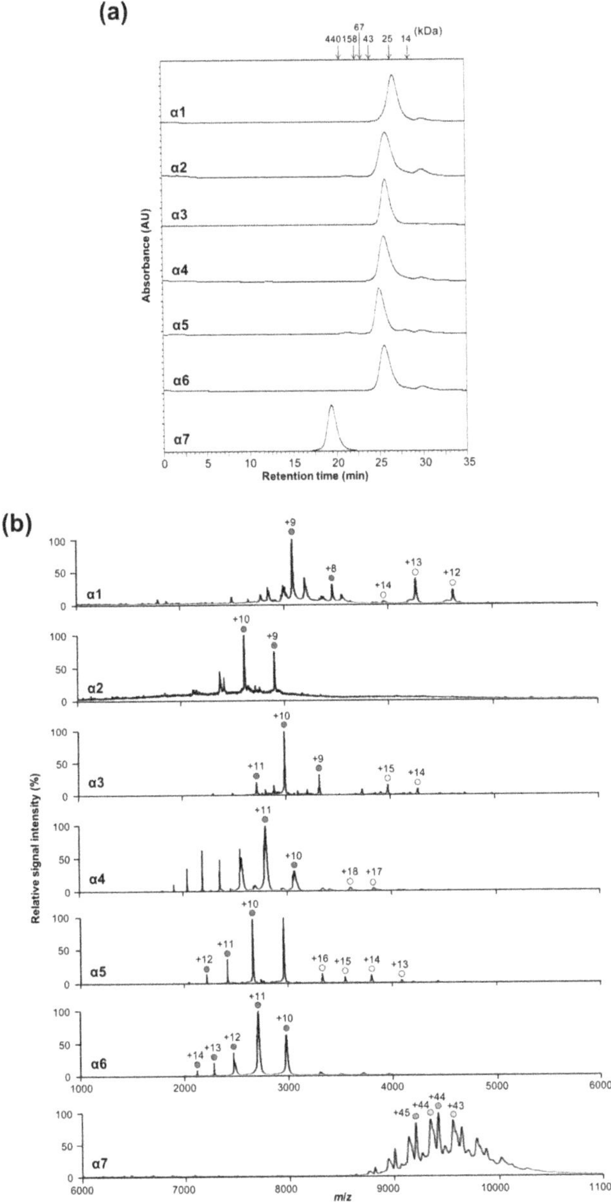

Figure 1. Characterization of the oligomeric states of proteasome α subunits. (**a**) Size-exclusion chromatogram of α1–α7 subunits. Arrows indicate eluted positions of the size markers. (**b**) Mass spectra of α1–α7 subunits under nondenaturing conditions. Blue and yellow circles indicate the ion series of α1–α6 monomers and dimers, respectively. Green and orange circles indicate the ion series of the α7 homotetradecamer. The estimated molecular masses of the α1–α7 subunits are as follows: 27,688.97 ± 9.91 and 55,398.59 ± 3.26 (α1 dimer and monomer, respectively); 26,034.50 ± 4.01 (α2); 59,650.89 ± 6.81 and 29,818.58 ± 0.72 (α3); 61,304.84 ± 9.52 and 30,669.77 ± 0.02 (α4); 53,162.75 ± 1.93 and 26,559.17 ± 1.39 (α5); 29,696.61 ± 9.90 (α6); and 414,406.97 ± 51.59 and 411,316.63 ± 48.20 (α7).

2.2. Mutational Disassembling of the α7 Homotetradecamer

To gain insight into the mechanisms of the assembly of α7, we inspected the crystal structure of its homotetradecamer [25]; our inspection highlighted five regions (regions 1–5) mediating intersubunit interactions, as shown in Figure 2. Regions 1, 2, and 3 are involved in intra-ring (*cis*) interactions, which are shared with the homoheptameric α-ring of archaeal 20S core particle [9,27], whereas regions 4 and 5 are involved in inter-ring (*trans*) interactions. Comparison of amino acid sequences across the human α subunits in terms of these regions revealed that region 1 involves highly conserved contacting pairs of residues (P16–Y25, F14–A29, and F14–P130), whereas most other pairs are unique for α7, which explains its specific self-assembling property, as shown in Figure S1 and Table S1. Intriguingly, mutational monomerization of an archaeal α subunit (*Thermoplasma acidophilum*) was achieved by truncating region 1 (residues 2–34), which was accompanied by alanine substitutions of arginine residues (Arg57, Arg86, and Arg130) in regions 2 and 3 [14]. Therefore, we tested whether the conserved residues in region 1 contribute to the formation of the human α7 homo-oligomer using an α7 mutant (α7*) in which the N-terminal segment in region 1 (residues 1–22, MSYDRAITVFSPDGHLFQVEYA) was replaced with a hexahistidine-containing segment (MGSSHHHHHHSSGLVPRGSHMGS). SEC and native MS indicated that α7* was monomeric in solution, as shown in Figure 3, thereby demonstrating that deletion of the N-terminal segment disrupts the homotetradecameric structure. Close inspection of region 1 in the crystal structure of the α7 homotetradecamer highlighted the hydrophobic ball-and-socket joint composed of Phe14 in one subunit fitting into a socket formed by Ala29 and Pro130 in the neighboring subunit, as shown in Figure 2c. A single-mutation F14A disassembled the α7 tetradecamer into the monomeric form, as shown in Figure 3, indicating that the N-terminal segment, particularly Phe14, is critically involved in intra-ring *cis* interaction as a prerequisite for inter-ring *trans* interaction. Because Phe14 as well as Ala29 and Pro130 in region 1 are all conserved across α1–α7, the remaining regions, i.e., regions 2–5, are likely to reinforce the α7-specific homophilic interaction, thereby stabilizing the tetradecameric complex.

The α7 tetradecamer involves inter-ring interactions mediated through regions 4 and 5. In contrast, despite the fact that like α7, *T. acidophilum* formed a double-ring structure in solution, the α subunit of the *Archaeoglobus fulgidus* proteasome exhibits a single-ring structure in the crystal [14,27]. We previously reported that the α7 single-ring structure can be stabilized on a mica surface and that the α7 tetradecameric double-ring structure is disassembled upon addition of the α6 subunit, thereby forming a 1:7 hetero-octameric α6/α7 complex [25,26]. These observations suggest that the inter-ring *trans* interaction is dispensable for stabilizing the heptameric ring structure. We examined this possibility via a mutational approach.

For this purpose, we attempted to introduce electrostatic repulsion at the double-ring interface by focusing on the three autologously contacting pairs Ser96–Ser96, Phe102–Phe102, and Tyr104–Tyr104. Ser96 was substituted with aspartate, whereas Phe102 and Tyr104 were both substituted with arginine, as shown in Figure 4a. In addition to these positions, because the Ser100 positions are spatially proximal to each other across the inter-ring interface, Ser100 was mutated into aspartate. In SEC, the great majority of this quadruple α7 mutant (α7^{SR}) eluted significantly later (at 21.1 min) than the wild-type α7 double ring (at 19.4 min), as shown in Figure 4b. Under nondenaturing conditions, the mass spectrum of this mutant exhibited a major ion series with molecular masses of 207,581 ± 78 and 209,044 ± 50 Da, corresponding to the heptameric α7 subunits (with a theoretical mass of 199,331 Da), as shown in Figure 4c. In addition, using atomic force microscopy (AFM), we confirmed the heptameric ring of the α7 mutant with a height of ~4 nm, which was half of that of the wild-type α7 double-ring (~9 nm) [26], as shown in Figure S2. All these data indicate that the double-ring tetradecamer of α7 is disassembled into a single heptameric ring by mutations bringing about electrostatic repulsion at the inter-ring interface, thereby demonstrating that the inter-ring interaction is dispensable for the formation of the homoheptameric ring of α7.

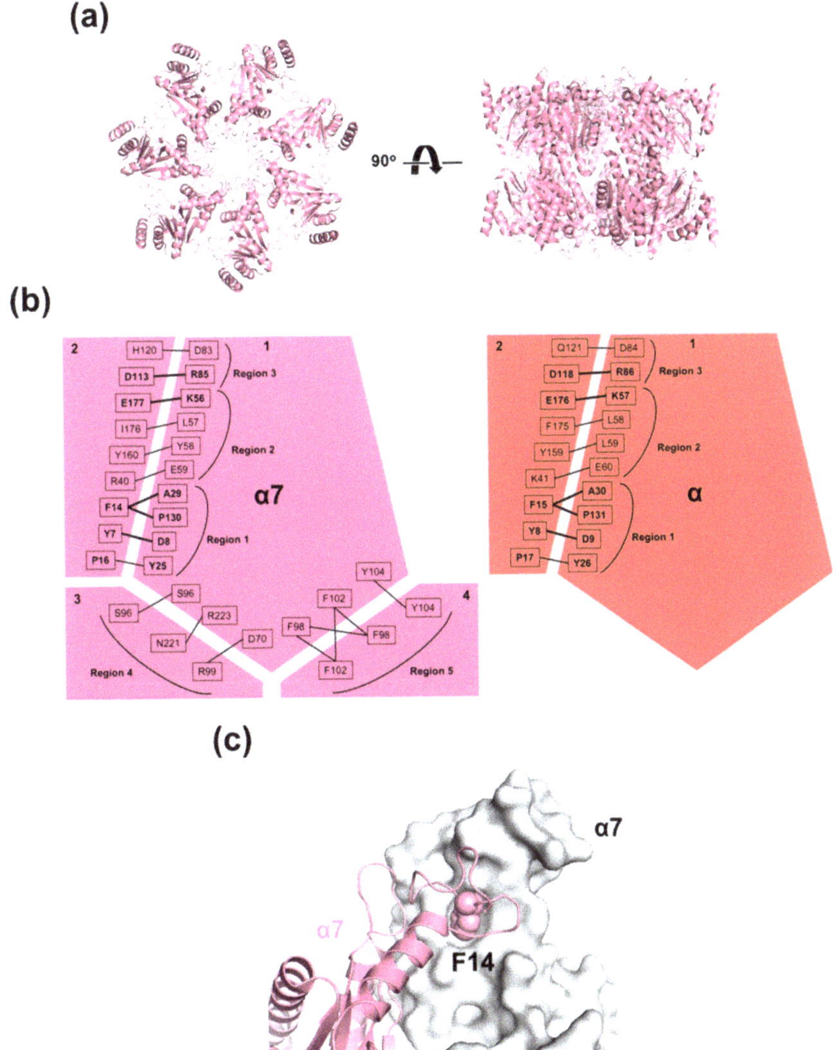

Figure 2. Schematic representation of the binding interfaces of the α7 homotetradecamer and the archaeal α homoheptamer. (**a**) Crystal structure of the α7 homotetradecamer (Protein Data Bank (PDB) code: 5DSV). The left and right structures are related by a rotation of 90° around the horizontal axis. (**b**) Schematic representations of intersubunit *cis* and *trans* interactions of the human α7 homotetradecamer (adjacent four molecules) together with *cis* interactions of the *Archaeoglobus fulgidus* proteasome α homoheptamer (adjacent two molecules, PDB code: 1J2P). Interaction pairs conserved between human α7 and *A. fulgidus* proteasome α subunits are highlighted by bold lines. (**c**) Surface and ribbon models of intersubunit *cis* interaction highlighting F14 residue in α7 homotetradecamer.

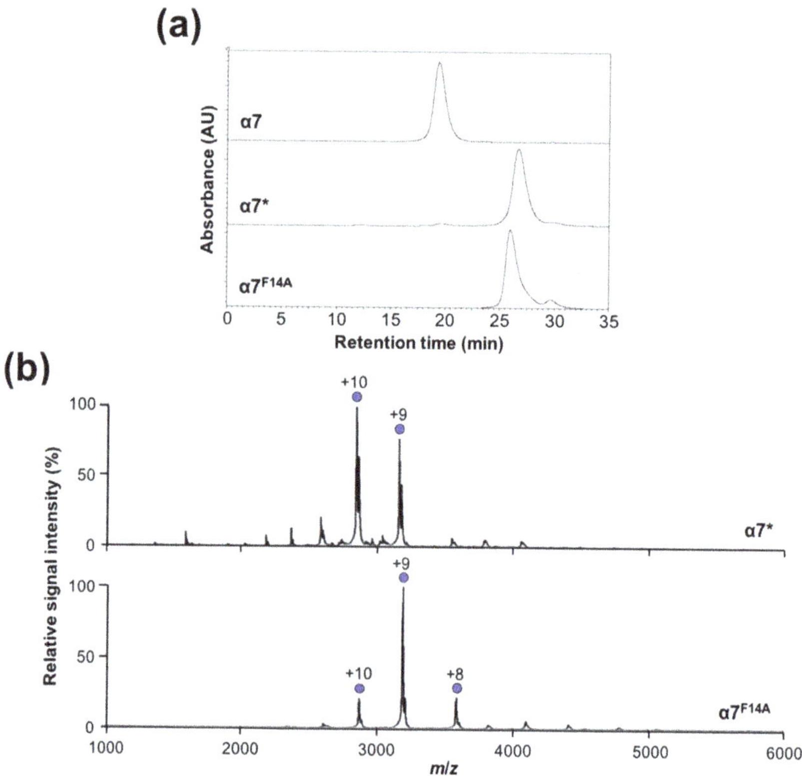

Figure 3. Characterization of the oligomeric states of α7 monomeric mutants. (**a**) Size-exclusion chromatogram of α7* and α7F14A together with wild-type α7. (**b**) Mass spectra of α7* and α7F14A under nondenaturing conditions. Blue circles indicate the ion series of the α7* or α7F14A monomer. The mass spectra of α7* and α7F14A mutants under nondenaturing conditions exhibited the major ion series with molecular masses of the monomer 28,456 ± 0 (with a theoretical mass: 28,594 Da) and 28,691 ± 0.05 (28,638 Da), respectively.

2.3. Disassembly of the α7 Double-Ring via Subunit Interactions

In addition to being disassembled by mutations, the α7 double-ring structure can be disassembled into its single-ring structure via interaction with the monomeric α6 subunit, thereby forming a 1:7 hetero-octameric α6/α7 complex [25,26]. We examined whether the other α subunits have such a disassembling capability by performing a comprehensive analysis of the oligomeric states of α7 in the presence of the α1–α6 subunits along with the monomerized α7 mutant α7*. The SEC data indicated that although neither α1, α2, α3, α5, nor α7* affected the tetradecameric structure of α7, approximately 20% and 50% of the α7 tetradecamer was disrupted into smaller complexes by α4 or α6, respectively, judging from the peak intensity reduction, as shown in Figure 5a. The almost identical elution times (19.7 min) between the resultant complexes generated in the α4/α7 and α6/α7 mixtures implied that like α6/α7, α4 and α7 also form a 1:7 hetero-octameric complex [25,26]. To confirm this, we performed native MS of the α4/α7 mixture; results indicated that like the α6/α7 complex [25], the α4/α7 mixture exhibits a major ion series corresponding to a 1:7 hetero-octameric α4/α7 complex with a molecular mass of 228,789 ± 9 Da (theoretical molecular mass: 228,984 Da) under nondenaturing conditions, as shown in Figure 5b. The native MS data also indicated that the higher-molecular–mass complexes observed for these mixtures under the present conditions corresponded to complexes composed of fourteen α7 subunits and one α4 or α6 subunit, indicating that α4 and α6 can bind to the α7 double ring.

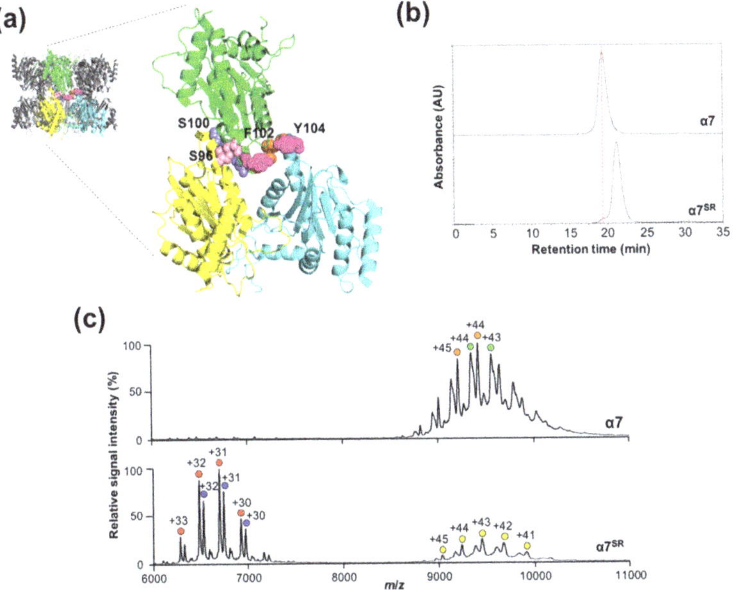

Figure 4. Generation of the single-ring α7 mutant. (**a**) Mutated positions of the single-ring mutant (α7^{SR}). The mutated resides shown in sphere models are labeled in the close-up view. (**b**) Size-exclusion chromatogram of α7^{SR}. Red dotted line indicates position of the SEC peak of α7 homotetradecamer. (**c**) Mass spectra of α7^{SR} under nondenaturing conditions. Green and orange circles indicate the ion series of the α7 homotetradecamer. Blue and red circles indicate the ion series of the homoheptameric complex of α7^{SR}. Yellow circles show those of the homotetradecameric complex of α7^{SR}.

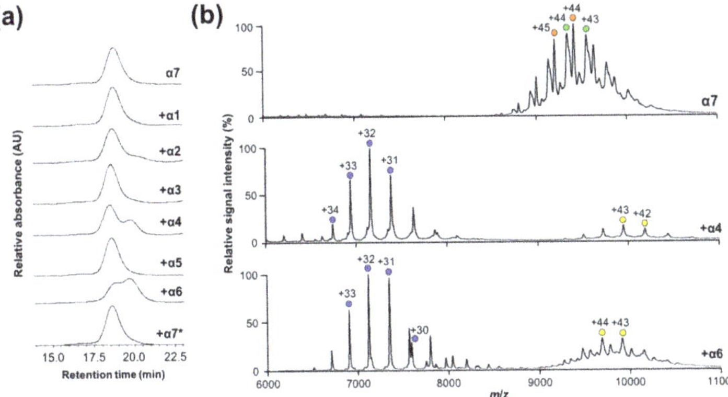

Figure 5. Evaluation of the α7-disassembling ability of the proteasome α subunits. (**a**) Size-exclusion chromatogram patterns of α7-containing fractions depending on the presence of the α1–α6 subunits and the α7* mutant (20 μM monomers). (**b**) Mass spectra of α4/α7 and α6/α7 mixtures under nondenaturing conditions. Green and orange circles indicate the ion series of the α7 homotetradecamer. Blue circles indicate the ion series of the 1:7 hetero-octameric complexes of α4/α7 and α6/α7, and yellow circles indicate those of the 1:14 heteropentadecameric complexes of α4/α7 and α6/α7.

Our previous AFM data indicated that the disassembly of the α7 double ring by α6 involves two steps—the α6 monomer initially cracks at the interface between two stacked α7 single rings and subsequently occupies the central pore of the α7 single ring [26]. Therefore, we examined possible interactions of the α7 single ring with α4 as well as α6. Using the α7^{SR} mutant, we performed SEC-based binding analysis with the α1–α6 subunits together with α7*, as shown in Figure 6a. Sodium dodecyl sulfate polyacrylamide gel electrophoresis (SDS-PAGE) analysis detected the α2, α4, and α6 but not the α1, α3, α5, and α7* in the fractions co-eluted with α7^{SR} at approximately 21 min. The mass spectra of the α7^{SR}/α2, α7^{SR}/α4, and α7^{SR}/α6 complexes under nondenaturing conditions showed a major ion series indicating their 1:7 hetero-octameric complexes with molecular masses of 225,164 $\pm$ 62, 229,435 $\pm$ 116, and 228,717 $\pm$ 124 Da (with theoretical masses of 225,230; 229,257; and 229,010 Da), respectively, as shown in Figure 6b. Regarding α2, MS peaks originating from the unbound α7 single ring were also observed, suggesting their weak interactions, as shown in Figure 6b. All these results indicate that α4 and α6, and to a lesser extent α2 but not the other α subunits, can interact with the disassembled α7 single ring in solution.

The cavitary surface of the homoheptameric ring of α7 exhibits unique charge distributions characterized by central negatively charged clusters surrounded by positively charged zones. The α2, α4, and α6 subunits commonly display positively and negatively charged clusters that are polarized near the vertex and at the base regions, respectively, of their triangle-shaped architecture; however, the negatively charged patch at the base region of α2 is smaller than those of α4 and α6, as shown in Figure S3. These charge complementarities may explain the selective accommodation of these α subunits in the cavity of the α7 homoheptameric ring. The double-ring cracking abilities of α4 and α6 may also be ascribed to their remarkably polarized charged clusters, although the structural basis of their transient interactions with the α7 double ring remains elusive.

It is plausible that the hetero-octameric complexes composed of the α7 and α4 (or α6) subunits cannot be re-associated into double rings because α4 as well as α6 occupies the central pore of the α7 homoheptameric ring, thereby sterically blocking inter-ring interactions. However, the hetero-octameric complexes are likely to be able to form complexes with unoccupied α7 homoheptameric rings, consequently encapsulating α4 and α6 within the α7 double-ring cage. One intriguing possibility is that these cages were detected as 14:1 α7/α4 and 14:1 α7/α6 complexes in the native mass spectra, as shown in Figure 5b.

2.4. Creation of Heterotetradecameric Double-Ring Structures of the Proteasomal Subunits

The self-assembling property of α7 suggests that proteasome formation involves some scrap-and-build mechanisms via which the α7 homotetradecamer is disassembled and the monomeric α7 subunit is integrated into the heteroheptameric α ring. To obtain insights into putative hetero-assembling processes involving α7, we used SEC to characterize possible interactions of the mutationally monomerized α7* with the other subunits, particularly α1 and α6, which flank α7 in the native heteroheptameric α-ring. Of note, the results indicated that α1, α4, and α6, but not the other α subunits, are co-assembled with α7*, giving rise to high–molecular-mass complexes of sizes comparable with that of the α7 homotetradecamer, as shown in Figure 7. The stoichiometry of the α7*/α1, α7*/α4, and α7*/α6 hetero-oligomeric complexes was estimated as 1:3.6, 1.2:1, and 2.5:1, respectively, based on the Coomassie Brilliant Blue (CBB)-staining SDS-PAGE results, as shown in Figure 7.

AFM data of the high-molecular-mass complex fraction of α7* with α4 or α6 identified tetradecameric particles of double-ring shape along with particles of single heptameric rings and oligomeric structures without ring shape, which presumably resulted from the disruption of the double-ring oligomers on the mica surface, as shown in Figure 8a. As for the α1/α7* complex, tetradecameric double rings and heptameric single rings were barely detected in the AFM observation possibly owing to their unstable structures on the mica surface. Instead, oligomeric structures were observed with a height of ~10 nm, which is comparable with those of the α7, α4/α7*, and α6/α7* tetradecamers (~9 nm). To obtain a higher-resolution structure, the α4/α7* hetero-oligomeric complex was subjected to negative-staining electron microscopy (EM), which also indicated that the complex had a double-ring tetradecameric structure, as shown in Figure 8b.

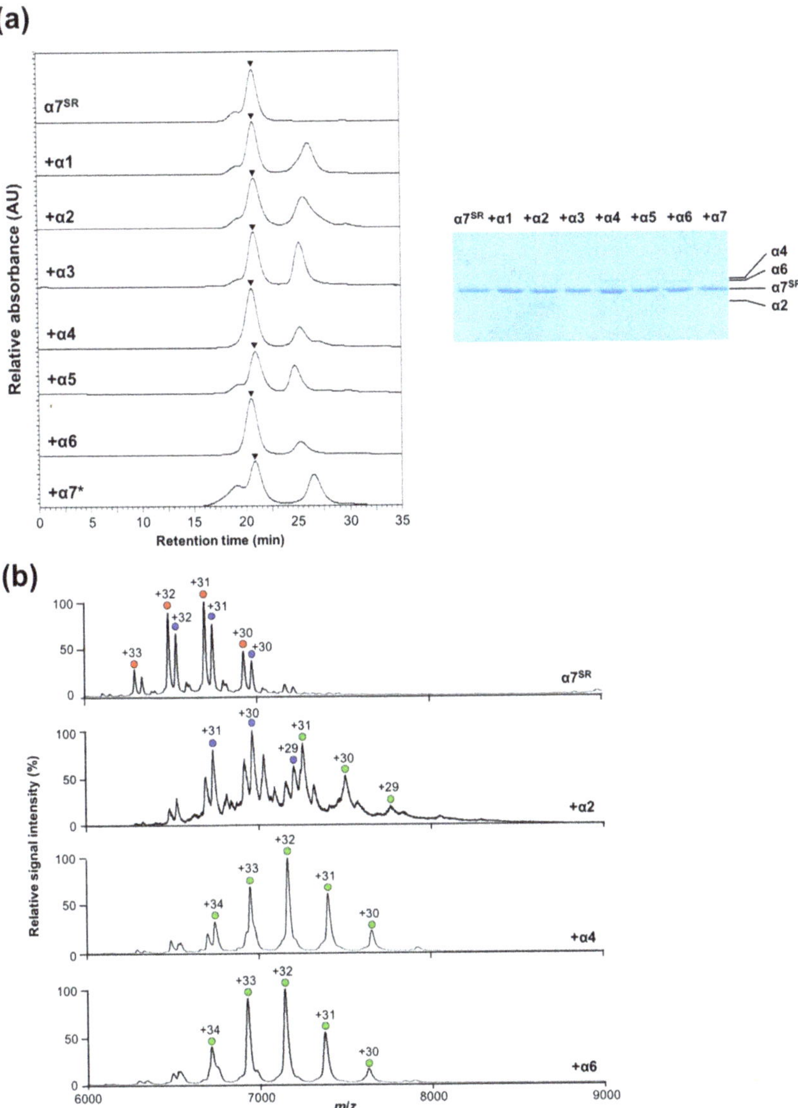

Figure 6. Examination of possible interactions of the α7 single ring and the proteasome α subunits. (**a**) Size-exclusion chromatogram of α7^{SR} in the presence of the α1–α6 subunits and the α7* mutant (20 μM monomers): SDS-PAGE of the size-exclusion chromatography (SEC) peaks (at approximately 21 min) originating from the α7^{SR}/α1–α6 and α7* complexes. The peak position is indicated by an inverted triangle. (**b**) Mass spectra of the α7^{SR}/α2, α7^{SR}/α4, and α7^{SR}/α6 mixtures under nondenaturing conditions. Blue and red circles indicate major ion series of the α7^{SR}. Green circles show the ion series of the 7:1 hetero-octameric complexes of α7^{SR}/α2, α7^{SR}/α4, and α7^{SR}/α6 complexes.

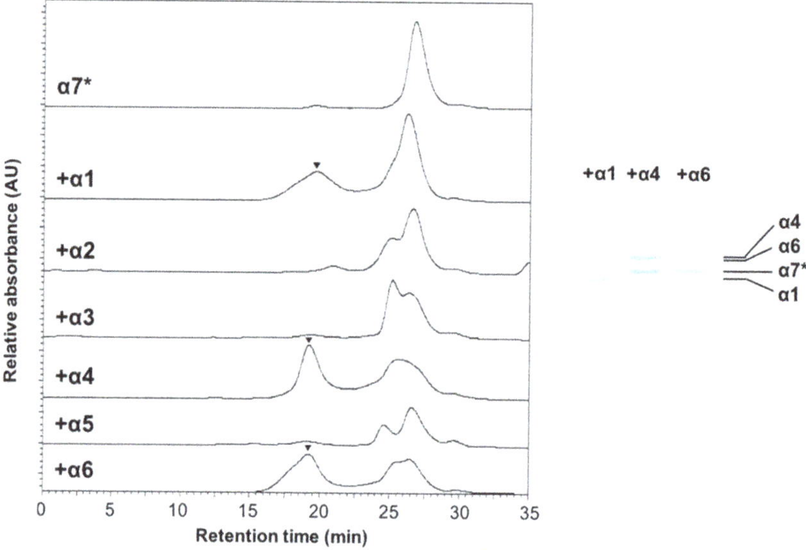

Figure 7. Exploration of possible formation of the α7 hetero-oligomeric complex mediated by proteasome α subunits. Size-exclusion chromatogram of α7* in the presence of α1–α6 subunits. The α1–α6 subunits (20 µM monomers) were mixed with an equimolar amount of α7* at 20 °C for 1 h, and the mixtures were subsequently analyzed by SEC. The peak position is indicated by an inverted triangle. SDS-PAGE of the purified α7*/α1, α7*/α4, and α7*/α6 complexes.

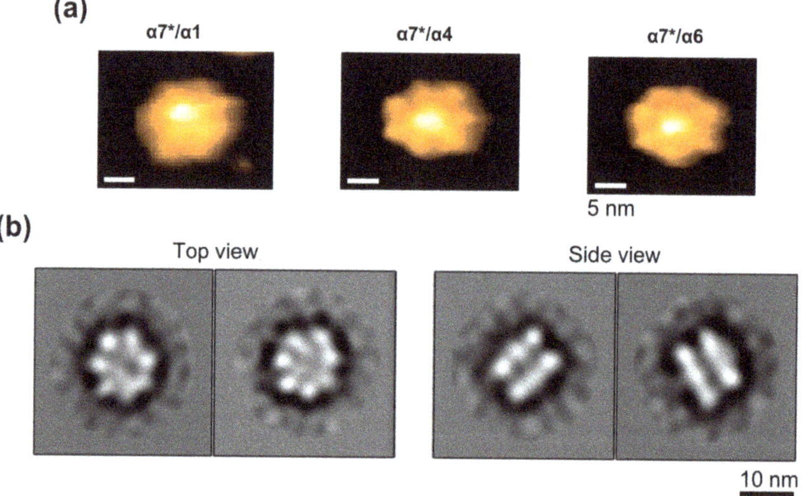

Figure 8. Structural characterization of the α7*/α4 and α7*/α6 hetero-oligomeric complexes. (**a**) AFM images of two typical orientations of the α7*/α1 (left), α7*/α4 (center), and α7*/α6 (right) complexes on bare mica. Scale bar: 5 nm. (**b**) Two-dimensional averaged image of an α7*/α4 particle subjected to single-particle negative-staining electron microscopy.

It is puzzling that mixture of the artificially monomerized subunit (α7*) and the intrinsically monomeric subunit (α4 or α6) yields the heterotetradecameric double-ring architecture. One might assume that, for example, the α6–α7* interaction is stronger than the α6–α6 and α7*–α7* interactions. However, such more favorable residue pair(s) could not be found by simply comparing the amino acid residues at the intersubunit interfaces, i.e., at regions 2–5, based on the crystal structure of the α7 homotetradecamer, as shown in Figure S1 and Table S1. Rather, the crystallographic data highlight a small but significant difference between the α7–α7 contacting mode [25] and the α6–α7 interaction in the crystal structure of the human 20S proteasome, as shown in Figure S4 [28]. Consequently, the overall quaternary structure of the α7 homoheptameric ring is markedly different from that of the native heteroheptameric α ring. This implies that accumulation of the slightly different contact mode at the subunit interface results in geometric frustration of the formation of the circular quaternary structure, thereby causing deformation or disruption of the ring structure. Based on the data obtained in this study, we suggest that not only local structural complementarity at the subunit interfaces but also geometric consistency and/or structural adjustability in terms of the formation of the circular structure are factors that determine the heptamerization of the proteasome α subunits. The present study demonstrates that geometric frustration can be compromised by combining structurally homologous protomers with potential but imperfect ability to self-assemble, thereby providing insights for controlling the assembly and disassembly of the proteasomal subunits. Furthermore, our findings would provide clues for the design and creation of circular hetero-oligomers based on the homo-oligomeric structural frameworks.

3. Materials and Methods

3.1. Preparation of Wild-Type and Mutated Proteasome α Subunits

Human proteasome α6 short isoform [*PSMA1* (P25786); residues 1–263] and α7- [*PSMA3* (P25788); residues 1–255] subunits were expressed and purified as described previously [23,24,26]. Genes encoding proteasome α1 [*PSMA6* (P60900); residues 1–246] and α4 [*PSMA7* (P60900); residues 1–248] were subcloned into the *Nde*I and *Sal*I sites of pET28b (Merck Millipore, Burlington, MA, USA), whereas the α2 gene [*PSMA2* (P25787); residues 1–234] was subcloned into the *Nde*I and *Xho*I sites of pRSFDuet-1 vector (Merck Millipore). In addition, genes encoding α3 [*PSMA4* (P25789); residues 1–261] and α5 [*PSMA5* (P28066); residues 1–241] were subcloned into the *Bam*HI and *Xho*I or *Sal*I sites of modified pCold-I and pCold-GST vectors (TaKaRa Bio Inc., Kusatsu, Japan), respectively, which contain the TEV protease cleavage site preceding the target genes. Monomeric mutants of α7 (designated as α7* and α7^{F14A}) were created via truncation of 22 N-terminal residues or introduction of the F14A mutation, respectively. The mutated α7 genes were subcloned into the *Nde*I and *Xho*I sites of the pET28b vector. In contrast, the single-ring mutant α7^{SR} was generated by introducing the S96D, S100D, F102R, and Y104R mutations using the wild-type construct in pRSFDuet-1. All expression plasmids were introduced into *Escherichia coli* BL21-CodonPlus (DE3)-RIL (Agilent Technologies, Santa Clara, CA, USA).

For producing recombinant proteins, the *E. coli* cells harboring the expression plasmids were grown in Luria–Bertani medium containing 15 µg/mL kanamycin or 50 µg/mL ampicillin. The α7^{SR} mutant was purified as employed for the wild-type α7. Briefly, except for α2, the recombinant proteins were purified from the soluble fraction obtained by sonication and centrifugation. The resultant cell lysates were subjected to affinity chromatography [Ni$^+$-charged Chelating Sepharose or Glutathione Sepharose 4B (GE Healthcare, Chicago, IL, USA)], and further purified using anion-exchange (RESOURCE Q, GE Healthcare) and size-exclusion (HiLoad 26/60 Superdex 75 or 200 pg; GE Healthcare) columns. The α2 was purified from the inclusion bodies and refolded according to standard dilution methods using a buffer containing 20 mM Tris-HCl (pH 8.0), 400 mM L-arginine, 250 mM NaSCN, 1 mM oxidized glutathione, and 5 mM reduced glutathione. The refolded protein was further purified using a HiLoad 26/60 Superdex 75pg column (GE Healthcare).

3.2. Determination of Molecular Mass

The molecular masses of the human wild-type proteasome α subunits ($\alpha 1$–$\alpha 7$) and the mutated $\alpha 7$ proteins ($\alpha 7^*$, $\alpha 7^{F14A}$, and $\alpha 7^{SR}$) were estimated using SEC and native MS. In SEC, the samples (0.3–10 μM) were loaded onto a Superose 6 increase 10/300 GL column (GE Healthcare) with 20 mM Tris-HCl (pH 8.0) and 150 mM NaCl at a flow rate of 0.75 mL/min. For calibrating the column, ribonuclease A (13.7 kDa), chymotrypsinogen A (25 kDa), ovalbumin (43 kDa), bovine serum albumin (67 kDa), conalbumin (75 kDa), aldolase (158 kDa), ferritin (440 kDa) (GE Healthcare), and blue dextran 2000 were used. The elution profiles were recorded as absorbance values at 280 nm.

In native MS, buffer exchange of the purified $\alpha 1$–$\alpha 7$ proteins (30–50 μM monomers) was performed using 150 mM ammonium acetate (pH 6.8–8.0) with a Bio-Spin 6 column (Bio-Rad, Hercules, CA, USA). The buffer-exchanged samples (5–20 μM monomers) were immediately subjected to nanoflow electrospray ionization MS analysis with gold-coated glass capillaries made in house. Approximately 2–5 μL samples were loaded for each measurement. Buffer-exchanged $\alpha 4$ and $\alpha 6$ (4 μM monomers) were mixed with $\alpha 7$ (2 μM tetradecamer) at 20 °C for 1 h and subsequently analyzed using native MS. In contrast, $\alpha 4$ and $\alpha 6$ (4 μM monomers) were mixed with $\alpha 7^{SR}$ mutant (4 μM heptamer) and incubated as employed for wild-type $\alpha 7$. As for $\alpha 2$, the sample was incubated with $\alpha 7^{SR}$ beforehand, and the buffer exchange was carried out for the mixture, and then subjected to native MS measurements. Spectra were acquired on a SYNAPT G2-S*i* HDMS mass spectrometer (Waters, Manchester, UK) in the positive ionization mode, as previously described [25]. Spectrum calibration was performed using 1 mg/mL of cesium iodide and analysis was performed using the Mass Lynx software (Waters).

3.3. AFM

For AFM sample preparation, the $\alpha 1$, $\alpha 4$, or $\alpha 6$ (20 μM monomers) was mixed with an equal molar amount of $\alpha 7^*$ at 20 °C for 1 h, and subsequently fractionated by SEC. The high-molecular-mass complex fractions were subjected to the AFM analysis. AFM was performed using a laboratory-constructed apparatus with cantilevers (7 μm long, 2 μm wide, and 90 nm thick) at room temperature [29]. Typical values of the spring constant, resonant frequency, and quality factor of the cantilever in an aqueous solution are approximately 0.2 N/m, 800 kHz, and 2, respectively. In AFM imaging, the free and set-point oscillation amplitudes were set to approximately 1.5 nm and 90% of the former, respectively. All samples were applied to either bare mica in 20 mM Tris-HCl (pH 8.0) with 150 mM NaCl, as previously described [26].

3.4. EM

The protein samples for EM measurement were prepared using the same protocol as in the AFM analysis. Negative-staining EM was performed using a conventional protocol, as previously described [30]. EM imaging of the $\alpha 4/\alpha 7^*$ hetero-oligomeric complex was performed at room temperature using a JEOL JEM 2200FS electron microscope (JEOL Ltd., Tokyo, Japan) equipped with a field emission gun operating at an acceleration voltage of 200 kV. A total of 40 images were obtained using a DE20 direct detection camera (Direct Electron LP, San Diego, CA, USA) at a detector magnification of 50,000 with an energy slit width of 20 eV using the low-dose mode. The image size was set to 1.09 Å per pixel on the camera. After subjection of motion collection with the DE_process_frames.py script, the obtained images were processed with Relion 2.0 software [31]. Subsequently, 1,534-particle images were extracted from the 40 images and subjected to two-dimensional classification after sorting with cross-correlation coefficients.

Supplementary Materials: Supplementary Materials can be found at http://www.mdpi.com/1422-0067/20/9/2308/s1.

Author Contributions: T.S. (Taichiro Sekiguchi), T.S. (Tadashi Satoh), E.K., and K.K. conceived and designed the study; T.S. (Taichiro Sekiguchi), T.S. (Tadashi Satoh), E.K., H.Y., and S.Y. performed protein designing and sample preparation; T.S. (Taichiro Sekiguchi), K.I., and S.U. performed native MS; T.K., H.W., and T.U. performed AFM; C.S. and K.M. performed EM; T.S. (Tadashi Satoh) and K.K. mainly drafted the manuscript.

Funding: This work was supported in part by the Grants-in-Aid for Scientific Research (Grant Numbers JP16H06280 to T.S., JP26460051 to E.K., JP17H05890 to H.Y., JP18H04512, JP18H01837 to T.U., and JP25102008, JP15H02491 to K.K.), by the Grants-in-Aid for Scientific Research on Innovative Areas-Platforms for Advanced Technologies and Research Resources, "Advanced Bioimaging Support" (JP16H06280), from the Ministry of Education, Culture, Sports, Science and Technology, Japan. This work was also supported by the Joint Studies Program in the Okazaki BIO-NEXT project of the Okazaki Institute for Integrative Bioscience (No. 303), by the Joint Research by Exploratory Research Center on Life and Living Systems (ExCELLS) (ExCELLS program No. 18-101 to T.U., and No. 18-402 to H.Y.), by Functional Genomics Facility, NIBB Core Research Facilities, and by SOKENDAI (The Graduate University for Advanced Studies).

Acknowledgments: The authors would like to thank Kumiko Hattori and Kiyomi Senda for their help in the preparation of recombinant proteins, Hiroki Kawamura for his contribution during the early stage of this study, and Michiko Nakano (Institute for Molecular Science) for constructing three-dimensional protein models useful for insightful discussion.

Conflicts of Interest: The authors declare no conflict of interest.

Abbreviations

AFM	atomic force microscopy
CBB	Coomassie Brilliant Blue
EM	electron microscopy
MS	mass spectrometry
m/z	mass-to-charge ratio
PDB	Protein Data Bank
SDS-PAGE	sodium dodecyl sulfate polyacrylamide gel electrophoresis
SEC	size-exclusion chromatography
SR	single-ring

References

1. Garcia-Bellido, A. Symmetries throughout organic evolution. *Proc. Natl. Acad. Sci. USA* **1996**, *93*, 14229–14232. [CrossRef]
2. Goodsell, D.S.; Olson, A.J. Structural symmetry and protein function. *Annu. Rev. Biophys. Biomol. Struct.* **2000**, *29*, 105–153. [PubMed]
3. Snider, J.; Houry, W.A. AAA+ proteins: Diversity in function, similarity in structure. *Biochem. Soc. Trans.* **2008**, *36*, 72–77.
4. Lin, Y.R.; Koga, N.; Vorobiev, S.M.; Baker, D. Cyclic oligomer design with de novo αβ-proteins. *Protein Sci.* **2017**, *26*, 2187–2194. [CrossRef] [PubMed]
5. Strauch, E.M.; Bernard, S.M.; La, D.; Bohn, A.J.; Lee, P.S.; Anderson, C.E.; Nieusma, T.; Holstein, C.A.; Garcia, N.K.; Hooper, K.A.; et al. Computational design of trimeric influenza-neutralizing proteins targeting the hemagglutinin receptor binding site. *Nat. Biotechnol.* **2017**, *35*, 667–671. [CrossRef]
6. Tanaka, K. The proteasome: Overview of structure and functions. *Proc. Jpn. Acad. Ser. B Phys. Biol. Sci.* **2009**, *85*, 12–36. [CrossRef] [PubMed]
7. Collins, G.A.; Goldberg, A.L. The Logic of the 26S Proteasome. *Cell* **2017**, *169*, 792–806. [CrossRef]
8. Kato, K.; Satoh, T. Structural insights on the dynamics of proteasome formation. *Biophys. Rev.* **2018**, *10*, 597–604. [CrossRef] [PubMed]
9. Löwe, J.; Stock, D.; Jap, B.; Zwickl, P.; Baumeister, W.; Huber, R. Crystal structure of the 20S proteasome from the archaeon *T. acidophilum* at 3.4 Å resolution. *Science* **1995**, *268*, 533–539.
10. Unno, M.; Mizushima, T.; Morimoto, Y.; Tomisugi, Y.; Tanaka, K.; Yasuoka, N.; Tsukihara, T. The structure of the mammalian 20S proteasome at 2.75 Å resolution. *Structure* **2002**, *10*, 609–618. [CrossRef]
11. Kish-Trier, E.; Hill, C.P. Structural biology of the proteasome. *Annu. Rev. Biophys.* **2013**, *42*, 29–49. [CrossRef] [PubMed]
12. Finley, D.; Chen, X.; Walters, K.J. Gates, Channels, and Switches: Elements of the Proteasome Machine. *Trends Biochem. Sci.* **2016**, *41*, 77–93. [CrossRef] [PubMed]
13. Budenholzer, L.; Cheng, C.L.; Li, Y.; Hochstrasser, M. Proteasome Structure and Assembly. *J. Mol. Biol.* **2017**, *429*, 3500–3524. [CrossRef] [PubMed]

14. Sprangers, R.; Kay, L.E. Quantitative dynamics and binding studies of the 20S proteasome by NMR. *Nature* **2007**, *445*, 618–622. [CrossRef] [PubMed]

15. Yashiroda, H.; Mizushima, T.; Okamoto, K.; Kameyama, T.; Hayashi, H.; Kishimoto, T.; Niwa, S.; Kasahara, M.; Kurimoto, E.; Sakata, E.; et al. Crystal structure of a chaperone complex that contributes to the assembly of yeast 20S proteasomes. *Nat. Struct. Mol. Biol.* **2008**, *15*, 228–236. [CrossRef]

16. Murata, S.; Yashiroda, H.; Tanaka, K. Molecular mechanisms of proteasome assembly. *Nat. Rev. Mol. Cell Biol.* **2009**, *10*, 104–115. [CrossRef] [PubMed]

17. Takagi, K.; Saeki, Y.; Yashiroda, H.; Yagi, H.; Kaiho, A.; Murata, S.; Yamane, T.; Tanaka, K.; Mizushima, T.; Kato, K. Pba3-Pba4 heterodimer acts as a molecular matchmaker in proteasome α-ring formation. *Biochem. Biophys. Res. Commun.* **2014**, *450*, 1110–1114. [CrossRef] [PubMed]

18. Kock, M.; Nunes, M.M.; Hemann, M.; Kube, S.; Dohmen, R.J.; Herzog, F.; Ramos, P.C.; Wendler, P. Proteasome assembly from 15S precursors involves major conformational changes and recycling of the Pba1-Pba2 chaperone. *Nat. Commun.* **2015**, *6*, 6123. [CrossRef]

19. Kurimoto, E.; Satoh, T.; Ito, Y.; Ishihara, E.; Okamoto, K.; Yagi-Utsumi, M.; Tanaka, K.; Kato, K. Crystal structure of human proteasome assembly chaperone PAC4 involved in proteasome formation. *Protein Sci.* **2017**, *26*, 1080–1085. [CrossRef]

20. Wu, W.; Sahara, K.; Hirayama, S.; Zhao, X.; Watanabe, A.; Hamazaki, J.; Yashiroda, H.; Murata, S. PAC1-PAC2 proteasome assembly chaperone retains the core α4-α7 assembly intermediates in the cytoplasm. *Genes Cells* **2018**, *23*, 839–848. [CrossRef]

21. Satoh, T.; Yagi-Utsumi, M.; Okamoto, K.; Kurimoto, E.; Tanaka, K.; Kato, K. Molecular and Structural Basis of the Proteasome α Subunit Assembly Mechanism Mediated by the Proteasome-Assembling Chaperone PAC3-PAC4 Heterodimer. *Int. J. Mol. Sci.* **2019**, *20*, 2231. [CrossRef]

22. Gerards, W.L.; Enzlin, J.; Häner, M.; Hendriks, I.L.; Aebi, U.; Bloemendal, H.; Boelens, W. The human α-type proteasomal subunit HsC8 forms a double ringlike structure, but does not assemble into proteasome-like particles with the β-type subunits HsDelta or HsBPROS26. *J. Biol. Chem.* **1997**, *272*, 10080–10086. [CrossRef] [PubMed]

23. Sugiyama, M.; Hamada, K.; Kato, K.; Kurimoto, E.; Okamoto, K.; Morimoto, Y.; Ikeda, S.; Naito, S.; Furusaka, M.; Itoh, K.; et al. SANS simulation of aggregated protein in aqueous solution. *Nucl. Instrum. Methods Phys. Res. A* **2009**, *600*, 272–274. [CrossRef]

24. Sugiyama, M.; Kurimoto, E.; Yagi, H.; Mori, K.; Fukunaga, T.; Hirai, M.; Zaccai, G.; Kato, K. Kinetic asymmetry of subunit exchange of homooligomeric protein as revealed by deuteration-assisted small-angle neutron scattering. *Biophys. J.* **2011**, *101*, 2037–2042. [CrossRef] [PubMed]

25. Ishii, K.; Noda, M.; Yagi, H.; Thammaporn, R.; Seetaha, S.; Satoh, T.; Kato, K.; Uchiyama, S. Disassembly of the self-assembled, double-ring structure of proteasome α7 homo-tetradecamer by α6. *Sci. Rep.* **2015**, *5*, 18167. [CrossRef] [PubMed]

26. Kozai, T.; Sekiguchi, T.; Satoh, T.; Yagi, H.; Kato, K.; Uchihashi, T. Two-step process for disassembly mechanism of proteasome α7 homo-tetradecamer by α6 revealed by high-speed atomic force microscopy. *Sci. Rep.* **2017**, *7*, 15373. [CrossRef]

27. Groll, M.; Brandstetter, H.; Bartunik, H.; Bourenkow, G.; Huber, R. Investigations on the maturation and regulation of archaebacterial proteasomes. *J. Mol. Biol.* **2003**, *327*, 75–83. [CrossRef]

28. Schrader, J.; Henneberg, F.; Mata, R.A.; Tittmann, K.; Schneider, T.R.; Stark, H.; Bourenkov, G.; Chari, A. The inhibition mechanism of human 20S proteasomes enables next-generation inhibitor design. *Science* **2016**, *353*, 594–598. [CrossRef]

29. Uchihashi, T.; Kodera, N.; Ando, T. Guide to video recording of structure dynamics and dynamic processes of proteins by high-speed atomic force microscopy. *Nat. Protoc.* **2012**, *7*, 1193–1206. [CrossRef]

30. Murata, K.; Nishimura, S.; Kuniyasu, A.; Nakayama, H. Three-dimensional structure of the α1-β complex in the skeletal muscle dihydropyridine receptor by single-particle electron microscopy. *J. Electron. Microsc. (Tokyo)* **2010**, *59*, 215–226. [CrossRef]

31. Scheres, S.H. RELION: Implementation of a Bayesian approach to cryo-EM structure determination. *J. Struct. Biol.* **2012**, *180*, 519–530. [CrossRef] [PubMed]

International Journal of
Molecular Sciences

Article

Molecular and Structural Basis of the Proteasome α Subunit Assembly Mechanism Mediated by the Proteasome-Assembling Chaperone PAC3-PAC4 Heterodimer

Tadashi Satoh [1], Maho Yagi-Utsumi [1,2,3], Kenta Okamoto [1], Eiji Kurimoto [4], Keiji Tanaka [5] and Koichi Kato [1,2,3,*]

[1] Graduate School of Pharmaceutical Sciences, Nagoya City University, 3-1 Tanabe-dori, Mizuho-ku, Nagoya 467-8603, Japan; tadashisatoh@phar.nagoya-cu.ac.jp (T.S.); mahoyagi@ims.ac.jp (M.Y.-U.); kenta.okamoto@icm.uu.se (K.O.)

[2] Exploratory Research Center on Life and Living Systems (ExCELLS), National Institutes of Natural Sciences, 5-1 Higashiyama, Myodaiji, Okazaki, Aichi 444-8787, Japan

[3] Institute for Molecular Science, National Institutes of Natural Sciences, 5-1 Higashiyama, Myodaiji, Okazaki, Aichi 444-8787, Japan

[4] Faculty of Pharmacy, Meijo University, Tempaku-ku, Nagoya 468-8503, Japan; kurimoto@meijo-u.ac.jp

[5] Laboratory of Protein Metabolism, Tokyo Metropolitan Institute of Medical Science, 2-1-6, Kamikitazawa, Setagaya-ku, Tokyo 156-8506, Japan; tanaka-kj@igakuken.or.jp

* Correspondence: kkatonmr@ims.ac.jp; Tel.: +81-564-59-5225; Fax: +81-564-59-5224

Received: 27 March 2019; Accepted: 3 May 2019; Published: 7 May 2019

Abstract: The 26S proteasome is critical for the selective degradation of proteins in eukaryotic cells. This enzyme complex is composed of approximately 70 subunits, including the structurally homologous proteins α1–α7, which combine to form heptameric rings. The correct arrangement of these α subunits is essential for the function of the proteasome, but their assembly does not occur autonomously. Assembly of the α subunit is assisted by several chaperones, including the PAC3-PAC4 heterodimer. In this study we showed that the PAC3-PAC4 heterodimer functions as a molecular matchmaker, stabilizing the α4-α5-α6 subcomplex during the assembly of the α-ring. We solved a 0.96-Å atomic resolution crystal structure for a PAC3 homodimer which, in conjunction with nuclear magnetic resonance (NMR) data, highlighted the mobility of the loop comprised of residues 51 to 61. Based on these structural and dynamic data, we created a three-dimensional model of the PAC3-4/α4/α5/α6 quintet complex, and used this model to investigate the molecular and structural basis of the mechanism of proteasome α subunit assembly, as mediated by the PAC3-PAC4 heterodimeric chaperone. Our results provide a potential basis for the development of selective inhibitors against proteasome biogenesis.

Keywords: assembly chaperone; molecular matchmaker; molecular modeling; proteasome; X-ray crystal structure

---

1. Introduction

The selective degradation of proteins in eukaryotic cells is essential for the maintenance of physiological homeostasis. Protein degradation is implemented primarily via the ubiquitin-proteasome system [1,2]. The proteasome is huge protein complex (26S), comprised of a 20S core particle (CP) and one or two 19S regulatory particles (RPs). The 20S CP, which has proteolytic activity, is composed of seven homologous α subunits, α1–α7, and seven homologous β subunits, β1–β7, which are assembled into a cylindrical structure with an $\alpha_{1-7}\beta_{1-7}\beta_{1-7}\alpha_{1-7}$ arrangement. The 19S RP is responsible for the collection of ubiquitinated substrates, the opening of the central gating pore of the 20S CP, and the de-ubiquitination and translocation-coupled unfolding of the substrates. Recent structural studies using cryo-electron microscopy have shed light upon the cooperative working mechanisms of this huge proteolytic machinery [3,4].

The correct arrangement of the proteasomal subunits is essential to the proper functioning of eukaryotic proteasomes. There is considerable evidence that the assembly of the eukaryotic 26S proteasome does not proceed spontaneously, but is mediated by several assembly chaperones [5–8]. The formation of the 20S CP is assisted by five proteasome-specific chaperones: PAC1–PAC4 and POMP in humans; Pba1–Pba4 and Ump1 in yeast. Four dedicated chaperones, p27 (Nas2), gankyrin (Nas6), PAAF1 (Rpn14), and S5b (Hsm3) are responsible for the formation of the base subcomplex of the 19S RP. Malfunctions of these assembly chaperones cause the accumulation of imperfectly assembled or mis-assembled complexes of the proteasomal subunits. For example, knock-down experiments involving PAC3 and PAC4 results in the accumulation of abnormal α-subunit oligomers lacking the $\alpha3$–$\alpha7$ subunits [9,10].

Because proteasome biogenesis is known to be significantly upregulated in cancer cells [11], the proteasome has potential as a target for therapeutic drugs for cancer treatment [12,13]. Bortezomib (Velcade) has been widely used as proteasome inhibitor for the treatment of patients with multiple myeloma [14]. The chaperones involved in proteasome assembly have also been considered as potential drug targets for anticancer treatments [12]. Selective inhibitors that specifically suppress proteasome biogenesis could be valuable for minimizing the undesired side effects which can occur when using compounds which target mature proteasomes.

Recently reported knock-out experiments indicated that $\alpha4$, $\alpha5$, $\alpha6$, and $\alpha7$ form a core assembly intermediate as part of the initial process of α-ring assembly, which is supported by PAC3-PAC4 [15]. However, most of the biochemical and structural data about the proteasome-assembly chaperones have been generated mainly from yeast proteins, which have only modest sequence identities with the human counterparts; less than 20%, for PAC3 when compared with Pba3. As with CP-assembly, yeast Pba3 and Pba4 have structural resemblance, and form a heterodimer [16] which functions as a matchmaker mediating the association between $\alpha4$ and $\alpha5$ [17]. It remains unclear, however, how the human PAC3-PAC4 complex functions in α-ring assembly through specific, direct interactions with cognate proteasomal subunits, although the crystal structures of human PAC3 and PAC4 have been solved for their homodimeric forms [16,18].

Structural insights into the chaperone-mediated formation of the human proteasome are important for the design and development of low-toxicity anticancer drugs which can inhibit the protein-protein interactions involved in the proteasome-assembly process. We performed a biochemical and biophysical study of the human PAC3-PAC4 heterodimer in order to understand the functional and structural mechanisms of α-ring formation mediated by the proteasome-assembling chaperones.

2. Results and Discussion

2.1. The PAC3-PAC4 Heterodimer Interacts Primarily with $\alpha5$

To study the biochemical processes involved in proteasome α-subunit assembly mediated by the PAC3-PAC4 heterodimer, we prepared all of the human proteasome α subunits as recombinant proteins. Although protocols to prepare PAC3 and PAC4 as individual recombinant proteins have been reported previously, their heterodimer is rather unstable, unlike the yeast orthologs Pba3 and Pba4 [16,18]. The recombinant PAC4 also has a tendency to form a domain-swapped homodimer [18]. To overcome these problems, we designed and prepared a PAC3-PAC4 heterodimer as a single-chain form, termed scPAC3/4, in which the C-terminus of PAC4 is connected to the N-terminus of PAC3 via a $(GGGS)_4$ liner. All of the recombinant proteins were produced using bacterial expression systems in *Escherichia coli*, and were successfully purified to homogeneity (Figure S1).

To determine which proteasomal α subunits interact with the PAC3-PAC4 heterodimer, we performed in vitro pull-down experiments using these recombinant proteins. In the pull-down assay, His$_6$-tagged scPAC3/4 was applied to Ni^{2+}-charged resin, and subsequently incubated with a mixture of all of the α-subunit proteins. Since α7 spontaneously forms an oligomer that is capable of capturing α6 [19,20], we carried out this experiment both in the absence and in the presence of α7. The pull-down experiments showed that scPAC3/4 reacted with several α subunits including α4, α5, and α6 (Figure 1). Addition of α7 had virtually no impact on the interaction of α6 with scPAC3/4, suggesting that it has a higher affinity for the PAC3-PAC4 heterodimer than for the α7 oligomer. To avoid ambiguity due to overlapping of the Coomassie Brilliant Blue (CBB)-stained bands, we performed the pull-down experiments using α1, α4, α5, and α6 individually. The pull-down assay showed that scPAC3/4 interacted most strongly with α5 and weakly with α4 and α6. By contrast, no interaction was detected between scPAC3/4 and α1. The interaction between α6 and scPAC3/4 appeared to be enhanced in the presence of the other α subunits. Since α5 and α6 occur consecutively in the native α ring, these data suggest that the PAC3-PAC4 heterodimer is important for α5-α6 subcomplex assembly during proteasome α-ring formation.

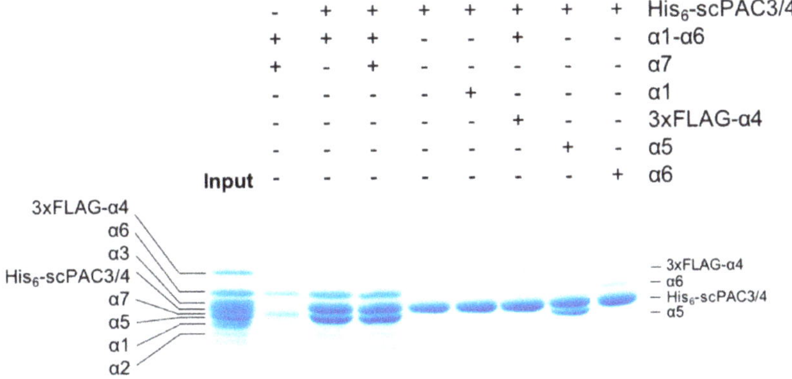

Figure 1. Pull-down experiments between the PAC3-PAC4 heterodimer and proteasome α subunits. The non-tagged α1–α3 and α5–α7 along with 3xFLAG-tagged α4 were mixed with His$_6$-tagged scPAC3/4 immobilized on Ni^{2+}-charged Chelating Sepharose beads. The 3xFLAG-tagged α4 was used to avoid the band overlap between α4 and scPAC3/4. After extensive washing, bound proteins were analyzed using CBB staining after sodium dodecyl sulfate polyacrylamide gel electrophoresis (SDS-PAGE). The 'Input' lane contained all α subunits and His$_6$-scPAC3/4 (0.5 μg each). The SDS-PAGE bands were assigned according to Figure S1a, and the bands originating from the His$_6$-scPAC3/4 and the bound α subunits are labeled.

2.2. The PAC3-PAC4 Heterodimer Acts as Molecular Matchmaker in α4-α5-α6 Assembly

In order to explore the functional mechanism of the PAC3-PAC4 heterodimer in proteasome assembly involving α4–α6, we investigated the inter-subunit interactions mediated by scPAC3/4. In a pull-down assay, glutathione *S*-transferase (GST)-fused α5 was used as a bait for probing its interactions with the other α subunits, both in the absence and in the presence of scPAC3/4. GST-α5 interacted weakly with α6 in the presence of scPAC3/4, while the other subunits were not reactive with α5 regardless of the presence or absence of scPAC3/4 (Figure 2). In contrast, GST-α5 weakly interacted with α4 regardless of the presence or absence of scPAC3/4 under this assay condition. The results were not influenced by the presence of α7 (Figures 1 and 2).

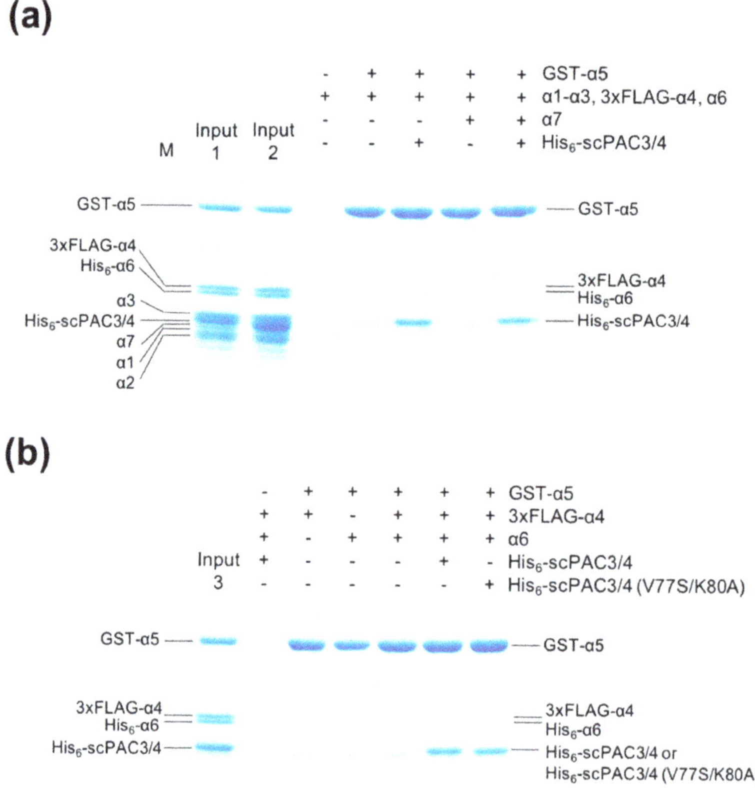

Figure 2. Pull-down experiments between α5 and the other α subunits. The α1–α4, α6, and α7 subunits were mixed with GST-tagged α5 immobilized on Glutathione Sepharose beads. The 3xFLAG-tagged α4 and His6-tagged α6 were used to avoid the overlap of their bands with those of the other α subunits or scPAC3/4. (**a**) Interactions between α5 and the other α subunits in the presence and absence of the PAC3-PAC4 heterodimer. The 'Input1' and 'Input2' lanes contained His6-scPAC3/4 and α1–α6 subunits in the absence and presence of α7, respectively (0.5 µg each). (**b**) Interaction between α5 and the adjacent α subunits, α4 and α6. The 'Input3' lane contained His6-scPAC3/4 and α4–α6 subunits. The pull-down experiment was also performed using an scPAC3/4 mutant with V77S and K80A substitutions in PAC3. Band assignments were carried out according to Figure S1b.

In yeast, the Pba3-Pba4 heterodimer acts as a matchmaker, reinforcing interactions between the α4 and α5 subunits [17]. The results of our pull-down analysis indicated that the PAC3-PAC4 heterodimer interacts with α4, α5, and α6, thereby acting as a molecular matchmaker for these proteasomal subunits. These findings suggest that the functional roles and interactions of this assembly chaperone complex with the proteasomal subunits are evolutionarily conserved between yeast and humans.

2.3. Structural Insights into the Mechanisms Underlying PAC3/PAC4-Dependent α4-α5-α6 Assembly

To investigate the structural mechanisms underlying the chaperone-dependent formation of the α4-α5-α6 subcomplex, we built a three-dimensional model of the putative quintet complex comprised of PAC3, PAC4, α4, α5 and α6, using previously-reported crystallographic data. Crystal structures for the PAC3 homodimer [16], domain-swapped PAC4 homodimer [18], and 20S proteasome [21] have been published. In addition, we newly determined a 0.96-Å high-resolution trigonal structure of the

PAC3 homodimer (Figure S2a). The overall structure of the trigonal form was very similar to that of the tetragonal structure we have previously published, except for a loop comprised of residues 51–61 (Figure S2b), suggesting that it is mobile. Loop flexibility was also observed in the corresponding segment of the yeast ortholog Pba3 in its heterodimer with Pba4 [16] (Figure S3). Our nuclear magnetic resonance (NMR) relaxation data from the human PAC3 homodimer confirmed that the loop is indeed mobile and disordered in solution (Figure S4).

In the quintet-complex models, in addition to the interactions between PAC3-PAC4 and α5, based on the crystal structure of the yeast counterparts [16], the assembly chaperone interacted with the neighboring α4 and α6 subunits (Figure 3a). When the PAC3-4/α4/α5/α6 quintet complex model was superimposed onto the crystal structure of 20S CP, PAC3 and PAC4 make steric hindrance with β6 and β5, respectively, which possibly triggers the release of PAC3-4 from the α-ring upon binding of the β subunits onto the α-ring. A complex model, model A, based on the 2.00-Å PAC3 structure showed that the mobile loop was turned toward the solvent. Another model based on the 0.96-Å structure, model B, showed that the corresponding loop contacts α6. Apart from interactions involving this mobile loop, intermolecular contacts between the PAC3-PAC4 heterodimer and the proteasomal subunits are almost identical in the two models. Therefore, in the rest of this paper, we discuss the structural basis of the PAC3/PAC4-dependent α4-α5-α6 subunit assembly using model B.

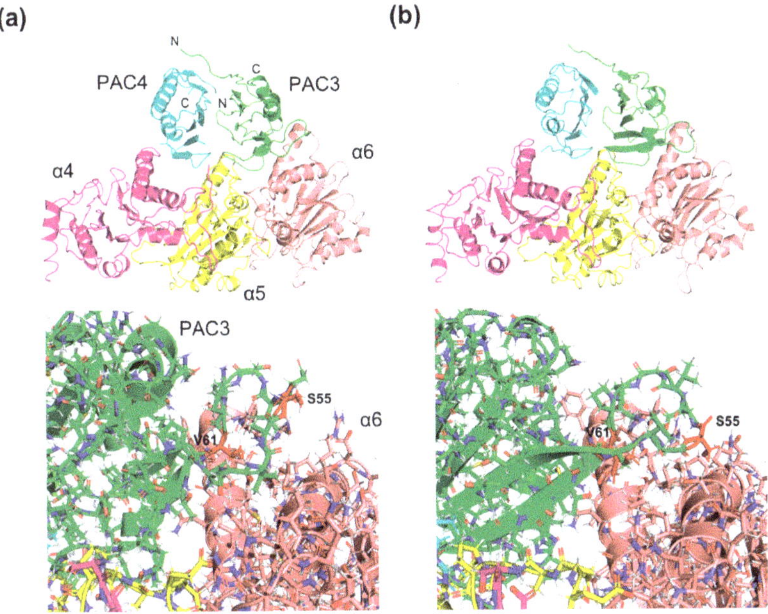

Figure 3. Three-dimensional model of the quintet complex comprising PAC3, PAC4, α4, α5, and α6. (**a**) Complex model A, based on the 2.00-Å PAC3 structure. (**b**) Complex model B, based on the newly-determined 0.96-Å structure. The positions of the N- and C-termini are indicted. Overall and close-up views between PAC3 and α6 of the quintet-complex models are shown in the upper and lower parts of the figure, respectively. Putative α6-binding residues of PAC3, Ser55 and Val61 (see also Figure 4b), are highlighted in red in both models to highlight the conformational differences of the loop between the two models.

In this model, the interaction of α5 with the PAC3-PAC4 heterodimer is mediated by Gln70, Glu72, and Lys104 in PAC3, Arg48 in PAC4, and Glu95, His99, Tyr103, and Asp129 in α5 through electrostatic interactions and hydrogen bonds (Figure 4). Our pull-down data indicated that scPAC3/4 interacted most strongly with α5 and weakly with α4 and α6 (Figure 1). The model predicted additional interactions between PAC3 and α6, and between PAC4 and α4 (Figure 4b,c). Specifically, Ser55, Lys80, and Asn81 of PAC3 form hydrogen bonds or electrostatic interactions with Ser110, Asp94, and Arg96 of α6, respectively. There are also predicted hydrophobic interactions of Val61, Phe85, and Val77 in PAC3 with Phe87, Phe97, and Leu93 in α6. Additionally, Asp70 and Arg85 of PAC4 have electrostatic interactions with Arg117 and Glu99 of α4, respectively (Figure 4c). Gln81 and Ile61 of PAC4 form hydrogen bonds and hydrophobic interactions with Ser93 and Val98 of α4, respectively.

To validate our docking model, we performed mutational experiments, especially focusing on the interaction between PAC3 and α6, which were specifically observed in the human proteins as compared with yeast counterparts. We constructed an scPAC3/PAC4 mutant in which putative α6-binding residues, Val77 and Lys80, of PAC3 are replaced with Ser and Ala, respectively. As expected, our mutational analysis indicated that mutations of Val77 and Lys80 of PAC3 impaired interaction with α6 but not with α4 and α5 (Figure 2b), confirming the validity of our docking model.

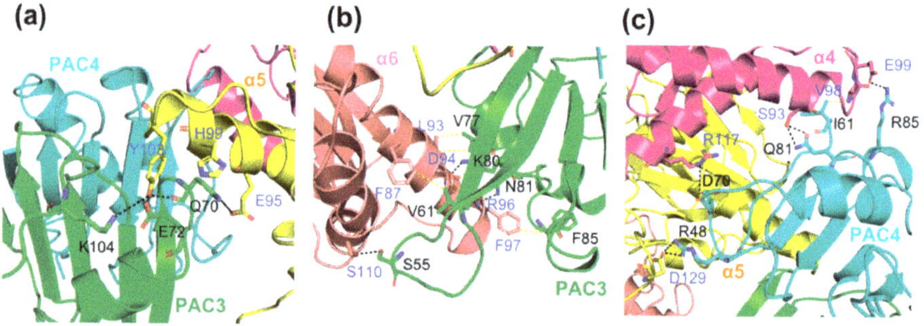

Figure 4. Predicted interaction interfaces between the PAC3-PAC4 heterodimer and proteasomal α4-α5-α6 subunits. (**a**) PAC3-α5. (**b**) PAC3-α6. (**c**) PAC4-α4 or α5 interfaces. Residues involved in the interactions are shown as stick representations. Potential hydrogen bonds and non-polar interactions are indicated as black and yellow dotted lines, respectively.

Although the overall structures of human PAC3 and PAC4 are similar to those of yeast Pba3 and Pba4 (RMSD = 1.9–2.1Å and 1.9–2.2 Å) respectively, their amino acid sequence similarities are low (PAC3 versus Pba3 11.0%; PAC4 versus Pba4 14.6%). The α-subunit contacting residues of human PAC3 and PAC4, as predicted by the model, are not well-conserved in the yeast orthologs Pba3 and Pba4 (Figure S5). Nevertheless, our model predicts that the complementarity at the interaction interfaces between the PAC3-PAC4 heterodimer and the proteasomal α4-α5-α6 subunits can be conserved in the yeast counterparts with a few exceptions. Perhaps the best example is the replacement of electrostatic interactions between Glu72 of PAC3 and His99 of α5 by non-polar contacts between Ala105 of Pba3 and Gln114 of α5. Therefore, despite the low sequence similarity, the overall interaction modes of the matchmaking chaperones with the proteasomal subunits appear to be conserved between humans and yeast. It is plausible that the conformational flexibility of the mobile 51–61 loop of PAC3, which carries the α6-contacting residues, contributes to the interaction adjustability.

In summary, we produced structural insights into the functional mechanisms of the PAC3-PAC4 heterodimer as a molecular matchmaker underpinning the α4-α5-α6 subcomplex during α-ring formation. These findings offer potential new approaches to the design of inhibitors against the protein-protein interactions involved in proteasome biogenesis.

3. Methods

3.1. Sample Preparation

Human proteasome α6 short isoform and α7 subunits were produced and purified as previously described [22,23]. Genes encoding the proteasome α1 and α4 subunits were subcloned into *Nde*I and *Sal*I sites in pET28b, and the α2 gene was inserted into the pRSFDuet-1 vector using *Nde*I and *Xho*I restriction enzyme sites (Merck Millipore, Burlington, MA, USA). As for α4, 3xFLAG sequence (DYKDHDGDYKDHDIDYKDDDDK) was added at the N-terminus. The α3 and α5 genes were subcloned into the *Bam*HI and *Xho*I or *Sal*I sites of modified pCold-I and pCold-GST vectors (TaKaRa Bio Inc., Kusatsu, Japan), respectively, in which a factor Xa cleavage site was replaced with that of TEV protease. The PAC3 and PAC4 genes were subcloned into *Nde*I and *Xho*I sites in pET28b, in which the C-terminus of PAC4 was connected to the N-terminus of PAC3 through a $(GGGS)_4$ liner. Standard polymerase chain reaction method was used to generate a V77S/K80A PAC3 mutant. *Escherichia coli* BL21-CodonPlus (DE3)-RIL (Agilent Technologies, Santa Clara, CA, USA) was used for all recombinant protein expression.

For the expression of recombinant proteins, the *E. coli* cells were grown in LB medium containing kanamycin or ampicillin. Briefly, the recombinant proteins were purified from the soluble fractions, except for α2, which was purified from the inclusion bodies and refolded using standard dilution methods.

Purification of these recombinant proteins was performed using affinity chromatography with Anti-FLAG M2 Affinity gel (Sigma-Aldrich, St. Louis, MO, USA), Ni^+-charged Chelating Sepharose, or Glutathione Sepharose 4B, anion-exchange chromatography with RESOURCE Q resin, and size exclusion chromatography with Superdex 75 pg or 200 pg resins (GE Healthcare, Chicago, IL, USA). For NMR analyses, the PAC3 homodimer was expressed in *E. coli* cells which were grown in M9 minimal medium containing $[^{13}C]$glucose (2.0 g/L) and/or $[^{15}N]NH_4Cl$ (1.0 g/L), and purified using a previously-described protocol [12].

3.2. Pull-Down Experiments

The 3xFLAG-tagged α4, GST-fused proteasome α5-subunit (GST-α5), non-tagged or His_6-tagged forms of proteasome α1, α2, α3, α6, and α7 subunits, and scPAC3/4 were used in the pull-down assays. For immobilization, 20 μg of His_6-tagged scPAC3/4 or GST-α5 was applied to Ni^{2+}-charged Chelating Sepharose or Glutathione Sepharose 4B (GE Healthcare) resins, respectively. The His_6-scPAC3/4-immobilized resins were incubated with 50 μg of α1–α7 subunits for 2 h at 4 °C in an incubation buffer (20 mM Tris-HCl (pH 8.0) and 150 mM NaCl). For α5, the GST-α5-immobilized resins were incubated with 50 μg of α1–α4, α6, and α7 in the presence and absence of 50 μg of scPAC3/4 as described above. Since α7 makes a stable complex with α6 [19,20], the pull-down experiments containing α7 were performed separately. The resins were washed four times with the incubation buffer, which contains 60 mM imidazole in the His_6-tag pull-down assays. Proteins bound to the Chelating or Glutathione Sepharose resins were eluted using 20 mM Tris-HCl (pH 8.0)/500 mM imidazole or 50 mM Tris-HCl (pH 8.0)/10 mM reduced glutathione, respectively, and analyzed by SDS-PAGE, stained with CBB.

3.3. Crystallization, X-ray Data Collection, and Structure Determination

For crystallization, purified non-tagged PAC3 homodimer was produced at a concentration of 8.0 mg/mL in 20 mM Tris-HCl (pH 7.5) and 150 mM NaCl. Crystals were obtained in a buffer containing 30% PEG2000 monomethyl ether and 0.1 M potassium thiocyanate with incubation at 20 °C for three to four days. Crystals were transferred into the reservoir solution and flash-cooled in liquid nitrogen. Diffraction intensities were integrated using XDS [24] and data scaling was carried out using AIMLESS [25]. The crystals of PAC3 belonged to space group $P3_121$ and diffracted up to a resolution of 0.96 Å.

The trigonal structure of PAC3 was solved by the molecular replacement method using MOLREP [26], using the previously-reported tetragonal structure (PDB code 2Z5E) [16] as a search model. Automated model building and manual model fitting to electron density maps were performed using ARP/wARP [27] and COOT [28], respectively. Model refinement was carried out using REFMAC5 [29], and structure validation was conducted using MolProbity [30]. The data collection and refinement statistics of the PAC3 homodimer are summarized in Table S1. The molecular graphics were prepared using PyMOL (Schrödinger, New York, NY, USA).

3.4. Computer-Aided Model Building

The quintet-complex model comprising PAC3, PAC4, α4, α5, and α6 was created by several rounds of superimpositions using the coordinates of the human PAC3 homodimer (PDB codes, 2Z5E [16], and 6JPT (from this study)), the human PAC4 homodimer (PDB code: 5WTQ) [18], the human 20S proteasome (5LE5) [21], the yeast Pba3-Pba4 heterodimer (2Z5B) [16], and the yeast Pba3-Pba4/α5 ternary complex (2Z5C) [16]. The human PAC3-PAC4 heterodimer was created by superimposition of PAC3 and PAC4 monomers onto yeast Pba3 and Pba4, respectively. The resulting PAC3-PAC4 model was superimposed onto the yeast Pba3-Pba4 structure complexed with α5. Finally, to make a quintet complex model, the PAC3-PAC4-α5 (yeast) model was superimposed onto the human α5 subunit of the 20S proteasome. Subsequent protonation and energy minimization was performed using the CHARMm force field with the Discovery Studio program suite [31] (BIOVIA, San Diego, CA, USA).

3.5. NMR Spectroscopy

^{13}C- and ^{15}N-labeled non-tagged PAC3 homodimer (0.3 mM) and ^{15}N-labeled non-tagged PAC3 homodimer (0.1 mM), dissolved in PBS (pH 6.8) containing 10% D_2O (v/v), 1 mM EDTA, and 0.01% NaN_3, were used for spectral assignment and relaxation experiments. All NMR data were acquired at 303 K using DMX-500, AVANCE-500, and AVANCE-800 spectrometers equipped with a 5-mm triple-resonance cryogenic probe (Bruker, Billerica, MA, USA). The NMR data were processed using TOPSPIN (Bruker) and NMRPipe [32]. Conventional 3D NMR experiments [33] were carried out for chemical shift assignments of the heteronuclear single-quantum correlation (HSQC) peaks originating from the PAC3 homodimer. Spectral assignments were carried out using SPARKY [34] and CCPNMR [35] software. ^{15}N relaxation parameters, T_1, T_2, and ^{15}N-^{1}H heteronuclear nuclear Overhauser effect (NOE) were obtained at 303 K using an AVANCE-800 spectrometer and analyzed using the Protein Dynamics software in the Dynamics Center (Bruker).

3.6. Accession Numbers

The coordinates and structural factors of the crystal structure of the PAC3 homodimer have been deposited in the Protein Data Bank under accession number 6JPT. Backbone ^{1}H and ^{15}N chemical shift data of the PAC3 homodimer have been deposited in the Biological Magnetic Resonance Data Bank under accession number 27844.

Supplementary Materials: Supplementary materials can be found at http://www.mdpi.com/1422-0067/20/9/2231/s1.

Author Contributions: T.S. and K.K. conceived and designed the study; T.S. and E.K. performed protein design and preparation. T.S. performed the pull-down and crystallographic experiments; M.Y.-U. and K.O. performed the NMR experiment; K.T. contributed the materials and reagents; T.S. and K.K. mainly wrote the manuscript.

Funding: This work was supported in part by the Grants-in-Aid for Scientific Research (Grant Numbers JP26460051 to E.K. and JP25102008, JP15H02491 to K.K.) and Nanotechnology Platform Program (Molecule and Material Synthesis) from the Ministry of Education, Culture, Sports, Science and Technology, Japan. This work was also supported by the Joint Research by Exploratory Research Center on Life and Living Systems (ExCELLS) (ExCELLS program No. 18-402).

Acknowledgments: We thank Kumiko Hattori, Kiyomi Senda and Yukiko Isono for their help in the preparation of recombinant proteins. We also thank Hirokazu Yagi (Nagoya City University) for his useful discussion. The diffraction data set was collected at Osaka University using BL44XU at SPring-8 and Nagoya University using BL2S1 at Aichi Synchrotron Radiation Center (Japan). We acknowledge the synchrotron beamline staff and Institute of Drug Discovery Science at Nagoya City University for providing the data collection and computational facilities.

Conflicts of Interest: The authors declare that they have no competing financial interests.

Abbreviations

CBB	Coomassie Brilliant Blue
CP	Core particle
GST	Glutathione *S*-transferase
HSQC	Heteronuclear single-quantum correlation
NMR	Nuclear magnetic resonance
NOE	Nuclear Overhauser effect
PAC	Proteasome-assembling chaperone
RP	Regulatory particle
SC	Single-chain
SDS-PAGE	Sodium dodecyl sulfate polyacrylamide gel electrophoresis

References

1. Tanaka, K. The proteasome: Overview of structure and functions. *Proc. Jpn. Acad. Ser. B Phys. Biol. Sci.* **2009**, *85*, 12–36. [CrossRef]
2. Baumeister, W.; Walz, J.; Zuhl, F.; Seemuller, E. The proteasome: Paradigm of a self-compartmentalizing protease. *Cell* **1998**, *92*, 367–380. [CrossRef]
3. Budenholzer, L.; Cheng, C.L.; Li, Y.; Hochstrasser, M. Proteasome Structure and Assembly. *J. Mol. Biol.* **2017**, *429*, 3500–3524. [CrossRef]
4. Collins, G.A.; Goldberg, A.L. The Logic of the 26S Proteasome. *Cell* **2017**, *169*, 792–806. [CrossRef] [PubMed]
5. Murata, S.; Yashiroda, H.; Tanaka, K. Molecular mechanisms of proteasome assembly. *Nat. Rev. Mol. Cell Biol.* **2009**, *10*, 104–115. [CrossRef]
6. Kish-Trier, E.; Hill, C.P. Structural biology of the proteasome. *Annu. Rev. Biophys.* **2013**, *42*, 29–49. [CrossRef]
7. Tomko, R.J., Jr.; Hochstrasser, M. Molecular architecture and assembly of the eukaryotic proteasome. *Annu. Rev. Biochem.* **2013**, *82*, 415–445. [CrossRef]
8. Kato, K.; Satoh, T. Structural insights on the dynamics of proteasome formation. *Biophys. Rev.* **2018**, *10*, 597–604. [CrossRef]
9. Hirano, Y.; Hayashi, H.; Iemura, S.; Hendil, K.B.; Niwa, S.; Kishimoto, T.; Kasahara, M.; Natsume, T.; Tanaka, K.; Murata, S. Cooperation of multiple chaperones required for the assembly of mammalian 20S proteasomes. *Mol. Cell* **2006**, *24*, 977–984. [CrossRef]
10. Le Tallec, B.; Barrault, M.B.; Courbeyrette, R.; Guerois, R.; Marsolier-Kergoat, M.C.; Peyroche, A. 20S proteasome assembly is orchestrated by two distinct pairs of chaperones in yeast and in mammals. *Mol. Cell* **2007**, *27*, 660–674. [CrossRef] [PubMed]
11. Almond, J.B.; Cohen, G.M. The proteasome: A novel target for cancer chemotherapy. *Leukemia* **2002**, *16*, 433–443. [CrossRef]
12. Doi, T.; Yoshida, M.; Ohsawa, K.; Shin-ya, K.; Takagi, M.; Uekusa, Y.; Yamaguchi, T.; Kato, K.; Hirokawa, T.; Natsume, T. Total synthesis and characterization of thielocin B1 as a protein–protein interaction inhibitor of PAC3 homodimer. *Chem. Sci.* **2014**, *5*, 1860–1868. [CrossRef]
13. Zhang, X.; Schulz, R.; Edmunds, S.; Kruger, E.; Markert, E.; Gaedcke, J.; Cormet-Boyaka, E.; Ghadimi, M.; Beissbarth, T.; Levine, A.J.; et al. MicroRNA-101 Suppresses Tumor Cell Proliferation by Acting as an Endogenous Proteasome Inhibitor via Targeting the Proteasome Assembly Factor POMP. *Mol. Cell* **2015**, *59*, 243–257. [CrossRef]
14. Chen, D.; Frezza, M.; Schmitt, S.; Kanwar, J.; Dou, Q.P. Bortezomib as the first proteasome inhibitor anticancer drug: Current status and future perspectives. *Curr. Cancer Drug Targets* **2011**, *11*, 239–253. [CrossRef]
15. Wu, W.; Sahara, K.; Hirayama, S.; Zhao, X.; Watanabe, A.; Hamazaki, J.; Yashiroda, H.; Murata, S. PAC1-PAC2 proteasome assembly chaperone retains the core α4-α7 assembly intermediates in the cytoplasm. *Genes Cells* **2018**, *23*, 839–848. [CrossRef]
16. Yashiroda, H.; Mizushima, T.; Okamoto, K.; Kameyama, T.; Hayashi, H.; Kishimoto, T.; Niwa, S.; Kasahara, M.; Kurimoto, E.; Sakata, E.; et al. Crystal structure of a chaperone complex that contributes to the assembly of yeast 20S proteasomes. *Nat. Struct. Mol. Biol.* **2008**, *15*, 228–236. [CrossRef]

17. Takagi, K.; Saeki, Y.; Yashiroda, H.; Yagi, H.; Kaiho, A.; Murata, S.; Yamane, T.; Tanaka, K.; Mizushima, T.; Kato, K. Pba3-Pba4 heterodimer acts as a molecular matchmaker in proteasome α-ring formation. *Biochem. Biophys. Res. Commun.* **2014**, *450*, 1110–1114. [CrossRef]

18. Kurimoto, E.; Satoh, T.; Ito, Y.; Ishihara, E.; Okamoto, K.; Yagi-Utsumi, M.; Tanaka, K.; Kato, K. Crystal structure of human proteasome assembly chaperone PAC4 involved in proteasome formation. *Protein Sci.* **2017**, *26*, 1080–1085. [CrossRef]

19. Ishii, K.; Noda, M.; Yagi, H.; Thammaporn, R.; Seetaha, S.; Satoh, T.; Kato, K.; Uchiyama, S. Disassembly of the self-assembled, double-ring structure of proteasome α7 homo-tetradecamer by alpha6. *Sci. Rep.* **2015**, *5*, 18167. [CrossRef]

20. Kozai, T.; Sekiguchi, T.; Satoh, T.; Yagi, H.; Kato, K.; Uchihashi, T. Two-step process for disassembly mechanism of proteasome α7 homo-tetradecamer by α6 revealed by high-speed atomic force microscopy. *Sci. Rep.* **2017**, *7*, 15373. [CrossRef]

21. Schrader, J.; Henneberg, F.; Mata, R.A.; Tittmann, K.; Schneider, T.R.; Stark, H.; Bourenkov, G.; Chari, A. The inhibition mechanism of human 20S proteasomes enables next-generation inhibitor design. *Science* **2016**, *353*, 594–598. [CrossRef]

22. Sugiyama, M.; Hamada, K.; Kato, K.; Kurimoto, E.; Okamoto, K.; Morimoto, Y.; Ikeda, S.; Naito, S.; Furusaka, M.; et al. SANS simulation of aggregated protein in aqueous solution. *Nucl. Instrum. Methods Phys. Res. A* **2009**, *600*, 272–274. [CrossRef]

23. Sugiyama, M.; Kurimoto, E.; Yagi, H.; Mori, K.; Fukunaga, T.; Hirai, M.; Zaccai, G.; Kato, K. Kinetic asymmetry of subunit exchange of homooligomeric protein as revealed by deuteration-assisted small-angle neutron scattering. *Biophys. J.* **2011**, *101*, 2037–2042. [CrossRef]

24. Kabsch, W. Xds. *Acta Crystallogr. D Biol. Crystallogr.* **2010**, *66*, 125–132. [CrossRef]

25. Evans, P.R. An introduction to data reduction: Space-group determination, scaling and intensity statistics. *Acta Crystallogr. D Biol. Crystallogr.* **2011**, *67*, 282–292. [CrossRef]

26. Vagin, A.; Teplyakov, A. MOLREP: An automated program for molecular replacement. *J. Appl. Crystallogr.* **1997**, *30*, 1022–1025. [CrossRef]

27. Langer, G.; Cohen, S.X.; Lamzin, V.S.; Perrakis, A. Automated macromolecular model building for X-ray crystallography using ARP/wARP version 7. *Nat. Protoc.* **2008**, *3*, 1171–1179. [CrossRef]

28. Emsley, P.; Lohkamp, B.; Scott, W.G.; Cowtan, K. Features and development of Coot. *Acta Crystallogr. D Biol. Crystallogr.* **2010**, *66*, 486–501. [CrossRef]

29. Murshudov, G.N.; Vagin, A.A.; Dodson, E.J. Refinement of macromolecular structures by the maximum-likelihood method. *Acta Crystallogr. D Biol. Crystallogr.* **1997**, *53*, 240–255. [CrossRef]

30. Chen, V.B.; Arendall, W.B., 3rd; Headd, J.J.; Keedy, D.A.; Immormino, R.M.; Kapral, G.J.; Murray, L.W.; Richardson, J.S.; Richardson, D.C. MolProbity: All-atom structure validation for macromolecular crystallography. *Acta Crystallogr. D Biol. Crystallogr.* **2010**, *66*, 12–21. [CrossRef]

31. Spassov, V.Z.; Yan, L. A fast and accurate computational approach to protein ionization. *Protein Sci.* **2008**, *17*, 1955–1970. [CrossRef] [PubMed]

32. Delaglio, F.; Grzesiek, S.; Vuister, G.W.; Zhu, G.; Pfeifer, J.; Bax, A. NMRPipe: A multidimensional spectral processing system based on UNIX pipes. *J. Biomol. NMR* **1995**, *6*, 277–293. [CrossRef] [PubMed]

33. Uekusa, Y.; Mimura, S.; Sasakawa, H.; Kurimoto, E.; Sakata, E.; Olivier, S.; Yagi, H.; Tokunaga, F.; Iwai, K.; Kato, K. Backbone and side chain ^{1}H, ^{13}C, and ^{15}N assignments of the ubiquitin-like domain of human HOIL-1L, an essential component of linear ubiquitin chain assembly complex. *Biomol. NMR. Assign.* **2012**, *6*, 177–180. [CrossRef] [PubMed]

34. Goddard, T.D.; Koeller, D.G. *Sparky*, Version 3.0; University of California: San Francisco, CA, USA, 1993.

35. Vranken, W.F.; Boucher, W.; Stevens, T.J.; Fogh, R.H.; Pajon, A.; Llinas, M.; Ulrich, E.L.; Markley, J.L.; Ionides, J.; Laue, E.D. The CCPN data model for NMR spectroscopy: Development of a software pipeline. *Proteins* **2005**, *59*, 687–696. [CrossRef] [PubMed]

International Journal of
Molecular Sciences

Communication

miR-7 Knockdown by Peptide Nucleic Acids in the Ascidian *Ciona intestinalis*

Silvia Mercurio [1], Silvia Cauteruccio [2,*], Raoul Manenti [1], Simona Candiani [3,*], Giorgio Scarì [4], Emanuela Licandro [2] and Roberta Pennati [1]

[1] Department of Environmental Science and Policy, Università degli Studi di Milano, 20133 Milano, Italy; sil.mercurio@gmail.com (S.M.); raoul.manenti@unimi.it (R.M.); roberta.pennati@unimi.it (R.P.)
[2] Department of Chemistry, Università degli Studi di Milano, 20133 Milano, Italy; emanuela.licandro@unimi.it
[3] Department of Earth Science, Environment and Life, Università degli Studi di Genova, 16126 Genova, Italy
[4] Department of Biosciences, Università degli Studi di Milano, 20133 Milano, Italy; nkiller@unimi.it
* Correspondence: silvia.cauteruccio@unimi.it (S.C.); simona.candiani@unige.it (S.C.); Tel.: +39-02-5031-4147 (S.C.); +39-01-0353-8051 (S.C.)

Received: 18 September 2019; Accepted: 14 October 2019; Published: 16 October 2019

Abstract: Peptide Nucleic Acids (PNAs) are synthetic mimics of natural oligonucleotides, which bind complementary DNA/RNA strands with high sequence specificity. They display numerous advantages, but in vivo applications are still rare. One of the main drawbacks of PNAs application is the poor cellular uptake that could be overcome by using experimental models, in which microinjection techniques allow direct delivery of molecules into eggs. Thus, in this communication, we investigated PNAs efficiency in miR-7 downregulation and compared its effects with those obtained with the commercially available antisense molecule, Antagomir (Dharmacon) in the ascidian *Ciona intestinalis*. Ascidians are marine invertebrates closely related to vertebrates, in which PNA techniques have not been applied yet. Our results suggested that anti-miR-7 PNAs were able to reach their specific targets in the developing ascidian embryos with high efficiency, as the same effects were obtained with both PNA and Antagomir. To the best of our knowledge, this is the first evidence that unmodified PNAs can be applied in in vivo knockdown strategies when directly injected into eggs.

Keywords: microRNA; *hnRNP K*; PNA; tunicates; LNA probe

1. Introduction

Peptide Nucleic Acids (PNAs) are artificial nucleic acids mimics [1], extensively used for the regulation of gene expression in cellular and molecular systems [2]. In PNAs, the neutral pseudo-peptide backbone, based on *N*-(2-aminoethyl)glycine units (Figure 1A), replaces the negatively charged sugar-phosphate chain of nucleic acids. PNAs can recognize and bind to DNA or RNA sequences according to regular Watson–Crick base pairing rules [3]. Unlike DNA or RNA, PNAs are chemically stable across a wide range of temperatures and pHs, and they are resistant to enzymatic degradation since they are not easily recognized by nucleases or proteases [4]. Moreover, one of the most remarkable properties of PNA is the excellent thermal stability of PNA/DNA and PNA/RNA duplexes, in comparison with DNA/DNA or DNA/RNA duplexes [5]. Indeed, the lack of charge repulsion between the neutral PNA strand and the DNA or RNA strand provides extremely stable complexes. For example, Shakeel et al. reported that the melting temperature (T_m) values for a 15-mer PNA/DNA or PNA/RNA duplex are generally 20 °C higher than the natural nucleic acid duplexes [6]. All these properties make PNAs excellent candidates for in vivo antisense and antigene therapies, targeting oncogenes, viruses, and bacteria [7]. Indeed, studies on the use of PNAs in knockdown technologies are accumulating, demonstrating PNAs potential for future therapeutic purposes, as well as for basic research. However, in vivo PNAs applications are still rare [8,9] due to some drawbacks, such as poor cellular uptake and low solubility in aqueous media [10], which can be improved either by conjugation with carrier molecules or by chemical modifications [11].

PNAs poor cellular uptake could be also overcome by using model organisms, such as ascidians, in which molecules can be directly injected in the target tissue/cell. Ascidians are marine invertebrates, closely related to vertebrates [12]. They develop through a swimming larva that shows the basic chordate features, comprising a notochord, which runs along the tail, and a dorsal tubular central nervous system (CNS) [13,14]. Particularly, the ascidian *Ciona intestinalis* is amenable to embryological manipulations. A variety of molecular tools were developed to perturb gene activity during its development, including microinjections of antisense molecules directly into unfertilized eggs [15,16].

To verify PNAs efficiency in gene downregulation during ascidian development, we chose one of the most evolutionarily conserved microRNAs (miRNAs), miR-7 [17], as PNAs target. miRNAs are a class of non-coding RNAs that regulate gene expression at post-transcriptional level. They are found in all animal lineages, where they modulate multiple biological processes [18,19]. A single miRNA has the potential to target a broad spectrum of mRNAs, possessing great regulatory potential [20]. However, in many cases, knockout of individual miRNA does not lead to critical effects, as the same pathway is often controlled by many of these molecules that collectively affect the pathway by exerting fine-tune functions and ensuring the correct progression of cellular and developmental programs [21].

In mammals, miR-7 is expressed predominantly in the pancreas, neural tissues and pituitary [22,23]. miR-7 is highly expressed in neurons with sensory or neurosecretory functions in fish and animals distantly related from vertebrates, such as annelids [24]. miR-7 expression in photoreceptors is similarly conserved, being reported in rodents [25], amphioxus [26], and even *Drosophila* [27]. A gene encoding for miR-7 is also present in the genome of the ascidian *C. intestinalis* (www.mirbase.org), but its expression has not been described yet.

Thus, in this communication, we aim to characterize miR-7 expression profile in the ascidian *C. intestinalis*, and then test PNAs in vivo knockdown efficiency in this species, comparing it with the commercial antisense molecule, Antagomirs (Dharmacon, USA) [28].

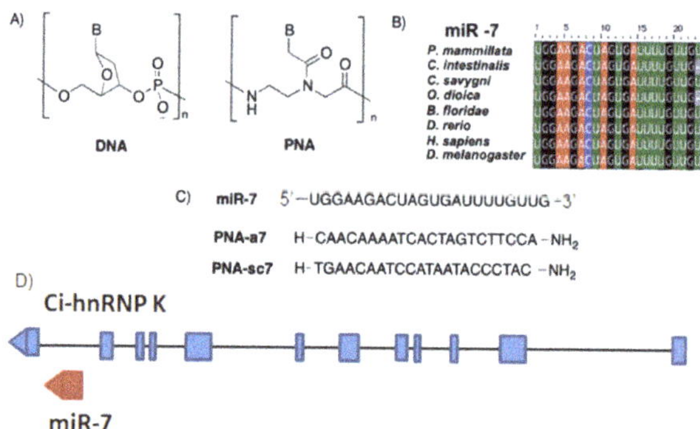

Figure 1. (**A**) DNA and Peptide Nucleic Acids (PNA) backbone; (**B**) multi-alignment of mature miR-7 sequences in different species: *Phallusia mammillata*, *Ciona intestinalis*, *Ciona savignyi*, *Oikopleura dioica*, *Branchiostoma floridae*, *Danio rerio*, *Homo sapiens*, *Drosophila melanogaster*; (**C**) miR-7 and PNAs sequences used in this study; (**D**) schematic representation of miR-7 genomic position inside the last Ci-hnRNP K intron (blue rectangles and interconnecting lines represent exons and introns, respectively; the red graph corresponds to the miR-7 sequence).

2. Results

2.1. miR-7 in Ciona intestinalis

Comparing miR-7 mature sequences in different animal models, we observed that miR-7 is highly conserved also in basal chordates: *C. intestinalis* miR-7 differs from that of *Homo sapiens*, only by the deletion of the terminal uracil. This feature is shared with another tunicate species, *Oikopleura dioica*, while in the available transcriptomes of two other ascidians, miR-7 mature sequences are completely conserved (Figure 1B).

In *C. intestinalis* genome, the miR-7 gene resides within the last intron of the heterogeneous nuclear ribonucleoprotein K (hnRNP K) gene, oriented in the same direction as the Ci-hnRNP K transcription unit (Figure 1D).

2.2. Genes Expression Profile

To determine the expression pattern of miR-7 mature transcripts during *C. intestinalis* development, standard in situ hybridization protocol [29] was ineffective. When hybridization with DIG-labeled Locked Nucleic Acid (LNA; Exiqon, Vedbaek, Denmark) probes were carried out overnight, unspecific stains were always detected in mesenchymal cells of late tailbud and larva trunk (Figure 2A). Extending the hybridization step to five days and increasing the hybridization temperature (5 °C more than the recommended temperature) was found to be optimal for miRNA detection with LNA probes, as confirmed by miR-124 results (Figure 2B). Performing this modified protocol, we found that miR-124 mature transcripts were abundantly present in all of the nervous system of *C. intestinalis* larva, as previously reported by Zeller and co-workers [30].

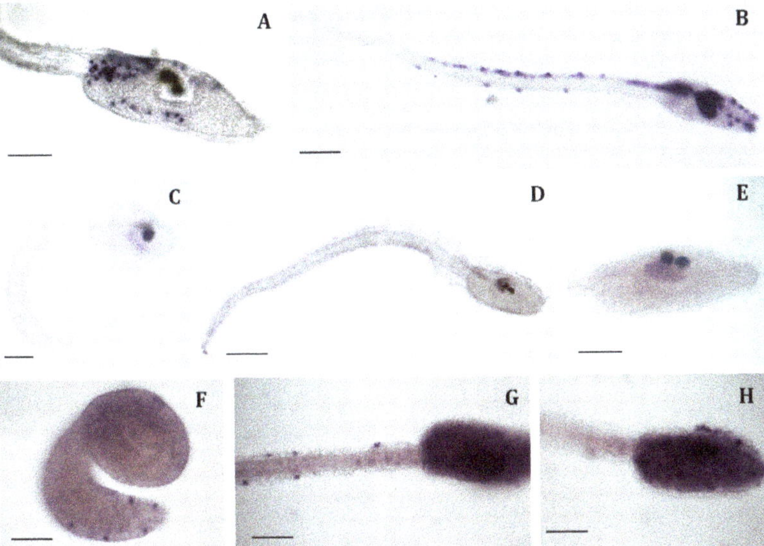

Figure 2. Whole mount in situ hybridization of *C. intestinalis* embryos. (**A**) Unspecific stain in mesenchymal cells of larva trunk, obtained when hybridization with LNA probes was performed overnight; (**B**) miR-124 expression in larval central and peripheral nervous system; (**C**) miR-7 expression at late tailbud stage: signal is clearly visible in the posterior ventral part of the sensory vesicle; (**D,E**) miR-7 expression at larva stage: the signal persists in the posterior ventral region and faintly extends in the neural ganglion; (**F**) hnRNP K expression at mid tailbud stage: signal appears more intense in the epidermal sensory neurons; (**G,H**) hnRNP K expression at larva stage: signal persists in epidermal sensory neurons and extends all over the trunk. (**A,E–H**) Scale bar = 60 μm; (**B–D**) Scale bar = 100 μm.

Using this protocol, we found that miR-7 expression started in the central nervous system at the late tailbud stage, but only in the ventral posterior part of the sensory vesicle (Figure 2C). At larval stage, the signal persisted in this region and faintly extended in the neural ganglion (Figure 2D,E).

hnRNP K, the gene hosting miR-7, was ubiquitously expressed at early developmental stages; but from mid tailbud stage, the signal was more intense in the epidermal sensory neurons, i.e., ascidian peripheral nervous system, of both trunk and tail (Figure 2F). This expression persisted at the late tailbud and larval stages, and a strong signal was also detectable all over the trunk (Figure 2G,H).

2.3. miR-7 Downregulation by PNAs

To evaluate PNAs effectiveness in miRNA knockdown, we designed a 22-mer PNA complementary to *C. intestinalis* miR-7 (PNA-a7, Figure 1C) and a PNA scrambled sequence with the same base composition of PNA-a7 (PNA-sc7, Figure 1C). Then, PNA-a7 and PNA-sc7 were microinjected in *C. intestinalis* eggs before in vitro fertilization. Moreover, we performed microinjections employing the commercial AntagomiR (AmiR-7), commonly used in miRNAs knockdown studies [28], and we compared the effects with those obtained with PNAs.

Preliminary trials revealed that the highest non-lethal concentrations were 0.7 mM for PNAs (PNA-a7 and PNA-sc7) and 0.3 mM for AmiR-7. Based on these results, PNAs solution seemed less toxic than AmiR-7, as embryos injected with concentrations higher than 0.3 mM AmiR-7 died before completing embryogenesis. All the following analyses were performed on embryos injected with 0.7 mM PNAs or 0.3 mM AmiR-7.

The developing rates of controls (injected with only the vital dye, Fast Green) and embryos injected with PNA-sc7, PNA-a7 or AmiR-7 were comparable, ranging from 69% (PNA-a7) to 72% (PNA-sc7) and no difference in sample morphology was recorded.

The specificity of PNA-a7 as well as AmiR-7 for miR-7 in *C. intestinalis* was previously checked by blast search in its genome by using cin-miR-7 (MIMAT0006091) as query. No identity with other microRNAs were found. Few mRNAs (for example: NLRC5-like or FAM192A-like mRNAs) having some sequence identity with miR-7 were obtained but the query coverage was lower, as there was some similarity but not for the entire sequence of miR-7.

To verify miRNA downregulation by PNAs, we first evaluated miR-7 expression by in situ hybridization. Results revealed that miR-7 expression was drastically reduced in embryos injected with PNA-a7 (Figure 3A), while miR-7 was normally expressed in embryos injected with PNA-sc7 (Figure 3B).

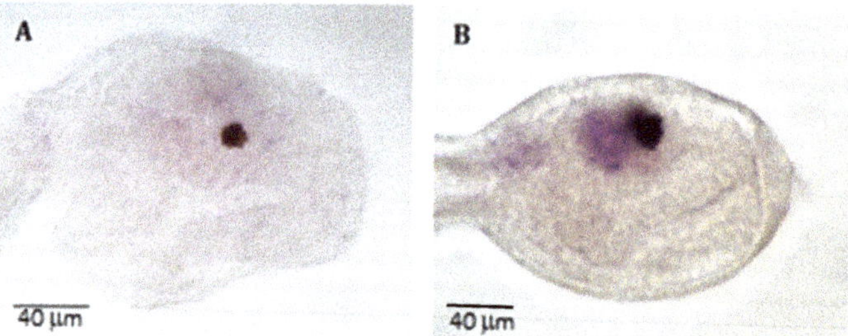

Figure 3. *C. intestinalis* miR-7 expression at the late tailbud stage in embryos injected with (**A**) PNA-a7 and (**B**) PNA-sc7. Scale bar = 40 µm.

Then, we checked the expression of some pan-neural genes: Ci-ETR [31] and Ci-Syn [32] in all the injected embryos. Ci-ETR expression was normal as its signal was observed throughout the central nervous system and in the epidermal sensory neurons of all injected samples (Figure 4A,C,E,G). Nevertheless, the expression of Ci-Syn was reduced in embryos injected with PNA-a7 (87%, $n = 28$) and AmiR-7 (91%, $n = 32$), compared to PNA-sc7 ($n = 41$). In control embryos and in embryos injected with PNA-sc7, the hybridization signal occurred in most of the sensory vesicle, in the motor ganglion, and extended into the posterior neural tube (Figure 4B,D). In embryos injected with PNA-a7, Ci-Syn transcripts were detected only in a subpopulation of neurons in the sensory vesicle and in the motor ganglion, and not detected in the posterior neural tube (Figure 4F). The same expression pattern was present in embryos injected with AmiR-7 (Figure 4H).

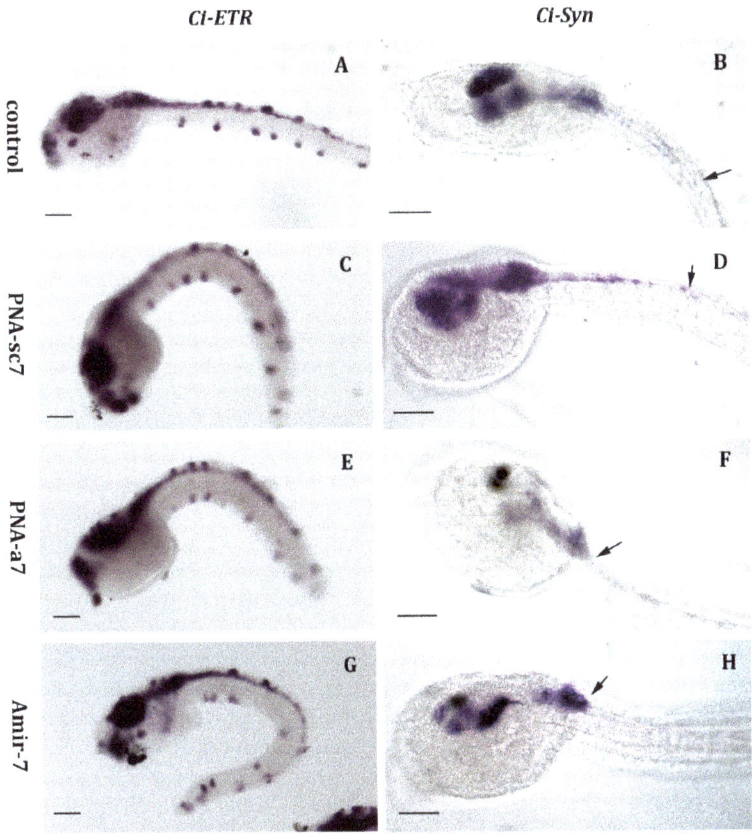

Figure 4. Whole mount in situ hybridization of *C. intestinalis* control embryos and embryos injected with PNA-sc7, PNA-a7, and AntagomiR (AmiR-7). (**A,C,E,G**) Ci-ETR expression in central and peripheral nervous system of late tailbud embryos; (**B,D,F,H**) Ci-Syn expression at late tailbud stage: in control and PNA-sc7 injected embryos, signal is detectable in most of the sensory vesicle, in the motor ganglion and along the posterior neural tube. In embryos injected with PNA-a7 and AmiR-7, transcripts are present only in a subpopulation of neurons in the sensory vesicle and in the motor ganglion, while no signal is recorded in posterior neural tube. Arrows indicate the posterior limit of the signal. (**A,C,E,G**) Scale bar = 15 μm. (**B,D,F,H**) Scale bar = 10 μm.

3. Discussion

PNAs are synthetic mimics of natural oligonucleotides, which bind complementary DNA/RNA strands with high sequence specificity [3]. In comparison with DNA/DNA or DNA/RNA duplexes, PNA/DNA and PNA/RNA duplexes show excellent thermal stability due to the neutral PNA backbone that lacks a repulsion charge when binding with DNA or RNA, providing more stable complexes [5]. Although they display numerous advantages, in vivo applications are still rare due to their poor cellular uptake. The latter, however, could be overcome by using experimental models, such as the ascidian

C. intestinalis. In this model organism, microinjections allow direct delivery of antisense molecules into eggs, perturbing gene activity during embryonic development [16]. To verify this hypothesis, we chose miR-7 as PNAs target. miRNAs are potent endogenous regulators of gene expression with fundamental roles in development [21].

In *C. intestinalis*, hundreds of miRNAs have been identified (www.mirbase.org) but expression data are reported only for miR-124 [30]. In fact, characterization of miRNAs expression is extremely challenging due to their tiny size and low level of expression. When performing in situ hybridization, the possibility to get nonspecific signals is rather high. Thus, we first optimized the hybridization protocol with DIG-labeled LNA (Exiqon) probes to obtain specific staining even when miRNA levels are particularly low, as in miR-7's case. In particular, modifications in hybridization temperature and incubation time were found to highly improve miRNAs detection.

Employing this protocol, we described for the first time miR-7 expression during ascidian development. In *C. intestinalis*, miR-7 mature transcripts were detected at late tailbud and larva stages in the central nervous system, particularly, in the ventral posterior part of the sensory vesicle (Figure 2C–E). miR-7 neural expression has been reported in different animal models, specifically in photoreceptors and/or neurosecretory tissues [24,26,33]. miR-7 is considered part of the evolutionary conserved fingerprint of neurosecretory cells [24]. In *C. intestinalis*, the ventral region of the larval sensory vesicle has been proposed to be homologous to vertebrate hypothalamus and retinal amacrine cells [34–36]. Thus, miR-7 expression also appears highly conserved in ascidians. Its expression domain, together with that already reported for Ci-Rx [37], Ci-Nk2, and Ci-Otp, [35] further supports the homology of this region to vertebrate hypothalamus [24,33,35]. Moreover, miR-7 was reported to be expressed in mammalian retina; and in ascidians, miR-7 was identified in the photoreceptive neuroepithelium [36], further indicating the striking evolutionary conservation of this miRNA among chordates.

In addition, miR-7 genomic position seems extremely conserved. In *C. intestinalis* genome, miR-7 was found within the last intron of the heterogeneous nuclear ribonucleoprotein K (hnRNP K) gene, similarly to those already reported for hsa-miR-7-1 in *Homo sapiens* [38] and in *Drosophila* [39].

hnRNP K is referred as a ubiquitously expressed gene involved in different aspects of RNA functions: transcription, editing, processing, and translation [40]. During vertebrate embryogenesis hnRNP K is uniformly expressed in all blastomeres; and only at later stages, transcripts appear more concentrated in specific tissues, such as central nervous system and mesodermal derivatives [41]. In *C. intestinalis*, Ci-hnRNP K displays a similar expression during early developmental stages; while at the late tailbud and larva stages, transcripts accumulate in the epidermal sensory neurons (Figure 2F–H), i.e., ascidian peripheral nervous system. Like in vertebrates, Ci-hnRNP K expression does not overlap with that of its host miR-7. Since genes are oriented in the same direction (Figure 1D), post-transcriptional regulation of miR-7 biogenesis is likely to occur, as demonstrated in both mice and humans [38].

Then, we designed a 22-mer PNA complementary to *C. intestinalis* miR-7 (PNA-a7, Figure 1C) as well as a PNA scrambled sequence (PNA-sc7, Figure 1C), to verify the specificity of the interaction between PNA-a7 and miR-7. Our hybridization analysis confirmed that PNA-a7 efficiently downregulates miR-7 and that PNAs interaction occurs in a sequence-specific manner, as samples injected with PNA-sc7 never showed a similar signal reduction (Figure 3). To further verify PNAs effectiveness, we compared the effects induced by miR-7 downregulation by injecting the commercial AntagomiR molecules, commonly used in miRNAs knockdown studies (AmiR-7) [28], and our PNAs. We checked the expression of two pan-neural genes, Ci-ETR [31] and Ci-Syn [32]. Ci-ETR signal was not affected by neither the PNAs nor AmiR-7 injection, suggesting that perturbation of miR-7 expression does not affect nervous system differentiation. On the contrary, Ci-Syn signal was reduced in the posterior neural tube of embryos injected with PNA-a7 and AmiR-7 (Figure 4F,H), but not in those injected with PNA-sc7. Synapsins are neuronal phosphoproteins that constitute a small family of synaptic molecules, specifically associated with synaptic vesicles. They exert a key role in neurite outgrowth and synapse formation [32]. In human neural embryonic stem cells, miR-7 overexpression during their neuronal differentiation increased synapsin expression. Synapsin mRNA is not a direct target of miR-7 but miR-7 levels are positively correlated with synapsin expression, suggesting that this miRNA could act upstream of the synapsin pathway, and play an important role in synaptic development [39]. Similarly, in *C. intestinalis* embryos, Ci-Syn expression decreased after miR-7 knockdown, indicating that miR-7 has a functional role in synaptic plasticity and neurite elongation in ascidians.

Comparable results were obtained with both PNA-a7 and AmiR-7, confirming the reliability of our results and the specificity of PNAs. These are in agreement with different in vitro research, which demonstrated PNAs specificity for their complementary miRNAs [42]. In vivo experiments using modified PNAs have been performed in mouse: miR-155 inhibition by PNAs was demonstrated to be sequence specific, not affecting levels of unrelated miRNAs and mainly recapitulating the effects of genetic deletion of miR-155 [8].

4. Materials and Methods

4.1. Animals and Embryos Culture

Adult *Ciona intestinalis* were collected along the coasts of Roscoff (France) by the fishing service of the Station Biologique de Roscoff. Animals were maintained in aquaria filled with artificial sea water (Instant Ocean; salinity ~32‰) and provided with a circulation system, as well as mechanical, chemical, and biological filters. Constant light conditions were preferred to promote gamete production [29]. For each experiment, gametes from three adults were obtained surgically from the gonoducts and in vitro cross-fertilization was performed. Embryos were reared in petri dishes in filtered artificial sea water buffered with 1 M HEPES (ASWH; pH 8.0) at 18 ± 1 °C up to the stages of interest (gastrula stage; neurula stage; initial, early, mid, and late tailbud stages; larva stage). Then, they were dechorionated in ASWH containing 1% sodium thioglycolate and 0.05% protease; fixed in 4% paraformaldehyde, 0.5 M NaCl, and 0.1 M 3-(*N*-morpholino)propanesulfonic acid (pH 7.5) for 90 min; dehydrated in ethanol series (30%, 50%, and 70%); and stored at −20 °C.

4.2. Reagents

Antagomirs are chemically modified, cholesterol-conjugated single-stranded RNA analogues complementary to the mature miRNA sequences, commonly applied in miRNAs knockdown research [28]. Based on *C. intestinalis* sequence, a specific anti-miR-7 antagomir (AmiR-7: 5′-C_SA_SACAAAAUCACUAGUCUU$_SC_SC_SA_S$-Chol-3′) was designed and synthetized by Dharmacon (USA).

4.3. PNAs Synthesis and Characterization

Two PNA oligomers were designed: PNA-a7 complementary to mature miR-7; PNA-sc7 with a scrambled sequence, but the same base composition of PNA-a7 as control (Figure 1C). They were synthesized by automated solid-phase synthesis using Boc/Z chemistry by means of the automated synthesizer, Applied Biosystems 433A Peptide Synthesizer (Monza, Milan, Italy), equipped with Synthassist 2.0 software. The commercially available Boc/Z-protected PNA monomers were purchased from ASM Research Chemicals GmbH (Hannover, Germany). The MBHA resin was purchased from VWR International, and it was loaded manually to 0.2 mmol/g with Boc/Z-adenine PNA monomer for PNA-a7, and with Boc/Z-cytosine PNA monomer for PNA-sc7 [43]. The PNA purification was performed using reverse phase high pressure liquid chromatography (RP-HPLC) with an Agilent 1200 Series system (Cernusco sul Naviglio, Milan, Italy), equipped with DAD analyzer (UV detection at 260 and 280 nm, Cernusco sul Naviglio). The purity of PNA-a7 and PNA-sc7 was checked by RP-HPLC analyses, and their identity was confirmed by electrospray-ionisation quadrupole time-of-flight mass spectrometry (ESI-Q-TOF MS) mass analysis (Q-Tof Micro, Waters). PNA-a7, calculated MW: 5875.4; ESI-MS: m/z found (calculated): 1470.1 (1469.9) $[MH_4^{4+}]$, 1176.2 (1176.1) $[MH_5^{5+}]$, 980.4 (980.2) $[MH_6^{6+}]$, 840.5 (840.3) $[MH_7^{7+}]$, 735.5 (735.4) $[MH_8^{8+}]$, 653.9 (653.8) $[MH_9^{9+}]$. PNA-sc7, calculated MW: 5875.4;

ESI-MS: m/z found (calculated): 1470.0 (1469.9) $[MH_4^{4+}]$, 1176.2 (1176.1) $[MH_5^{5+}]$, 980.3 (980.2) $[MH_6^{6+}]$, 840.4 (840.3) $[MH_7^{7+}]$, 735.5 (735.4) $[MH_8^{8+}]$, 653.9 (653.8) $[MH_9^{9+}]$.

The melting temperature (T_m) of PNA-a7/DNA duplex was calculated according to the linear model for the melting temperature prediction of PNA/DNA duplexes [44]. In particular, taking into account the following formula:

$$T_{m,\,\mathrm{pred}} = c_0 + c_1 \times T_{m,\,\mathrm{nnDNA}} + c_2 \times f_{\mathrm{pyr}} + c_3 \times length \tag{1}$$

in which $T_{m,\,\mathrm{nnDNA}}$ is the melting temperature as calculated using the nearest neighbor model for the corresponding DNA/DNA duplex, applying ΔH^0 and ΔS^0 values as described by SantaLucia et al. [45], f_{pyr} denotes the fractional pyrimidine content, *length* is the PNA sequence length in bases, and the constants were determined to be $c_0 = 20.79$, $c_1 = 0.83$, $c_2 = -26.13$, $c_3 = 0.44$. The calculated $T_{m,\,\mathrm{pred}}$ of PNA-a7 was found to be 65.2 °C.

4.4. Microinjections

For microinjections, only batches in which 90% or more of the embryos developed normally were used. Concentrations of injected solutions were determined by preliminary experiments. We tested the following concentrations: 0.3, 0.5, and 0.7 mM of PNAs (PNA-a7 and PNA-sc7); and 0.3 and 0.5 mM of AmiR-7. For each molecule, the maximum non-lethal concentration was chosen. Dechorionated eggs were microinjected with a solution of 0.7 mM PNAs (PNA-a7 or PNA-sc7) in distilled water or 0.3 mM AmiR-7 plus 5 µg/µL Fast Green as vital dye, as previously described [15]. Embryos were reared at 18 ± 1 °C until they reached late tailbud stage [46].

4.5. Whole Mount In Situ Hybridization

To describe gene expression during development and evaluate microinjection effects, a standard protocol for whole mount in situ hybridization (WISH) was employed [29] with some modifications. Dechorionated embryos and larvae were permeabilized with 2 µg/mL proteinase K in PBS + 0.1% Tween20 for 5 min at 37 °C. To detect miR-7 mature transcripts (MIMAT0003552), a hybridization step was carried out with a DIG-labeled Locked Nucleic Acid (LNA) probe (cin-miR-7-5p: 5′-UGGAAGACUAGUGAUUUUGUUG; RNA T_m = 76 °C) for 5 days at 50 °C. The specificity of the miR-7 signal was confirmed by results obtained using the LNA probe against *C. intestinalis* miR-124, whose expression pattern is well known [30]. The riboprobe specific for hnRNP K was obtained from a GC27a23 plasmid contained in the *C. intestinalis* gene collection release I [47]. DIG-labelled riboprobes were transcribed with Sp6 (antisense) and T7 (sense) RNA polymerase, using a DIG RNA labelling kit (Roche, Monza, Italy). Microinjection effects were explored employing riboprobes against the pan-neural marker Ci-ETR [31] and the gene Ci-Syn, encoding for synapsin, a protein specifically associated with synaptic vesicles [32]. For each probe, at least 40 injected and control embryos were analyzed.

5. Conclusions

Overall, our results demonstrate the in vivo biological activity of PNA oligomers directed against miR-7 in *C. intestinalis* embryos. This animal model allowed direct injection of the anti-miR PNA in eggs, overcoming the typical drawbacks associated with the PNAs poor cellular uptake [48]. One still open problem in antisense approaches is the way of delivering antisense molecules to their target cells in a complex organism. Our results suggest that PNA-a7 is able to reach its specific target in the developing ascidian embryos with high efficiency, as underlined by the lack of effects induced by the scrambled sequence PNA-sc7. To the best of our knowledge, this is the first evidence that unmodified PNA can be successfully used in knockdown strategies in a multicellular organism.

Moreover, our results could be the basis for future quantitative analyses investigating in detail the effect of PNAs.

Author Contributions: Conceptualization, S.M. and S.C. (Silvia Cauteruccio); ascidian manipulation and investigation, S.M., R.M., and R.P.; methodology, G.S.; PNA synthesis, S.C. (Silvia Cauteruccio); writing—original draft preparation, S.M.; writing—review and editing, S.C. (Silvia Cauteruccio), S.C. (Simona Candiani), and R.P.; supervision, E.L., R.P.; project administration, E.L., R.P.; funding acquisition, E.L., R.P.

Funding: This research was funded by FONDAZIONE CARIPLO, grant number 2013-0752.

Conflicts of Interest: The authors declare no conflict of interest.

Abbreviations

ASWH	Artificial Sea Water with HEPES
AmiR-7	AntagomiR anti-miR-7
Boc	*tert*-Butyloxycarbonyl
ESI-Q-TOF MS	Electrospray-ionisation quadrupole time-of-flight mass spectrometry
LNA	Locked Nucleic Acid
MBHA	4-Methylbenzhydrylamine hydrochloride
PNA	Peptide Nucleic Acid
PNA-a7	Peptide Nucleic Acid anti-miR-7
PNA-sc7	Peptide Nucleic Acid scrambled sequence
RP-HPLC	Reverse phase-high pressure liquid chromatography
Z	Benzyloxycarbonyl

References

1. Nielsen, P.E.; Egholm, M.; Berg, R.; Buchardt, O. Sequence-selective recognition of DNA by strand displacement with a thymine-substituted polyamide. *Science* **1991**, *254*, 1497–1500. [CrossRef] [PubMed]
2. Peter, E.N. Gene Targeting and Expression Modulation by Peptide Nucleic Acids (PNA). *Curr. Pharm. Des.* **2010**, *16*, 3118–3123.
3. Egholm, M.; Buchardt, O.; Christensen, L.; Behrens, C.; Freier, S.M.; Driver, D.A.; Berg, R.H.; Kim, S.K.; Norden, B.; Nielsen, P.E. PNA hybridizes to complementary oligonucleotides obeying the Watson–Crick hydrogen-bonding rules. *Nature* **1993**, *365*, 566–568. [CrossRef] [PubMed]
4. Karkare, S.; Bhatnagar, D. Promising nucleic acid analogs and mimics: Characteristic features and applications of PNA, LNA, and morpholino. *Appl. Microbiol. Biotechnol.* **2006**, *71*, 575–586. [CrossRef]
5. Hyrup, B.; Nielsen, P.E. Peptide nucleic acids (PNA): Synthesis, properties and potential applications. *Bioorg. Med. Chem.* **1996**, *4*, 5–23. [CrossRef]
6. Shakeel, S.; Karim, S.; Ali, A. Peptide nucleic acid (PNA)—A review. *J. Chem. Technol. Biotechnol.* **2006**, *81*, 892–899. [CrossRef]
7. Hatamoto, M.; Ohashi, A.; Imachi, H. Peptide nucleic acids (PNAs) antisense effect to bacterial growth and their application potentiality in biotechnology. *Appl. Microbiol. Biotechnol.* **2010**, *86*, 397–402. [CrossRef]
8. Fabani, M.M.; Abreu-Goodger, C.; Williams, D.; Lyons, P.A.; Torres, A.G.; Smith, K.G.C.; Enright, A.J.; Gait, M.J.; Vigorito, E. Efficient inhibition of miR-155 function in vivo by peptide nucleic acids. *Nucleic Acids Res.* **2010**, *38*, 4466–4475. [CrossRef]
9. Yan, L.X.; Wu, Q.N.; Zhang, Y.; Li, Y.Y.; Liao, D.Z.; Hou, J.H.; Fu, J.; Zeng, M.S.; Yun, J.P.; Wu, Q.L. Knockdown of miR-21 in human breast cancer cell lines inhibits proliferation, in vitro migration and in vivo tumor growth. *Breast Cancer Res.* **2011**, *13*, R2. [CrossRef]
10. Nielsen, P.E. Addressing the challenges of cellular delivery and bioavailability of peptide nucleic acids (PNA). *Q. Rev. Biophys.* **2006**, *38*, 345–350. [CrossRef]
11. Rozners, E. Recent Advances in Chemical Modification of Peptide Nucleic Acids. *J. Nucleic acids* **2012**, *2012*, 518162. [CrossRef] [PubMed]
12. Delsuc, F.; Brinkmann, H.; Chourrout, D.; Philippe, H. Tunicates and not cephalochordates are the closest living relatives of vertebrates. *Nature* **2006**, *439*, 965–968. [CrossRef] [PubMed]
13. Passamaneck, Y.J.; Di Gregorio, A. Ciona intestinalis: Chordate development made simple. *Dev. Dyn.* **2005**, *233*, 1–19. [CrossRef] [PubMed]
14. Satoh, N. The ascidian tadpole larva: Comparative molecular development and genomics. *Nature Rev. Genet.* **2003**, *4*, 285–295. [CrossRef] [PubMed]

15. Kari, W.; Zeng, F.; Zitzelsberger, L.; Will, J.; Rothbächer, U. Embryo Microinjection and Electroporation in the Chordate Ciona intestinalis. *J. Vis. Exp.* **2016**, *116*, e54313. [CrossRef] [PubMed]

16. Stolfi, A.; Gandhi, S.; Salek, F.; Christiaen, L. Tissue-specific genome editing in Ciona embryos by CRISPR/Cas9. *Development* **2014**, *141*, 4115–4120. [CrossRef] [PubMed]

17. Prochnik, S.E.; Rokhsar, D.S.; Aboobaker, A.A. Evidence for a microRNA expansion in the bilaterian ancestor. *Dev. Genes Evol.* **2007**, *217*, 73–77. [CrossRef] [PubMed]

18. Alberti, C.; Cochella, L. A framework for understanding the roles of miRNAs in animal development. *Development* **2017**, *144*, 2548–2559. [CrossRef]

19. Paul, P.; Chakraborty, A.; Sarkar, D.; Langthasa, M.; Rahman, M.; Bari, M.; Singha, R.S.; Malakar, A.K.; Chakraborty, S. Interplay between miRNAs and human diseases. *J. Cell. Physiol.* **2018**, *233*, 2007–2018. [CrossRef]

20. Ambros, V. The functions of animal microRNAs. *Nature* **2004**, *431*, 350–355. [CrossRef]

21. Smibert, P.; Lai, E.C. Lessons from microRNA mutants in worms, flies and mice. *Cell Cycle* **2008**, *7*, 2500–2508. [CrossRef] [PubMed]

22. Bravo-Egana, V.; Rosero, S.; Molano, R.D.; Pileggi, A.; Ricordi, C.; Domínguez-Bendala, J.; Pastori, R.L. Quantitative differential expression analysis reveals miR-7 as major islet microRNA. *Biochem. Biophys. Res. Commun.* **2008**, *366*, 922–926. [CrossRef] [PubMed]

23. Sanek, N.A.; Young, W.S. Investigating the In Vivo Expression Patterns of miR-7 microRNA Family Members in the Adult Mouse Brain. *MicroRNA* **2012**, *1*, 11–18. [CrossRef] [PubMed]

24. Tessmar-Raible, K.; Raible, F.; Christodoulou, F.; Guy, K.; Rembold, M.; Hausen, H.; Arendt, D. Conserved Sensory-Neurosecretory Cell Types in Annelid and Fish Forebrain: Insights into Hypothalamus Evolution. *Cell* **2007**, *129*, 1389–1400. [CrossRef]

25. Huang, K.M.; Dentchev, T.; Stambolian, D. MiRNA expression in the eye. *Mamm. Genome* **2008**, *19*, 510–516. [CrossRef]

26. Candiani, S.; Moronti, L.; De Pietri Tonelli, D.; Garbarino, G.; Pestarino, M. A study of neural-related microRNAs in the developing amphioxus. *EvoDevo* **2011**, *2*, 15. [CrossRef]

27. Li, X.; Cassidy, J.J.; Reinke, C.A.; Fischboeck, S.; Carthew, R.W. A MicroRNA Imparts Robustness against Environmental Fluctuation during Development. *Cell* **2009**, *137*, 273–282. [CrossRef]

28. Mattes, J.; Yang, M.; Foster, P.S. Regulation of MicroRNA by Antagomirs. *Am. J. Respir.Cell Mol. Biol.* **2007**, *36*, 8–12. [CrossRef]

29. Messinetti, S.; Mercurio, S.; Pennati, R. Bisphenol A affects neural development of the ascidian Ciona robusta. *J. Exp. Zool.* **2019**, *331*, 5–16. [CrossRef]

30. Chen, J.S.; Pedro, M.S.; Zeller, R.W. miR-124 function during Ciona intestinalis neuronal development includes extensive interaction with the Notch signaling pathway. *Development* **2011**, *138*, 4943–4953. [CrossRef]

31. Satou, Y.; Takatori, N.; Yamada, L.; Mochizuki, Y.; Hamaguchi, M.; Ishikawa, H.; Chiba, S.; Imai, K.; Kano, S.; Murakami, S.D.; et al. Gene expression profiles in Ciona intestinalis tailbud embryos. *Development* **2001**, *128*, 2893–2904. [PubMed]

32. Candiani, S.; Moronti, L.; Pennati, R.; De Bernardi, F.; Benfenati, F.; Pestarino, M. The synapsin gene family in basal chordates: Evolutionary perspectives in metazoans. *BMC Evol. Biol.* **2010**, *10*, 32. [CrossRef] [PubMed]

33. Christodoulou, F.; Raible, F.; Tomer, R.; Simakov, O.; Trachana, K.; Klaus, S.; Snyman, H.; Hannon, G.J.; Bork, P.; Arendt, D. Ancient animal microRNAs and the evolution of tissue identity. *Nature* **2010**, *463*, 1084–1088. [CrossRef] [PubMed]

34. Hamada, M.; Shimozono, N.; Ohta, N.; Satou, Y.; Horie, T.; Kawada, T.; Satake, H.; Sasakura, Y.; Satoh, N. Expression of neuropeptide- and hormone-encoding genes in the Ciona intestinalis larval brain. *Dev. Biol.* **2011**, *352*, 202–214. [CrossRef]

35. Moret, F.; Christiaen, L.; Deyts, C.; Blin, M.; Joly, J.-S.; Vernier, P. The dopamine-synthesizing cells in the swimming larva of the tunicate Ciona intestinalis are located only in the hypothalamus-related domain of the sensory vesicle. *Eur. J. Neurosci.* **2005**, *21*, 3043–3055. [CrossRef]

36. Razy-Krajka, F.; Brown, E.R.; Horie, T.; Callebert, J.; Sasakura, Y.; Joly, J.-S.; Kusakabe, T.G.; Vernier, P. Monoaminergic modulation of photoreception in ascidian: Evidence for a proto-hypothalamo-retinal territory. *BMC Biol.* **2012**, *10*, 45. [CrossRef]

37. D'Aniello, S.; D'Aniello, E.; Locascio, A.; Memoli, A.; Corrado, M.; Russo, M.T.; Aniello, F.; Fucci, L.; Brown, E.R.; Branno, M. The ascidian homolog of the vertebrate homeobox gene Rx is essential for ocellus development and function. *Differentiation* **2006**, *74*, 222–234. [CrossRef]

38. Choudhury, N.R.; de Lima Alves, F.; de Andrés-Aguayo, L.; Graf, T.; Cáceres, J.F.; Rappsilber, J.; Michlewski, G. Tissue-specific control of brain-enriched miR-7 biogenesis. *Genes Dev.* **2013**, *27*, 24–38. [CrossRef]

39. Liu, J.; Githinji, J.; Mclaughlin, B.; Wilczek, K.; Nolta, J. Role of miRNAs in Neuronal Differentiation from Human Embryonic Stem Cell—Derived Neural Stem Cells. *Stem Cell. Rev.* **2012**, *8*, 1129–1137. [CrossRef]

40. Bomsztyk, K.; Denisenko, O.; Ostrowski, J. hnRNP K: One protein multiple processes. *Bioessays* **2004**, *26*, 629–638. [CrossRef]

41. Liu, Y.; Gervasi, C.; Szaro, B.G. A crucial role for hnRNP K in axon development in Xenopus laevis. *Development* **2008**, *135*, 3125–3135. [CrossRef] [PubMed]

42. Piva, R.; Spandidos, D.A.; Gambari, R. From microRNA functions to microRNA therapeutics: Novel targets and novel drugs in breast cancer research and treatment. *Int. J. Oncol.* **2013**, *43*, 985–994. [CrossRef] [PubMed]

43. Prencipe, G.; Maiorana, S.; Verderio, P.; Colombo, M.; Fermo, P.; Caneva, E.; Prosperi, D.; Licandro, E. Magnetic peptide nucleic acids for DNA targeting. *Chem. Commun.* **2009**, *40*, 6017–6019. [CrossRef] [PubMed]

44. Giesen, U.; Kleider, W.; Berding, C.; Geiger, A.; Orum, H.; Nielsen, P.E. A formula for thermal stability (Tm) prediction of PNA/DNA duplexes. *Nucleic Acids Res.* **1998**, *26*, 5004–5006. [CrossRef] [PubMed]

45. SantaLucia, J.; Allawi, H.T.; Seneviratne, P.A. Improved nearest-neighbor parameters for predicting DNA duplex stability. *Biochemistry* **1996**, *35*, 3555–3562. [CrossRef]

46. Hotta, K.; Mitsuhara, K.; Takahashi, H.; Inaba, K.; Oka, K.; Gojobori, T.; Ikeo, K. A web-based interactive developmental table for the ascidian Ciona intestinalis, including 3D real-image embryo reconstructions: I. From fertilized egg to hatching larva. *Dev. Dyn.* **2007**, *236*, 1790–1805. [CrossRef]

47. Satou, Y.; Yamada, L.; Mochizuki, Y.; Takatori, N.; Kawashima, T.; Sasaki, A.; Hamaguchi, M.; Awazu, S.; Yagi, K.; Sasakura, Y.; et al. A cDNA resource from the basal chordate Ciona intestinalis. *Genesis* **2002**, *33*, 153–154. [CrossRef]

48. Urtishak, K.A.; Choob, M.; Tian, X.; Sternheim, N.; Talbot, W.S.; Wickstrom, E.; Farber, S.A. Targeted gene knockdown in zebrafish using negatively charged peptide nucleic acid mimics. *Dev. Dyn.* **2003**, *228*, 405–413. [CrossRef]

MDPI

St. Alban-Anlage 66

4052 Basel

Switzerland

Tel. +41 61 683 77 34

Fax +41 61 302 89 18

www.mdpi.com

International Journal of Molecular Sciences Editorial Office

E-mail: ijms@mdpi.com

www.mdpi.com/journal/ijms